Nanoscience and Engineering

Nanoscience and Engineering

Edited by **Rich Falcon**

NY RESEARCH
P R E S S

New York

Published by NY Research Press,
23 West, 55th Street, Suite 816,
New York, NY 10019, USA
www.nyresearchpress.com

Nanoscience and Engineering
Edited by Rich Falcon

International Standard Book Number: 978-1-63238-497-3 (Hardback)

Printed in the United States of America.

Contents

Preface

Over the recent decade, advancements and applications have progressed exponentially. This has led to the increased interest in this field and projects are being conducted to enhance knowledge. The main objective of this book is to present some of the critical challenges and provide insights into possible solutions. This book will answer the varied questions that arise in the field and also provide an increased scope for furthering studies.

This book unravels the recent studies in the field of nanoscience. It includes some of the vital pieces of work being conducted across the globe, on various topics related to this field. Nanoscience refers to study of the structures that have nanoscale dimensions. It studies the operations and characteristics of these structures. With the help of engineering, appropriate tools are designed to study nanostructures like specialized microscopes, etc. The principles of nanoscience are applied in diverse areas such as medicine, computer science, energy, information technology, etc. Most of the topics introduced in this book cover new techniques and their different applications in this discipline. The ever growing need for advanced technology is the reason that has fueled the research in this subject. This text will serve as a beneficial reference guide for engineers, researchers, professionals, academicians and students engaged in the field of nanoscience.

I hope that this book, with its visionary approach, will be a valuable addition and will promote interest among readers. Each of the authors has provided their extraordinary competence in their specific fields by providing different perspectives as they come from diverse nations and regions. I thank them for their contributions.

Editor

Interaction between Kaolin and Urea in Organoclay and Its Impact on Removing Methylene Blue from Aqueous Solution

Sabri M. Husssein[1], Omar H. Shihab[2], Sattar S. Ibrahim[1], Naser M. Ahmed[3*]

[1]Department of Chemistry, College of Science, University of Anbar, Anbar, Iraq
[2]Department of Chemistry, College of Women Education, University of Anbar, Anbar, Iraq
[3]Nano-Optoelectronics Research and Technology Laboratory, School of Physics, University Sains Malaysia, Penang, Malaysia
Email: *naser@usm.my

Abstract

Interaction between kaolin (particle size 53 and 106 μm) and urea was studied by infrared spectroscopy and powder X-ray diffraction. Interaction was found to be dependent on the particle size of kaolin raw material. Nature of interaction achieved through the formation of hydrogen bonds between urea and both AlOH and Si-O surface of kaolinite. Effect of temperature on equilibrium adsorption of methylene blue (MB) from aqueous solution using kaolin also studied, the results were analyzed by Langmuir and frendlich isotherms. Thermodynamic parameters such as ΔG, ΔH and ΔS were calculated. Results suggested that the MB adsorption on kaolin was spontaneous and exothermic process.

Keywords

Kaolin, Urea, Intercalation, Thermodynamic, Methylene Blue and Adsorption

1. Introduction

Kaolin is one of the clay materials widely used in a large number of applications such as in ceramics, paper coating, paper filling, paint extender rubber filler, cracking catalyst or cements, oil refinery and water treatment (adsorption of dyes and other pollutant) [1]-[4] with the chemical composition $Al_2Si_2O_5(OH)_4$. For each application the engineering properties of the clays must be carefully designed to obtain the desired result. Clays are

---

*Corresponding author.

usually defined as natural materials presenting fine granulometry. Often, these materials exhibit a lamellar structure as a consequence of the crystalline arrangement formed by the silicon and aluminum oxides, which are the main components of clays. These structures are displayed by these materials. Kaolinite is a common 1:1 dioctahedral phyllosilicate (clay) mineral found throughout the world in highly-weathered environments. Being a 1:1 mineral, it has one silica tetrahedral layer and one aluminum octahedral layer combine to form a unique structural arrangement in which sheets of tetrahedral and octahedral overlap each other, leading to structural changes such as 2:1 (one octahedral sheet between two tetrahedral sheets) and 1:1 (one tetrahedral sheet to one octahedral sheet) that characterize the various clay minerals [5] [6].

Kaolinite is a 1:1 tetrahedral aluminosilicate with two distinct basal cleavge faces. One of them consist of tetrahedral siloxane surface formed by very chemically inert Si-O-Si bonds, while the other constituted by an Octahedral sheet $Al(OH)_3$ can be distributed and broken bands have the ability to accommodate OH group. The layers are bonded by hydrogen bonds. Hydrogen bonds occur between oppositely charged ends of a permanent dipole [7] [8].

Interaction between clays and organic compounds have received increase attention due to the wide ranges of applications especially in chromatography separations [9], to remove organic pollutants from air [10], and water [11], and to develop improved formulation for pesticides and as chemical sensor and molecular sieves [12].

This research primarily studies the nature intercalations between kaolinite and urea by using FTIR and XRD. Furthermore, it also studies the impacts to adsorption capacities made by the interaction, the kineticsod adsorption and application to remove dye from aqueous solution.

2. Experimental

Kaolinite used in this study was hydrated aluninum silicate, which was provided from general company for the manufacture of glass and ceramic (ceramic factory) in Ramadi. Chemical analysis of kaolin is shown in **Table 1**. Urea powder with a melting point of 132°C - 135°C, and density of 1.33 g/ml was obtained from sigma Aldrich.

2.1. Preparation of Kaolin-Urea Organoclay (Granular Size 53 µm and 106 µm)

1. 70 gm of grinded kaolin of granular size 53 µm was weighed and placed in a Beaker (capacity of 500 ml).
2. 35 gm of Urea was weighed and then added to the clay on the same Beaker.
3. The mixture was mixed by an electrical mixer in its dry form.
4. Suitable amount of water then added to the mixture with keeping continuous stirring, till getting a solution of kaolin-urea.
5. The mixture then placed at a porcelain crucible and heat in an oven at 90°C till dryness.
6. The products, finally was grinded and became ready to the required tests (FTIR, XRD and Adsorption of methylene blue (MB).
7. Same procedure was used on kaolin (partical size 106 µm).

2.2. Preparation of Methylen Blue Solution

1 gm of MB dye was dissolved in one liter of double distilled water to obtain 1000 ppm MB dye solution. UV-Vis spectra of this solution appeared an absorption band at λ_{max} = 660 nm.

Table 1. Chemical analysis of kaolin.

Al_2O_3	>23%
SiO_2	45% - 50%
Fe_2O_3	<3%
CaO	3%
MgO	<2%
L.O.I	12% - 13%

2.3. Steps of Adsorption

- 0.5 gm of the prepared organoclay was weighed, each alone and placed at 25 ml volumetric flask.
- 10 ml of methylen blue solution dye of the required concentration was added and stirred, very well to the clay.
- The flasks were placed at shaker water bath at different temperatures (10°C, 30°C, 40°C and 50°C) and stirred for 1 hour each.
- The solutions were filtered.
- The absorption was measured for each filtrate at 660 nm.
- The adsorption required calculation according to (Langmuir and Freundlich isotherms), from which the thermodynamic constant can be obtained (ΔG, ΔH and ΔS).

2.4. Preparation of Nano Organoclays

1) Weigh 5 g of prepared organoclay and placed in a glas Baker 250 ml.

2) Added 200 ml of a solution of urea concentration 2 M and shake well.

3) The solution is placed on the ultrasonic (probe ultrasonic) and placed the amount of ice around the beaker for one hour.

4) Separating the precipitate from the filtrate using a centrifuge.

5) Dried the pricipitate and then grinds it and conducted the tests required.

6) Returned the same previous steps for (Thiourea, Acetamide, DMSO and DMF).

3. Results and Discussion

3.1. FTIR Results

Vibrational spectroscopy is a key technique in the study of formation and structural characterization of kaolinite intercalates [13].

FTIR of kaolin, urea and kaolin-urea complex are shown in **Figure 1** and **Figure 2**. From these figures, one could observed that kaolin show two sharp bands at 3694 cm^{-1} and 3625 cm^{-1}. The literature however shows conflicting assignment of these bands [14], band at 3694 cm^{-1} belong to hydroxyl group in specific lattice sites in the layer and resulting from vibrational coupling of three surface of hydroxyl in the primitive cell and the dipole oscillation in perpendicular to the layer, while band at 3625 cm^{-1} in belong to hydroxyl group lie within lamellae

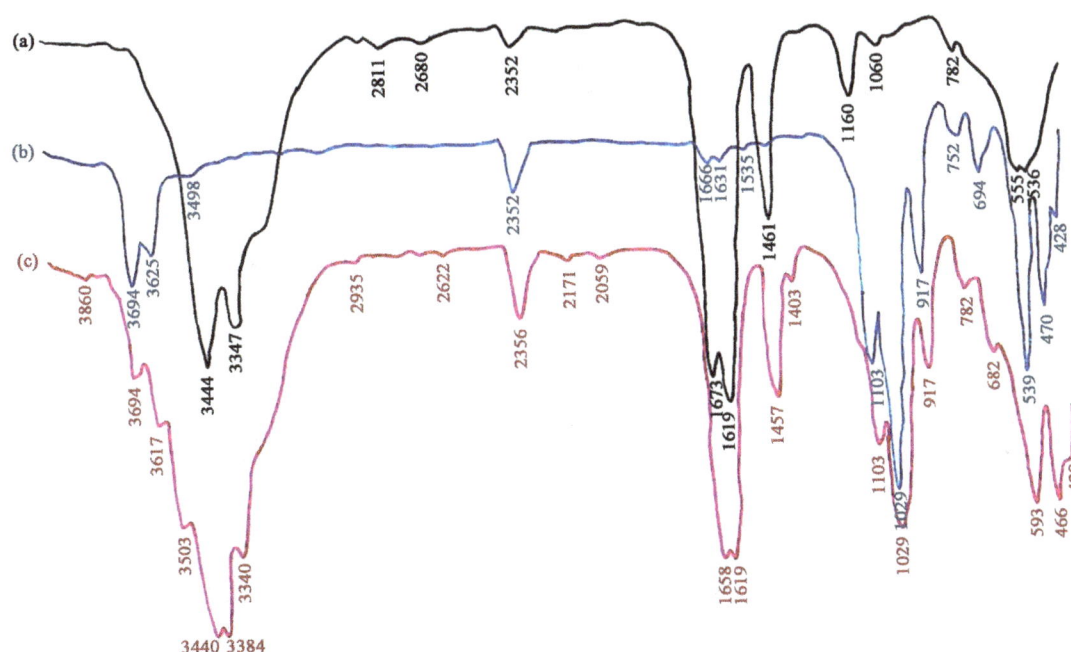

Figure 1. FTIR spectra of (a) urea; (b) kaolin 53 µm and (c) kaolin 53 µm-urea complex.

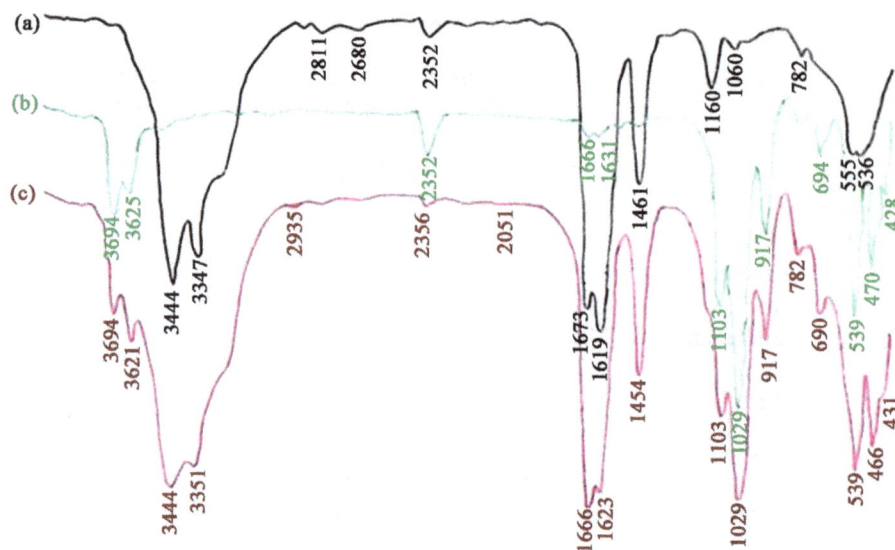

Figure 2. FTIR spectra of (a) kaolin 106 μm; (b) urea and (c) kaolin 106 μm-urea complex.

in plane common to both the tetrahedral and octahedral sheets. Upon intercalation with urea, the intensity of these two bands decrease and shifted to lower frequency, Also a new bands at 3503 cm^{-1} appeared due to the breaking of some hydrogen bonds between the kaolinite layers and formation of new band, which usually involve the inner surface OH group and change are observed in the intensities of bands assigned to vibrations of these groups [13].

Bands at 3440 and 3444 cm^{-1} in the **Figure 1** and **Figure 2** (Chart C) which appear in the results intercalation of kaolinite 53 and 106 μm with urea respectively are attributed to formation H-bond between NH$_2$ group from urea and Oxygen group of tetrahedral sheet for kaolinite.

The newly formed bands at 3384 and 3503 cm^{-1} in the intercalation of kaolinite 53 μm with urea confirmed the asymmetric and symmetric NH$_2$ stretching frequencies involved in weak H-bonding with the inner hydroxyls [15]-[18].

Band at 2352 cm^{-1} in urea chart and kaolinite 53 and 106 μm started disappear when intercalated urea with kaolinite 106 μm and happened shifted in this band to the 2356 cm^{-1} when intercalate urea with kaolinite 53 μm.

Also same effect appeared for the band at 1673 cm^{-1}, which assigned for the C=O group of urea, upon interaction with kaolin, formation a bond between C=O and OH group in Gibbsite-like layer so it shifted to 1658 and 1666 cm^{-1} whene kaolinite 53 and 106 μm interactions with urea respectinely (Chart C in **Figure 1** and **Figure 2**). CN stretching of free urea appeared at 1461 cm^{-1}, upon interaction with kaolinite shifted to 1457 and 1454 cm^{-1} in the **Figure 1** and **Figure 2** Chart C respectively, and a new band at 1403 cm^{-1} appeared. This suggest urea in this system would then be considered to exist in two forms anionic and complex (ion dipole) as shown in **Figure 3**.

3.2. XRD Results

The XRD curves of raw kaolin chart (A), and kaolin-Urea complexes charts (B and C) are shown in **Figure 4**.

From this figure one could observe that the strongest three peaks and their values are recorded in **Table 2**.

From this table:

Peaks at $2\theta = 12.3044$, d(Å) = 7.18765, intensity = 403 and $2\theta = 24.9208$, d(Å) = 3.57009, intensity = 362 are attributed to kaolinite and $2\theta = 26.6345$, d(Å) = 3.34415, intensity = 286 is due to SiO$_2$.

Peak in Chart B at $2\theta = 22.4669$, d(Å) = 3.95417, intensity = 1019 is attributed to urea, and peaked at $2\theta =$ 268558, d(Å) = 3.31709, intensity = 247 is due to SiO$_2$. Band at $2\theta = 25.1508$, d(Å) = 3.53796, intensity = 227 is due to kaolinite.

These peaks in Chart C at $2\theta = 22.3047$, d(Å) = 3.98256, intensity = 2249, $2\theta = 29.3554$, d(Å) = 3.04008, intensity = 371 and $2\theta = 24.6676$, d(Å) = 3.60616, intensity = 370 are assigned to urea, SiO$_2$ and kaolinite respectively.

From the results in **Table 2** and make comparison between these values, on could concluded that strong intercalation between kaolinite layers and urea as a result of appearance high intensity of peaks are due to urea and

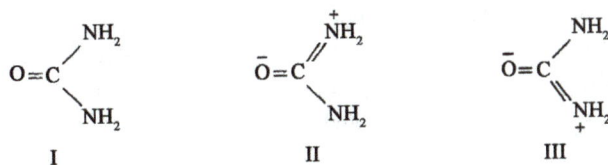

Figure 3. Anionicformsin ureamolecule.

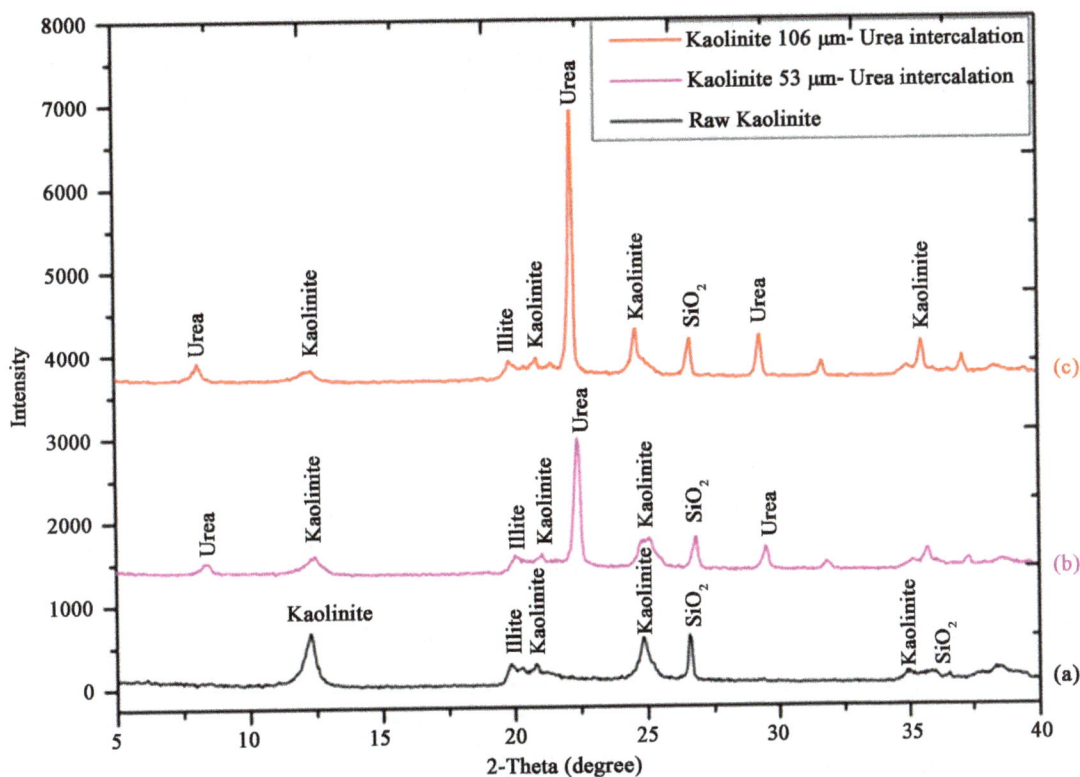

Figure 4. The XRD pattern of raw kaolinite (a); kaolinite 53 μm-urea intercalation (b); and kaolinite 106 μm-urea intercalation (c).

Table 2. Values of XRD for strong peaks in **Figure 4** Chart A.

Assignment	Kaoline	Kaolinite 53 μm-Thiourea Complex	Kaolinite 53 μm-Thiourea Complex
	12.3044	22.4669	22.3047
2θ	24.9208	26.8558	29.3554
	26.6345	25.1508	24.6676
	7.18765	3.95417	3.98256
d-spacing d(Å)	3.57009	3.31709	3.04008
	3.34415	3.53796	3.60616
	403	1019	2249
Intensity (counts)	362	247	371
	286	227	370

in the same time happened shifted and decrease in the intensity of kaolinite and SiO_2 when the intercalation is event. The intercalation caused the destruction of the hydrogen bonding between the kaolinite layers [14]. And from results in this table show decreasing in intensity of peaks when the kaolin 53 μm-urea intercalated with urea compared with other complex this indicates that this kaoline a granular size 53 μm is the best.

3.3. Adsorption Results

Effects of temperature on the equilibrium adsorption of methylene blue from aqueous solution using kaolin (partical size 53 and 106 μm) and kaolin-urea complex were studied.

The equilibrium adsorption data were analyzed using two widely applied isotherms: Langmuir and Freundlich. The results were shown in **Table 3** and **Table 4**. Non-linear method was used for comparing the best fit of the isotherms. Best fit was found to be Langmuir isotherm.

3.3.1. Thermodynamic Parameters

Thermodynamic parameters such as ΔG, ΔH and ΔS were calculated using adsorption equilibrium constant obtained from Langmuir isotherm and shown in **Table 5**.

Results suggested that methylene blue adsorption on kaolin was spontaneous and exothermic process.

Decrease a negative value of ΔG with increase the value of ΔH (-ve) indicate that the adsorption reaction was exothermic.

Percentage of adsorption (Q%) for kaolin and kaolin-urea at conc. 100 ppm of methylen blue are shown in **Table 6**.

3.3.2. Transmission Electron Microscopy (TEM)

TEM is a microscopy technique in which a beam of electrons is transmitted through an ultra-thin specimen, interacting with the specimen as it passes through. An image is formed from the interaction of the electrons transmitted through the specimen; the image is magnified and focused onto an imaging device, such as a fluorescent screen, on a layer of photographic film, or to be detected by a sensor such as a CCD camera.

Figure 5 and **Figure 6** show the TEM photographs of Kaolin (53 and 106 μm)-urea complexes.

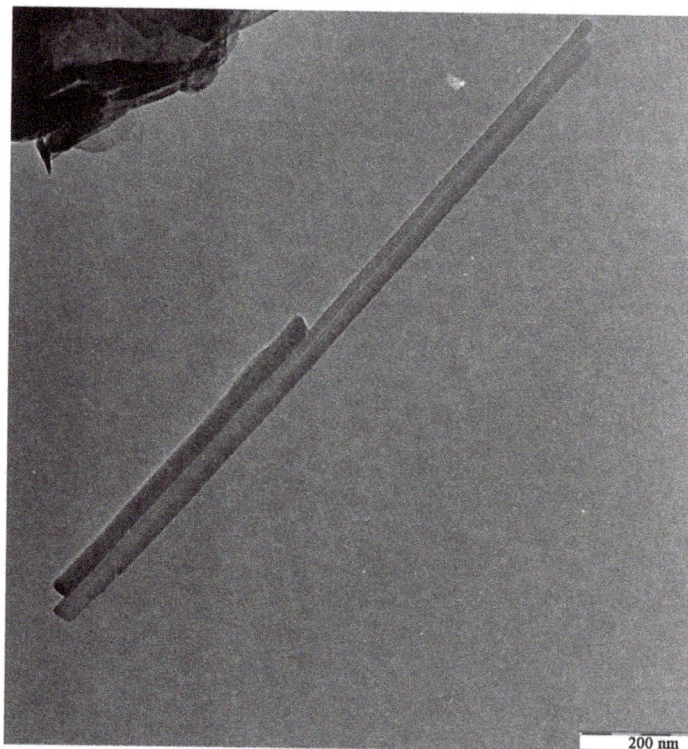

Figure 5. TEM image of kaolinite 53 μm urea complexes.

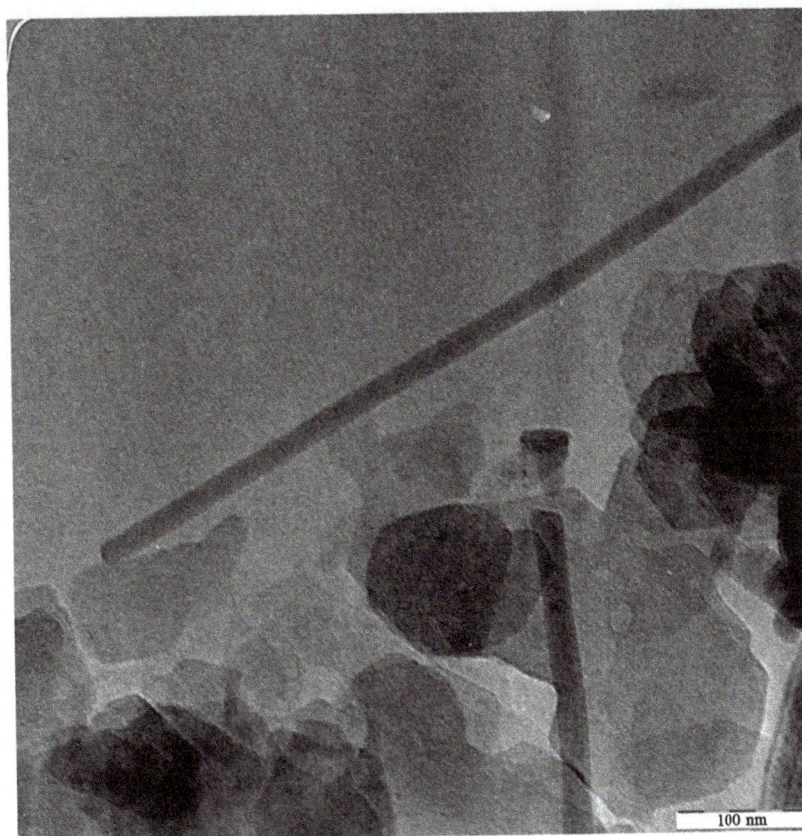

Figure 6. TEM image of kaolinite 106 µm urea complexes.

Table 3. Langmuir constant for adsorption at conc. 100 ppm of methylene blue.

Sample	Particle Size µm	Langmuir Constant	Temperature K			
			283	303	313	322
Kaolin	53	K_f	1000	1000	1000	500
		a	1	1	1	0.5
		R^2	0.535	0.411	0.504	0.64
Kaolin-Urea	53	K_f	0	0	0	0
		a	0	0	0	0
		R^2	0.758	0.879	0.944	0.957
Kaolin	106	K_f	0	250	142.857	250
		a	0	−1.5	−1.714	−0.5
		R^2	0.817	0.933	0.933	0.345
Kaolin + Urea	106	K_f	1000	1000	500	500
		a	2.0	3.0	1.5	0.5
		R^2	0.65	0.737	0.808	0.640

Table 4. Freundlich constant for adsorption at conc. 100 ppm of methylene blue.

Sample	Particle Size μm	Freundlich Constant	Temperature K			
			283	303	313	322
Kaolin	53	K_f	419.75	404.57	309.2	285.759
		n	1.315	1.207	1.331	1.360
		R^2	0.913	0.938	0.897	0.864
Kaolin-Urea	53	K_f	371.53	297.85	229.08	186.638
		n	1.680	1.980	2.624	2.923
		R^2	0.860	0.876	0.867	0.877
Kaolin	106	K_f	319.15	1127.19	1879.31	434.51
		n	1.751	0.536	0.379	1.360
		R^2	0.880	0.983	0.970	0.864
Kaolin-Urea	106	K_f	263.02	224.38	207.01	202.301
		n	1.633	1.908	1.754	1.481
		R^2	0.885	0.874	0.924	0.920

Table 5. Thermodynamic parameters at conc. 100 ppm Methylen blue.

Sample	Particle Size μm	ΔH KJ/mol	ΔS KJ/mol·k	ΔG KJ/mol			
				283 K	303 K	313 K	322 K
Kaolin	53	−9.877	0.01858	−15.0757	−15.7056	−15.644	−15.8984
Kaolin-Urea	53	−6.59965	0.0325576	−15.661	−16.668	−17.119	−18.742
Kaolin	106	−16.9356	−0.006187	−15.3088	−14.8637	−14.8618	−15.1733
Kaolin-Urea	106	−9.935	0.01738	−14.6689	−15.6398	−15.904	−16.2948

Table 6. Percentage of adsorption (Q%) for kaolin and kaolin-urea at conc. 100 ppm of methylen blue.

Sample	Particle Size μm	Q% at Different Temp.			
		283 K	303 K	313 K	322 K
Kaolin	53	99.853	99.803	99.7521	99.726
Kaolin-Urea	53	99.8714	99.8662	99.861	99.803
Kaolin	106	99.8506	99.726	99.699	99.648
Kaolin-Urea	106	99.8039	99.7987	99.7313	99.6639

From this figures show formation of nanotubeit is also very clearly in the images. The average sizes of particles are in the range of 20.2 - 24.5 nm in **Figure 5** and from 20.8 - 27.7 nm in **Figure 6**.

References

[1]   Belver, C., Munor, M.A. and Vicente, M.A. (2002) Chemical Activation of a Kaolinite under Acid and Alkaline Conditions. *Chemistry of Materials*, **14**, 2033-2043.
http://dx.doi.org/10.1021/cm0111736

[2]   Vaga, G. (2007) Effect of Acid Treatments on the Physicochemical Properties of Kaolin Clay. *Epitoanyag*, **59**, 4-8.

Interaction between Kaolin and Urea in Organoclay and Its Impact on Removing Methylene Blue...

9

[3] Caulcante, A.M., Torres, L.G. and Welho, G.L.V. (2005) Effect of Acid Treatments on the Physicochemical Properties of Kaolin Clay. *Journal of Chemical Engineering*, **22**, 2682-2865.

[4] Salawudeen, T.O., Dada, E.O. and Alagbe, S.O. (2007) Performance Evaluation of Acid Treated Clays for Palm Oil Bleaching. *Journal of Engineering and Applied Sciences*, **2**, 1677-1680.

[5] Grim, R.E. (1962) Clay Mineralogy. McGraw Hill, New York.

[6] Valenzuela-Díaz, F.R., Souza-Santos, P. and Souza-Santos, H. (1992) A importância das argilas industriais brasileiras II. *Quimica Industrial*, **44**, 31-35.

[7] Fell, J.R., MacGregor, P., Stapledon, D. and Bell, G. (2005) Geotechnical Engineering of Dams. A. A. Balkema, Leiden.

[8] Mitchell, J.K. and Soga, K. (2005) Fundamentals of Soil Behavior. 3rd Edition, John Wiley & Sons, Hoboken.

[9] Wang, Y., Chem, F.B. and Wu, K.C. (2004) Twin-Screw Extrusion Compounding of Polypropylene/Organoclay Nanocomposites Modified by Maleated Polypropylenes. *Journal of Applied Polymer Science*, **93**, 100-112. http://dx.doi.org/10.1002/app.20407

[10] Sonawane and Meshram, S. (2011) Photo Catalytic Dehydration of Phenol Using ZnO Nanoclay under UV Irradiation in CSTR. *Chemical Engineering Journal*, **72**, 632-637.

[11] Xiang, Y.B., Wang, N., Song, J.M., Cai, D.Q. and Wu, Z.Y. (2013) Micro-Nanopores Fabricated by High-Energy Electron Beam Irradiation: Suitable Structure for Controlling Pesticide Loss. *Journal of Agricultural and Food Chemistry*, **61**, 5215-5219.

[12] Al-Marsoumi Sabri, M.H. and Farouk, K. (2010) Improving the Properties Iraqi Kaoline as an Alternative to the Plastic Clay. Patent No. 2143.

[13] Farmer, V.C. (2000) Transverse and Longitudinal Crystal Modes Associated with OH Stretching Vibrations in Single Crystals of Kaolinite and Dickite. *Spectrochimica Acta Part A*, **56**, 927-930. http://dx.doi.org/10.1016/S1386-1425(99)00182-1

[14] Frost, R., Kristof, J., Rintoul, L. and Kloprogge, J. (2000) Raman Spectroscopy of Urea and Urea-Intercalation Kaolinite at 77 K. *Spectrochimica Acta Part A*, **56**, 1681-1691. http://dx.doi.org/10.1016/S1386-1425(00)00223-7

[15] Orzechowski, K., Stonka, T. and Glowinski, J. (2006) Dielectric Properties of Intercalated Kaolinite. *Journal of Physics and Chemistry of Solids*, **67**, 915-919. http://dx.doi.org/10.1016/j.jpcs.2006.03.001

[16] Ledoux, R.L. and White, J.L. (1966) Infrared Studies of Hydrogen Bonding Interaction between Kaolinite Surfaces and Intercalated Potassium Acetate, Hydrazine, Formamide, and Urea. *Journal of Colloid and Interface Science*, **21**, 127-152. http://dx.doi.org/10.1016/0095-8522(66)90029-8

[17] Zhu, X.Y., Yan, C.J. and Chen, J.Y. (2012) Application of Urea-Intercalated Kaolinite for Paper Coating. *Applied Clay Science*, **55**, 114-119. http://dx.doi.org/10.1016/j.clay.2011.11.001

[18] Valaskova, M., Barabaszova, K., Hundakova, M., Ritz, M. and Plevova, E. (2011) Effects of Brief Milling and Acid Treatment on Two Ordered and Disordered Kaolinite Structures. *Applied Clay Science*, **54**, 70-76. http://dx.doi.org/10.1016/j.clay.2011.07.014

2

The Casimir Topological Effect and a Proposal for a Casimir-Dark Energy Nano Reactor*

Mohamed S. El Naschie

Department of Physics, University of Alexandria, Alexandria, Egypt
Email: Chaossf@aol.com

Abstract

A basically topological interpretation of the Casimir effect is given as a natural intrinsic property of the geometrical topological structure of the quantum-Cantorian micro spacetime. This new interpretation compliments the earlier conventional interpretation as vacuum fluctuation or as a Schwinger source and links the Casimir energy to the so called missing dark energy density of the cosmos. We start with a general outline of the theoretical principle and basic design concepts of a proposed Casimir dark energy nano reactor. In a nutshell the theory and consequently the actual design depends crucially upon the equivalence between the dark energy density of the cosmos and the faint local Casimir effect produced by two sides boundary condition quantum waves. This Casimir effect is then colossally amplified as a one sided quantum wave pushing from the inside on the one sided Möbius-like boundary with nothing balancing it from the non-existent outside. In view of the present theory, this one sided Möbius-like boundary of the holographic boundary of the universe is essentially what leads to the observed accelerated expansion of the cosmos. Thus in principle we will restructure the local topology of space using material nanoscience technology to create an artificial local high dimensionality with a Dvoretzky theorem like volume measure concentration. Needless to say the entire design is based completely on the theory of quantum wave dark energy proposed by the present author. The quintessence of the present theory is easily explained as the ϕ^3 intrinsic Casimir topological energy where $\phi = \left(\sqrt{5}-1\right)\big/2$ produced from the zero set ϕ of the quantum particle when we extract the empty set quantum wave ϕ^2 from it and find $\phi - \phi^2 = \phi^3$ by restructuring space via plates similar to that of the classical Casimir experiments but with some modification.

Keywords

Casimir Effect, Dark Energy, *E*-Infinity, Cantorian Spacetime, Nano Reactor, Free Energy

---

*Dedicated to President Abdul Fattah Al-Sisi.

1. Introduction and Motivation

It is difficult, if not near impossible to give even a glimpse into the past work done in a vibrant and vast field like the zero point vacuum energy which spans the entirety of fundamental theoretical and experimental physics. So rather than attempting the unfeasible, we have included in the present paper a very large amount of references and the reader is referred to these publications and the references therein [1]-[58].

The present paper has two different messages to communicate, a scientific one centred around the quantum vacuum as a source of energy [1]-[58] and a socio-economical, political message that we must invest in this new revolutionary source of energy [59]-[61]. The idea of zero point energy and the fluctuation of vacuum may seem at first glance to be more science fiction than science fact. However, there are, and since quite some time, a host of hard core experimental evidence that the vacuum may be more real ad fundamental than most of what we habitually consider the materialistic reality of physical phenomenon [1]-[58]. We just need to mention in this context the Lamb shift, Schwinger correction [62]-[66] and the van der Waals forces to realize how physical and real the vacuum is [22]-[24] [28]. None the less, and we do not think it is a minority opinion, nothing could be more impressive and inspiring as the Casimir effect [22]-[24]. This effect is a natural consequence and fundamental aspect of quantum field theory. There are at least two fundamental interpretations of this miraculous effect [63] [64]. The first is loosely connect to boundary conditions and the zero point quantum vacuum fluctuation which may be the common way of looking at the Casimir effect within the working physicists community. The second, which may be more theoretical and fundamental, is to see Casimir as a source in the mould of J. Schwinger's way of thinking and not far from the Casimir operators of quantum field theory [62] [63] [66]. Thus we could look upon the Casimir effect as a cousin of Hawking's negative energy fluctuation around a black hole or as Unruh's temperature for an accelerated, observed in a Rindler wedge, universe. Alternatively we could follow Schwinger's ideas and see it as something related to a fundamental mathematical scenario such as the Banach-Tarski theorem advanced for the first time in the cosmology of the big bang by the present author [67] [68].

In the present paper however we opted for a rather different point of viewing the Casimir effect as a natural topological necessity of a Cantorian spacetime fabric which was woven from an infinite number of zero Cantor sets and empty Cantor sets [37]. The zero set is taken following von Neumann-Connes dimensional function to model the quantum particles while the empty set models the quantum wave. Following this road we come we come to realize that the Casimir latent energy is nothing but the universal fluctuation ϕ^3 which gives birth to

the core of Cantorian-fractal spacetime by inversion $1/\phi^3 = 4 + \phi^3 = 4.23606797$ where $\phi = \left(\sqrt{5}-1\right)/2$ [52] [53]. This is nothing but the difference between the Hausdorff dimension of the particle zero set ϕ and the wave empty set ϕ^2. The result not surprisingly is almost equal to double the value found using imaginative modifycation of the classical Casimir experiment by Zee [62] who found the dimensionless Casimir energy to be $\pi/24 \simeq 0.1308$ [62]. Using E-infinity methodological reasoning, the exact value of Zee in the limit must be the ratio of the dimensionality of a Calabi-Yau transfinite manifold $6+k = 6.18033889$ and the transfinite dimension of bosonic string theory, i.e. 26.18033989. That means $\left(6+k\right)/\left(26+k\right) = \phi^3/2$. Needless to say, the division by 2 is due to the subdivision of the "vacuum" of E-infinity theory and is analogous to dividing Hardy's entanglement $P(H) = \phi^5$ by 2 to obtain the density of the ordinary measurable energy of the cosmos $E(O) = \left(\phi^5/2\right)mc^2$. The dimensional quantity analogous to mc^2 for the Casimir effect is trivially clear to be $\hbar c$ where c is the speed of light and $\hbar$ is the Planck quantum. From this new topological interpretation it becomes obvious that Casimir ϕ^3 is the counterfactual or global part of Hardy's entanglement $P(H) = \left(\phi^3\right)\left(\phi^n\right)$ where n is the number of quantum particles and is found for $n = 0$. It is therefore closely related to the Unruh temperature where $n = 1$, the Immirzi parameter $n = 3$ as well as Hardy's generic quantum-topological entanglement $n = 2$. These insights are not only simple mathematical insights. It goes far beyond that and suggests that Casimir energy and dark energy are two sides of the same coin, differing only with regard to exo and endo boundary conditions [6] [7] which will be made clearer in the main body of the present work. Second, by manipulating the local dimensionality of spacetime using an elaborate and complex set up of Casimir plates system we could build a nano universe and extract its dark energy concentrated at its boundary. The way to do this economically may be five, ten or more years of experimental work using the modern developments of cutting edge nanotechnology [1]. Never the less, the promise of near to infinite, clean, free energy is a goal worth any effort and the financial risks are minimal compared to the possible gains, so let the present modest steps be the first into this new world of a nano, Casimir-dark energy reactor.

The paper is structured as follows: In Section 2 immediately after the introduction we give vital information regarding the theory used, namely E-infinity Cantorian spacetime theory [30]-[40]. In Section 3 the vital idea of a one sided Möbius-like boundary of the holographic boundary of the universe is introduced. Sections 4 and 5 discuss the experimental role of nano technology in converting a mathematical model into a real, useful device to produce clean energy. Section 6 gives our main theoretical results and finally section 7 is the general conclusion.

2. Background Information

Based on his E-infinity Cantorian spacetime theory [1]-[21], it was recently argued by the author that the Casimir effect is a local manifestation of the quantum wave while dark energy is the global manifestation of the same [1]. The only difference is that of the details of the boundary conditions [1] [2]. It was further reasoned by the author that the universe as a whole has a one sided boundary akin to that of higher dimensional Möbius band and consequently the "local" Casimir effect ramifies at this one sided boundary located at infinity to produce the negative gravity pressure of the conjectured dark energy [1]-[3]. In other words, three rather mysterious physical notions are tied together and explained in terms of each other. At the top resides the quantum wave [4], which is not a mathematical artifact [4]-[8] but according to E-infinity theory of dark energy, a real physical entity fully described by the empty set fixed by Connes-El Naschie bi dimension $\left(-1,\phi^2\right)$ where $\phi = 1/\left(\sqrt{5}+1\right)$ [1]-[8].

On the other hand the gradient caused by different wave energy density in different bounded regions of space compared to the unbounded outside of the same space is behind the Casimir forces which in the limit can be show to be equal to the difference between the quantum zero set $(0,\phi)$ and the wave empty set $\left(-1,\phi^2\right)$ leading to $\phi - \phi^2 = \phi^3$ topological energy pressure [1] [5]-[8]. Finally at the edge of the universe there is only internal Casimir quantum wave pressure not balanced by outside pressure which is the dark energy concentration of 96 percent as per the consequences of Dvoretzky's theorem and the present author's dissection of Einstein's $E = mc^2$ to $E(O) = mc^2/22$ for ordinary energy of the quantum particle and $E(D) = mc^2\left(21/22\right)$ for the dark energy of the quantum wave [7] [8].

The sources of the ideas contained in the present work go back to many years ago when we attempted to improve on the traditional fast and slow fission reactors using the modern mathematics of fractals and nonlinear dynamics [9]-[17]. The second source is our recent reinterpretation of Einstein's $E = mc^2$ and finally the third source is the unexpected results of the earlier mentioned Dvoretzky's theorem of Banach spaces [3] [6]. However er in the final analysis building an actual reactor could not be possible, not even in principle, without first a sound theory [1]-[56] and second the combination of modern nanotechnology and state of the art Casimir effect experimentation [18]-[24]. In addition a reasonable amount of imaginative thinking similar to that of the man who is famed for inventing the 20th century is also recommended [49] [57].

As we mention at the beginning of our introduction, to keep the present paper short and yet to cover the large amount of the needed prerequisites we opted for a condensed presentation coupled to a large number of references. We recommend to start by reading Ref. 1 and Ref. 60, then it is a personal choice of how to proceed after that.

3. The Boundary of the Holographic Boundary of the Universe and the Empty Set

The holographic boundary theory goes back to the pioneering work of 'tHooft and Susskind [25]-[27]. On the other hand the principle that the boundary of a boundary is zero goes back to the out of the box thinking of J. A. Wheeler [28]. Pushing their ideas further still, it became obvious to the present author that the boundary of the holographic boundary is not only a zero limit set but actually a hierarchy of empty and emptier still sets ramifying at a most general form of a one sided higher dimensional Möbius band [28]-[33]. This limit set resembles a fundamental polyhedron group or better still, a Schottky-Kleinian group [29]-[33] which changes the topology of our conventional Casimir experiment to that of a sphere with internal Casimir pressure inflating the balloon-like universe and makes it expand into the surrounding "nothingness" fixed by the well known E-infinity formula $d_c^{(-\infty)} = \phi^\infty = 0$ where $\phi = 1/\left(\sqrt{5}+1\right)$ [34]. From the preceding elementary reasoning it is clear that Casimir-effect and dark energy have the same cause, namely the topology of a Banach-spacetime like manifold and the only difference is the difference of local exophysics and global endophysics and the respective associated boundary conditions [1] [2]. There is already a vast body of literature on the subject published in the last three years alone by the present author and his associates [1]-[56]. However, what we are aiming at in the present paper is to point out the way to move from theory to useful, practical application of which nothing could be more important and pressing than building a free energy reactor, based on real science rather than wishful thinking. Thus we will combine the dreams of visionaries like N. Tesla with hard nosed modern mathematics and physics which were not yet available in the time of Tesla [49].

4. Enter Nanotechnology

There has been no want of imaginative experimental set ups for measuring, testing and visualizing the Casimir effect since it was proposed by Dutch physicist, H. Casimir [22]-[24]. In recent years nanotechnology invaded all scientific fields and played a significant role in Casimir effect experiments. Thanks to E-infinity we now know that the true physical-mathematical connection between dark energy and the Casimir effect. A natural consequence of this discovered reality of the quantum wave, is rendering it a relatively simple task to find a way to harness dark energy or Casimir energy. Of course this "simple" is extremely difficult but no longer impossible.

We can start with a highly complex sub-structuring of space using nano tubes and nano particles and create that way nanosphere packing modelling the moonshine conjecture which relates superstrings to other fields of theoretical physics. We presently have, in embryonic form, the main idea of constructing a nano universe and extracting dark energy from its nano boundary of its holographic boundary. Our program to actually extract energy from such a nano reactor may still need five or more years but the road is marked and reasonably clear. It is only at the edge of the universe that 96% of the energy resides as dark energy. However we could create many nano universes from which its 96% energy concentration could be extracted without actually reaching to the boundary of our universe [3]-[8].

5. Laboratory Work between Real and Gedanken Experiments

In non-commutative geometry as well as E-infinity theory, the Penrose universe plays a significant role as a generic concrete model for both theories [50]-[53]. On the other hand Penrose universe or Penrose fractal tiling is basically a quasi-crystal mathematical model with the forbidden 5-fold symmetry [53] [54]. This form of matter not found naturally on earth, was produced experimentally by the great Israeli engineer D. Schechtman, who after facing a long period of fierce opposition from high profile scientists, for instance Nobel Laureate Linus Pauling, was rehabilitated and bestowed with a Nobel Prize. The 5 fold symmetry could be thought of theoretically as five Kaluza-Klein dimensions and using nano particles and nano tubes combinations we could build in the lab a nano holographic universe [5]-[8] akin to our own from which energy could be experimented with and extracted. For sure it will be a journey in unchartered seas with many trials and errors but sooner or later we will find out the right road to a Casimir dark energy nano reactor [1] [22]. There are other conceivable ways of producing artificial nano universes with high dimensionality for Dvoretzky's theorem to be applicable. For instance we could use Ji-Huan He's ten dimensional polytope [42] as a skeleton to grow on it a hierarchy of nano particles using the methods applied in the clustering of diffusion limited aggregation. In other words, we can let our scientific imagination run free but checked with E-infinity mathematical rigor and nanotechnological facts.

6. The Topological Casimir-Dark Energy Density Using E-Infinity Theory

It may come as a pleasant mild surprize that exact limits could easily be established for Casimir-dark energy using nothing more than the topology of our E-infinity Cantorian spacetime [56] [60]. We can do this in a variety of ways which are essentially tautologies leading to the same basic conclusion in the limit. Thus we could view the energy density of the space outside the two Casimir plates as that of Einstein's $E = mc^2$ density, *i.e.* $\gamma(\text{Einstein}) = 1$. Inside the plate the energy density in the limit could only be a statical, quasi potential energy of the quantum particle, *i.e.* $E = mc^2/22$ and consequently $\gamma(0) = 1/22$. It follows then that the net pressure of the Casimir plates must be $1-(1/22)=21/22$ which is, in the meantime rather well known, as the dark energy density of spacetime. A second way to interpret the same situation and reach the same result is to argue that within the Casimir plates there is no "space" except for the empty set with a Hausdorff dimension ϕ^2 where $\phi=1/(\sqrt{5}+1)$. Outside on the other hand we have the zero set. The difference is a net $\phi-\phi^2=\phi^3$ which is the universal fluctuation of spacetime and simply the reciprocal value of its Hausdorff dimension $(1/\phi^3)=4+\phi^3$ [3] [52] [56]. Finally we could see the situation as the difference of the completely empty set in the limit, *i.e.* zero between the Casimir plates and the spacetime fluctuation ϕ^3 [60]. That way the Casimir effect could be set in the limit equal to ϕ^3 and may easily be seen to be a relative to the Immirzi parameter ϕ^6 and the Unruh temperature ϕ^4 apart of Hardy's entanglement ϕ^5, *i.e.* a member of a generalized quantum-topological entanglement family [60].

7. Discussion, Conclusion and a Plea for a Peaceful Future

It would be a gross error to place the present nano reactor proposal within the context of science fiction. There is definitely a trivial element of speculation and trial and error but that is all. Exploding stars and galaxies are scientific facts. Consequently to presume that these are only topological defects in to near infinitely large space-time is not outlandish nor science fiction [54]-[56]. In fact the near identity of the Casimir effect and dark energy and the fact that both originate from the quantum wave aspect of quantum mechanics clearly shows to any open minded scientific thinker that to pursue clean free energy is a scientific real and reachable aim. The 4.5% of ordinary energy in the universe is nothing but the multiplicative volume of a five dimensional K-K zero set while the 95.5% dark energy is the additive volume of the same 5D Kaluza-Klein empty set [34]. Seen that way we think that making humanity free from oil and traditional sources of energy is a higher and moral aim worth investing heavily in for what is a million or even billion dollar research grant funding compared to the three trillion dollar Iraq war [59]. In fact the highly enlightened rules of the United Arab Emirates are already looking towards a future free of oil based energy [61]. It was Nobel Laureate in Economics, Prof. J. Stiglitz who calculated with Prof. L. Bilmes the true cost of the Iraq war for the USA. The staggering three trillion dollars do not actually include the loss and destruction for the economy of the entire world. The author dares to say with a tongue in cheek, that the mere sight of only one trillion dollars funding for our nano Casimir-dark energy reactor is sufficient to make this reactor spontaneously pop out of spacetime like virtual particles!

The author, who was born and raised in the Middle East with its unrivalled rich history and unparalleled che-quered present day politics feels morally obliged to call all the governments of the region to participate in a new dawn of science and life.

References

[1] El Naschie, M.S. (2015) Three Quantum Particles Hardy Entanglement from the Topology of Cantorian-Fractal Space-time and the Casimir Effect as Dark Energy—A Great Opportunity for Nanotechnology. *American Journal of Nano Research and Applications*, **3**, 1-5.

[2] El Naschie, M.S. (2014) Casimir-Like Energy as a Double Eigenvalue of Quantumly Entangled System Leading to the Missing Dark Energy Density of the Cosmos. *International Journal of High Energy Physics*, **1**, 55-63.

[3] El Naschie, M.S. (2014) The Measure Concentration of Convex Geometry in a Quasi Banach Spacetime behind the Supposedly Missing Dark Energy of the Cosmos. *American Journal of Astronomy & Astrophysics*, **2**, 72-77.

[4] Slezak, M. (2015) Quantum Wave Function Gets Real. *New Scientist*, 7 February, 14. http://dx.doi.org/10.1016/S0262-4079(15)60242-1

[5] El Naschie, M.S. (2015) Dark Energy and Its Cosmic Density from Einstein's Relativity and Gauge Fields Renormalization Leading to the Possibility of a New 'tHooft Quasi Particle. *The Open Astronomy Journal*, **8**, 1-17. http://dx.doi.org/10.2174/1874381101508010001

[6] El Naschie, M.S. (2015) Banach Spacetime-Like Dvoretzky Volume Concentration as Cosmic Holographic Dark Energy. *International Journal of High Energy Physics*, **2**, 13-21.

[7] El Naschie, M.S. (2014) From $E=mc^2$ to $E=mc^2/22$—A Short Account of the Most Famous Equation in Physics and Its Hidden Quantum Entanglement Origin. *Journal of Quantum Information Science*, **4**, 284-291. http://dx.doi.org/10.4236/jqis.2014.44023

[8] El Naschie, M.S. (2014) The Hidden Quantum Entanglement Roots of $E = mc^2$ and Its Genesis to $E = mc^2/22$ plus

$mc^2(21/22)$ Confirming Einstein's Mass-Energy Formula. *American Journal of Electromagnetics and Applications*, **2**, 39-44. http://dx.doi.org/10.11648/j.ajea.20140205.11

[9] El Naschie, M.S. (1999) From Implosion to Fractal Spheres. A Brief Account of the Historical Development of Scientific Ideas Leading to the Trinity Test and Beyond. *Chaos, Solitons & Fractals*, **10**, 1955-1965. http://dx.doi.org/10.1016/S0960-0779(99)00030-2

[10] El Naschie, M.S. and Al Athel, S. (2000) Estimating the Eigenvalue of Fast Reactors and Cantorian Space. *Chaos, Solitons & Fractals*, **11**, 1957-1961. http://dx.doi.org/10.1016/S0960-0779(99)00069-7

[11] El Naschie, M.S. (2000) On Nishina's Estimate of the Critical Mass for Fission and Early Nuclear Research in Japan. *Chaos, Solitons & Fractals*, **11**, 1809-1818. http://dx.doi.org/10.1016/S0960-0779(99)00172-1

[12] El Naschie, M.S. (2000) Remarks on Heisenberg's Farm-Hall Lecture on the Critical Mass of Fast Neutron Fission. *Chaos, Solitons & Fractals*, **11**, 1327-1333. http://dx.doi.org/10.1016/S0960-0779(99)00136-8

[13] El Naschie, M.S. and Hussein, A. (2000) On the Eigenvalue of Nuclear Reaction and Self-Weight Buckling. *Chaos, Solitons & Fractals*, **11**, 815-818. http://dx.doi.org/10.1016/S0960-0779(99)00106-X

[14] El Naschie, M.S. (2000) Elastic Buckling Loads and Fission Critical Mass as an Eigenvalue of a Symmetry Breaking Bifurcation. *Chaos, Solitons & Fractals*, **11**, 631-629. http://dx.doi.org/10.1016/S0960-0779(99)00063-6

[15] El Naschie, M.S. (2000) On the Zel'dovich-Khuriton Critical Mass for Fast Fission. *Chaos, Solitons & Fractals*, **11**, 819-824. http://dx.doi.org/10.1016/S0960-0779(99)00113-7

[16] El Naschie, M.S. (2000) On the Eigenvalue of Transport Reaction Involving Fast Neutrons. *Chaos, Solitons & Fractals*, **11**, 929-934. http://dx.doi.org/10.1016/S0960-0779(99)00066-1

[17] El Naschie, M.S. (2000) Heisenberg's Critical Mass Calculations for an Explosive Nuclear Reaction. *Chaos, Solitons & Fractals*, **11**, 987-997. http://dx.doi.org/10.1016/S0960-0779(99)00110-1

[18] El Naschie, M.S. (1998) Chaos and Fractals in Nano and Quantum Technology. *Chaos, Solitons & Fractals*, **9**, 1793-1802.

[19] El Naschie, M.S. (2006) Nanotechnology for the Developing World. *Chaos, Solitons & Fractals*, **30**, 769-773. http://dx.doi.org/10.1016/j.chaos.2006.04.037

[20] El Naschie, M.S. (2007) The Political Economy of Nanotechnology and the Developing World. *International Journal of Electrospun Nanofibers and Applications*, **1**, 41-50.

[21] El Naschie, M.S. (1998) Some Tentative Proposals for the Experimental Verification of Cantorian Micro Spacetime. *Chaos, Solitons & Fractals*, **9**, 143-144. http://dx.doi.org/10.1016/S0960-0779(97)00175-6

[22] Johnston, H. (2012) Physicists Solve Casimir Conundrum. Physicsworld.com, July 18, 2012.

[23] Rencroft, S. and Swain, J. (1998) What Is the Casimir Effect? Scientific American, June 22, 1998.

[24] Wongjun, P. (2015) Casimir Dark Energy, Stabilization and the Extra Dimensions and Gauss-Bonnet Term. *The European Physical Journal C*, **75**, 6.

[25] El Naschie, M.S. (2007) A Review of Application and Results of *E*-Infinity. *International Journal of Nonlinear Science & Numerical Simulation*, **8**, 11-20. http://dx.doi.org/10.1515/IJNSNS.2007.8.1.11

[26] Smolin, L. (2001) The Strong and the Weak Holographic Principles. *Nuclear Physics B*, **601**, 209-247. http://dx.doi.org/10.1016/S0550-3213(01)00049-9

[27] El Naschie, M.S. (2006) Holographic Dimensional Reduction: Centre Manifold Theorem and *E*-Infinity. *Chaos, Solitons & Fractals*, **29**, 816-822. http://dx.doi.org/10.1016/j.chaos.2006.01.013

[28] Misner, C., Thorne, K. and Wheeler, J.A. (1973) Gravitation. Freeman, New York.

[29] El Naschie, M.S. (2003) Kleinian Groups in *E*-Infinity and Their Connection to Particle Physics and Cosmology. *Chaos, Solitons & Fractals*, **16**, 637-649. http://dx.doi.org/10.1016/S0960-0779(02)00489-7

[30] El Naschie, M.S. (2005) A Guide to the Mathematics of *E*-Infinity Cantorian Spacetime Theory. *Chaos, Solitons & Fractals*, **25**, 955-964. http://dx.doi.org/10.1016/j.chaos.2004.12.033

[31] El Naschie, M.S. (2004) The Concepts of *E*-Infinity: An Elementary Introduction to the Cantorian-Fractal Theory of Quantum Physics. *Chaos, Solitons & Fractals*, **22**, 495-511. http://dx.doi.org/10.1016/j.chaos.2004.02.028

[32] El Naschie, M.S. (2003) Complex Vacuum Fluctuation as a Chaotic "Limit" Set of Any Kleinian Group Transformation and the Mass Spectrum of High Energy Particle Physics via Spontaneous Self Organization. *Chaos, Solitons & Fractals*, **17**, 631-638. http://dx.doi.org/10.1016/S0960-0779(02)00630-6

[33] El Naschie, M.S. (2003) Modular Groups in Cantorian *E*-Infinity High Energy Physics. *Chaos, Solitons & Fractals*, **16**, 353-366. http://dx.doi.org/10.1016/S0960-0779(02)00440-X

[34] El Naschie, M.S. (1994) On Certain "Empty" Cantor Sets and Their Dimensions. *Chaos, Solitons & Fractals*, **4**, 293-

296. http://dx.doi.org/10.1016/0960-0779(94)90152-X

[35] He, J.-H., Xu, L., Zhang, L.-N. and Wu, X.-H. (2007) Twenty Six Dimensional Polytope and High Energy Spacetime Physics. *Chaos, Solitons & Fractals*, **33**, 5-13. http://dx.doi.org/10.1016/j.chaos.2006.10.048

[36] El Naschie, M.S. (1994) Is Quantum Space a Random Cantor Set with a Golden Mean Dimension at the Core? *Chaos, Solitons & Fractals*, **4**, 177-179. http://dx.doi.org/10.1016/0960-0779(94)90141-4

[37] El Naschie, M.S. (2008) Mathematical Foundation of *E*-Infinity via Coxeter and Reflection Groups. *Chaos, Solitons & Fractals*, **37**, 1267-1268. http://dx.doi.org/10.1016/j.chaos.2008.02.001

[38] El Naschie, M.S. (1995) Banach-Tarski Theorem and Cantorian Micro Spacetime. *Chaos, Solitons & Fractals*, **5**, 1503-1508. http://dx.doi.org/10.1016/0960-0779(95)00052-6

[39] El Naschie, M.S. (1995) On the Initial Singularity and the Banach-Tarski Theorem. *Chaos, Solitons & Fractals*, **5**, 1391-1392. http://dx.doi.org/10.1016/0960-0779(95)99645-2

[40] El Naschie, M.S. (1998) COBE Satellite Measurement, Hyper Spheres, Superstrings and the Dimension of Spacetime. *Chaos, Solitons & Fractals*, **9**, 1445-1471. http://dx.doi.org/10.1016/S0960-0779(98)00120-9

[41] El Naschie, M.S. (2001) Infinite Dimensional Branes and the *E*-Infinity Topology of Heterotic Superstrings. *Chaos, Solitons & Fractals*, **12**, 1047-1055. http://dx.doi.org/10.1016/S0960-0779(00)00130-2

[42] El Naschie, M.S. (2007) Ji-Huan He's Ten Dimensional Polytope and High Energy Particle Physics. *International Journal of Nonlinear Science & Numerical Simulation*, **8**, 475-476. http://dx.doi.org/10.1515/IJNSNS.2007.8.4.475

[43] El Naschie, M.S. (1999) Hyper-Dimensional Geometry and the Nature of Physical Spacetime. *Chaos, Solitons & Fractals*, **10**, 155-158. http://dx.doi.org/10.1016/S0960-0779(98)00235-5

[44] Finkelstein, D. (1982) Quantum Sets and Clifford Algebras. *International Journal of Theoretical Physics*, **21**, 489-503.

[45] El Naschie, M.S. (2002) Derivation of the Threshold and Absolute Temperature T_c=273.16 k from the Topology of Quantum Spacetime. *Chaos, Solitons & Fractals*, **14**, 1117-1120. http://dx.doi.org/10.1016/S0960-0779(02)00053-X

[46] El Naschie, M.S. (2008) Quarks Confinement via Kaluza-Klein Theory as a Topological Property of Quantum-Classical Spacetime Phase Transition. *Chaos, Solitons & Fractals*, **35**, 825-829. http://dx.doi.org/10.1016/j.chaos.2007.08.057

[47] El Naschie, M.S. (2002) On a Class of General Theories for Higher Energy Particle Physics. *Chaos, Solitons & Fractals*, **14**, 649-668. http://dx.doi.org/10.1016/S0960-0779(02)00033-4

[48] He, J.-H. (2009) Hilbert Cube Model for Fractal Spacetime. *Chaos, Solitons & Fractals*, **42**, 2754-2759. http://dx.doi.org/10.1016/j.chaos.2009.03.182

[49] Lomas, R. (1999) The Man Who Invented The Twentieth Century, Nicola Tesla, Forgotten Genius of Electricity. Headline Books, London.

[50] Helal, M., Marek-Crnjac, L. and He, J.-H. (2013) The Three Page Guide to the Most Important Results of M.S. El Naschie's Research in *E*-Infinity Quantum Physics. *Open Journal of Microphysics*, **3**, 141-145. http://dx.doi.org/10.4236/ojm.2013.34020

[51] El Naschie, M.S. (2004) A Review of *E*-Infinity Theory and the Mass Spectrum of High Energy Physics. *Chaos, Solitons & Fractals*, **19**, 209-236. http://dx.doi.org/10.1016/S0960-0779(03)00278-9

[52] El Naschie, M.S. (2009) The Theory of Cantorian Spacetime and High Energy Particle Physics (An Informal Review). *Chaos, Solitons & Fractals*, **41**, 2635-2646. http://dx.doi.org/10.1016/j.chaos.2008.09.059

[53] Marek-Crnjac, L. and He, J.-H. (2013) An Invitation to El Naschie's Theory of Cantorian Spacetime and Dark Energy. *International Journal of Astronomy and Astrophysics*, **3**, 464-471.

[54] El Naschie, M.S. (2013) A Resolution of Cosmic Dark Energy via Quantum Entanglement Relativity Theory. *Journal of Quantum Information Science*, **3**, 23-26. http://dx.doi.org/10.4236/jqis.2013.31006

[55] El Naschie, M.S. (2013) What Is the Missing Dark Energy in a Nutshell and the Hawking-Hartle Quantum Wave Collapse. *International Journal of Astronomy & Astrophysics*, **3**, 205-211. http://dx.doi.org/10.4236/ijaa.2013.33024

[56] El Naschie, M.S. (2013) Topological-Geometrical and Physical Interpretation of the Dark Energy of the Cosmos as a "Halo" Energy of the Schrodinger Quantum Wave. *Journal of Modern Physics*, **4**, 591-596. http://dx.doi.org/10.4236/jmp.2013.45084

[57] Peat, F.D. (1983) In Search of Nikola Tesla. Ashgrove Publications, London & Bath.

[58] Susskind, L. and Lindesay, J. (2005) The Holographic Universe. World Scientific, Singapore.

[59] Stiglitz, J. and Bilmes, L. (2008) The Three Trillion Dollar War: The True Cost of the Iraq Conflict. Allen-Lane, Penguin Books, London.

[60] El Naschie, M.S. (2013) A Unified Newtonian-Relativistic Quantum Resolution of Supposedly Missing Dark Energy

of the Cosmos and the Constancy of the Speed of Light. *International Journal of Modern Nonlinear Theory & Application*, **2**, 43-54. http://dx.doi.org/10.4236/ijmnta.2013.21005

[61] Malek, C. (2015) Abu Dhabi Crown Prince Details UAE Leaders' Vision of Future without Oil. The National Newspaper, UAE, February 10th, 2015.
http://www.thenational.ae/uae/government/abu-dhabi-crown-prince-details-uae-leaders-vision-of-future-without-oil?utm_content="%20vision%20of%20future%20without%20oil

[62] Zee, A. (2003) Quantum Field Theory in a Nutshell. Princeton University Press, Princeton.

[63] Duplantier, B. and Rivasseau, V., Eds. (2003) Vacuum Energy-Renormalization. Birkhauser, Basel.

[64] Milonni, P.W. (1994) The Quantum Vacuum. Academic Press, Boston.

[65] Parsegian, V.A. (2006) Van der Waals Forces. Cambridge University Press, Cambridge.

[66] Huang, K. (2007) Fundamental Forces of Nature. World Scientific, Singapore.

[67] Wapner, L.M. (2005) The Pea and the Sun. A.K. Peters Ltd., Wellesley.

[68] El Naschie, M.S. (1995) Banach-Tarski Theorem and Cantorian Micro Spacetime. *Chaos, Solitons & Fractals*, **5**, 1503-1508. http://dx.doi.org/10.1016/0960-0779(95)00052-6

3

Effect of Aluminium Doping on Structural and Magnetic Properties of Ni-Zn Ferrite Nanoparticles

K. Vijaya Kumar[1]*, D. Paramesh[2], P. Venkat Reddy[2]

[1]Department of Physics, Jawaharlal Nehru Technological University Hyderabad College of Engineering, Nachupally (Kondagattu), Karimnagar-Dist, Telangana State, India
[2]Sreenidhi Institute of Science and Technology (Autonomous), Hyderabad, India
Email: *kvkphd@gmail.com

Abstract

Aluminium doped Ni-Zn ferrite nanoparticles of general formula of $Ni_{0.5}Zn_{0.5}Al_xFe_{2-x}O_4$ (x = 0.0, 0.2, 0.4, 0.6, 0.8, 1.0, 1.2, 1.4, 1.6, 1.8, 2.0) have been synthesized by sol-gel auto combustion method and characterized using X-ray diffraction (XRD), scanning electron microscopy (SEM), energy dispersive X-ray (EDX), Fourier transform spectroscopy (FTIR) and vibrating sample magneto meter (VSM). XRD studies confirm that all compositions show single phase cubic spinel structure. The crystallite size was calculated using the Debye-Scherrer formula and found in the range of 17 - 52 nm. The lattice parameter "a" is found to decrease with increasing Al^{3+} content. The SEM images clearly show the crystalline structure and EDX patterns confirm the compositional formation of the synthesized compositions. The results of FTIR analysis indicated that the functional groups of Ni-Zn spinel ferrite were formed during the sol-gel synthesis process. The IR spectra showed two main absorption bands, the high frequency band v_1 around 600 cm^{-1} and the low frequency band v_2 around 400 cm^{-1} arising from tetrahedral (*A*) and octahedral (*B*) interstitial sites in the spinel lattice. As doping is increased the magnetic behavior is found to decrease and the composition x = 2.0 ferrite appears to be exhibiting superparamagnetism as the coercive field and retentivity are found near zero.

Keywords

Al-Ni-Zn Ferrite Nanoparticles, XRD, EDX, SEM, FTIR, VSM

---

*Corresponding author.

1. Introduction

Since few decades, the ferrites have attained a good position of economic, engineering and magnetic importance due to their excellent physical and chemical properties. Ferrites have a wide range of applications in microwave absorbance, number of electronic devices as radio, TV sets, integrated non-reciprocal circuits, high frequency transformers, memory core devices, rod antennas and telecommunication applications. Furthermore, they are used for gas and ethanol sensors. Due to wide range of applications, the different compositional substitutions into the ferrites have gained importance. The most common soft ferrites are Ni-Zn and Mn-Zn ferrites. The crystal structure of Ni-Zn ferrites spinel configuration is based on a face centered cubic lattice of oxygen ions. The unit cell consists of eight formula units of the type $[ZnFe_{1-x}]A[Ni_{1-x}Fe_{1+x}]BO_4$, where A and B represent tetrahedral and octahedral sites, respectively [1]-[8]. They have a low coercivity and are called soft ferrites. The physical properties of such nanoferrites are highly sensitive to the method of preparation, grain size, chemical composition, sintering temperature, atmosphere, type of substituents and the distribution of cations among tetrahedral and octahedral sites [9]. The conventional methods for the preparation of ferrites have certain limitations such as long heating schedule at high temperatures, higher grain size, higher time consumption etc. The experimental conditions used in the preparation of these materials play an important role in the properties and the particle size of the ferrite nano particles produced. For this reason, a great variety of experimental methods have been used in the production of nano particles, like the sol-gel auto combustion technique. Among several synthesis methods including sol-gel auto-combustion [10], co-precipitation [11], hydrothermal [12], high-energy ball milling [13] and micro-emulsion [14] are developed to make nickel ferrite nano particles. Sol-gel auto combustion technique has many advantages over other methods such as the effect of minimal contamination, processing simplicity, low cost, high level of reactivity, easy control of the particle size and the efficiency of more homogeneous mixing of the component materials that lead to the formation of nanocrystallites.

Nickel-zinc ferrites are soft magnetic materials having low coercivity and high electrical resistivity, which makes them an excellent core material for power transformer in electronic and telecommunication applications [15]. Nickel and zinc have strong occupational preference for tetrahedral and octahedral sites respectively, which makes nickel ferrite a model inverse spinel and zinc ferrite a model normal one [16]. However, in Ni-Zn ferrites, the compositional variation can be resulted in the formation of mixed spinel structure, due to the redistribution of metal ions over the tetrahedral and octahedral sites, which can be drastically modified the ferrites properties [17] [18].

Aluminum substituted Ni-Co ferrites have high electrical resistivity, low eddy current losses, square nature of hysteresis loops, high stability and high value of saturation magnetization, hence they have enormous technological application over wide range of frequencies [19]. Recently, the diamagnetic substitution in mixed ferrites has received special attention. The role played by the substituents in modifying the physical properties of basic ferrites and the mechanism behind enhanced magnetic response is not widely studied. Fabrication of ferrite materials of high quality, low cost and low loss at high frequency for power applications is ever demanding. In this aspect, we have chosen Al doped Ni-Zn ferrites with a composition of $Ni_{0.5}Zn_{0.5}Al_xFe_{2-x}O_4$ (x = 0.0, 0.2, 0.4, 0.6, 0.8, 1.0, 1.2, 1.4, 1.6, 1.8, 2.0) to obtain novel behavior of ferrites on the nanoscale.

2. Experimental

The starting materials were analytical grade (AR) with 99% of purity. The materials were zinc nitrate hexahydrate-$Zn(NO_3)_2 \cdot 6H_2O$(AR), nickel nitrate-$Ni(NO_3)_2 \cdot 6H_2O$(AR), aluminium nitrate nonahydrate-$Al(NO_3)_3 \cdot 9H_2O$ (GR), Ferric Nitrate-$Fe(NO_3)_3 \cdot 9H_2O$(GR) & citric acid monohydrade-$C_6H_8O_7 \cdot H_2O$(GR). Citric acid helps for the homogenous distribution and segregation of the metal ions. During water dehydration, it suppresses the precipitation of metal nitrates because it has electronegative oxygen atoms interacting with electropositive metal ions. Therefore, at a relative low temperature the precursors can form a homogenous single phase ferrite. The mixed solution was neutralized to pH 7 by adding ammonia, it helps for the well formation of gel and improves the solubility of metal ions. Metal nitrates taken in the required stoichiometric ratio were dissolved in a optimum amount of distilled water and mixed together. Then the citric acid was added to the nitrate solution in 3:1 molar ratio. The analytical grade liquid ammonia was added drop by drop to the nitrate solution under constant stirring to maintain the pH value 7. The resulting solution was constantly heated on the magnetic stirrer around 300°C to allow gel formation.

The resultant gel was kept in open air environment to remove the absorbed water [20] and the precursor powder was sintered under the constant heating conditions at 600°C for 5 hours to obtain the final product. The resultant powder was grinded into fine particles by an agate mortar and pestle. Finally, the fine power was pressed into pellets with the help of hydraulic press by applying the 5 tons pressure. The structural properties of samples was studied by Rigaku X-ray diffractometer (Rigaku Miniflex II) using the CuK radiation (wavelength = 1.5406 Å). Scanning electron microscopy (SEM) images were obtained using a TESCAN, MIRA II LMH microscope. The composition was determined by energy dispersive X-ray spectroscopy (EDX, Inca Oxford, attached to the SEM). FTIR analysis carried out including the magnetic properties by using vibrating sample magnetometer (EZ VSM model) at room temperature.

3. Results and Discussions

The XRD patterns of $Ni_{0.5}Zn_{0.5}Al_xFe_{2-x}O_4$ (x = 0.0, 0.2, 0.4, 0.6, 0.8, 1.0, 1.2, 1.4, 1.6, 1.8, 2.0) ferrite nanoparticles presented in **Figure 1** confirmed that all the calcined samples at 600°C are in crystalline state with cubic spinel crystal structure [21]. Broadening of the XRD peaks says the nano-crystalline behaviour of ferrite. The grain size of the nano-particles were calculated from the most intense peak (3 1 1) of XRD data using Debye-Scherer equation.

$$t = \frac{0.9\lambda}{\beta \cos\theta} \tag{1}$$

Lattice parameter, X-ray density were calculated by the following equations.

$$a = d\sqrt{h^2 + k^2 + l^2} \tag{2}$$

$$\rho_x = \frac{8M}{Na^3} \tag{3}$$

The calculated particle size was found in the range of 17 nm to 52 nm, **Figure 2** shows the lattice parameter values were found to decrease from 8.472 Å to 8.205 Å with increasing the Al doping [22]. The lattice parameter values observed in $Ni_{0.5}Zn_{0.5}Al_xFe_{2-x}O_4$ (x = 0.0, 0.2, 0.4, 0.6, 0.8, 1.0, 1.2, 1.4, 1.6, 1.8, 2.0) are in good agreement with the reported values of cubic spinel ferrites [23] [24]. The decrease in the value of lattice parameters with increase in Al^{3+} doping can be explained on the basis of difference in their ionic radius of Fe and Al ions. (Ionic radius of Al^{3+} ion is 0.57 Å and Fe^{3+} ion 0.67 Å). The decrease in the lattice parameter is due to the replacement of Fe^{3+} ion by Al^{3+} ion of a smaller ionic radius [25]. Similar behavior of lattice constant was reported in the literature [26].

Using the values of lattice constant (a) and the distance between magnetic ions (ion jump lengths) available in tetrahedral (A-site), octahedral (B-site) i.e. "L_A" and "L_B" respectively was calculated by using the following relations [27].

$$L_A = \left(\sqrt{\frac{3}{4}}\right)a \tag{4}$$

$$L_B = \left(\sqrt{\frac{2}{4}}\right)a \tag{5}$$

where, a = lattice constant.

Calculated values of ion jump lengths (L_A and L_B) are given in **Table 1**, which shows that ion jump lengths decreased by the increasing the Al content. The ion jump lengths (L_A and L_B) are directly proportional to lattice constant (a) value. Hence, it was observed that the ion jump lengths decreased with the increasing of the Al concentration.

Figure 3 represents the complete samples (total = 11) SEM images. SEM images of compositions (x = 1.6 and x = 2.0) have somehow been reversed.

Figure 4 shows the EDX pattern of $Ni_{0.5}Zn_{0.5}Al_xFe_{2-x}O_4$ (x = 1.0), the peaks of the elements Ni, Zn, Al, Fe and O were observed on the EDX image and all samples confirmed the homogeneous mixing of the Ni, Zn, Fe, Al and O atoms in pure and doped ferrites. The observed composition is almost equal to that of the sample produced by stoichiometric calculations without precipitating cations [28].

FTIR spectra for the compositions of $Ni_{0.5}Zn_{0.5}Al_xFe_{2-x}O_4$ (x = 0.0, 0.2, 0.4, 0.6, 0.8, 1.0, 1.2, 1.4, 1.6, 1.8, 2.0) have been taken using in the range of 500 - 4000 cm^{-1}. **Figure 5** shows the higher frequency band and lower frequency band are assigned to the tetrahedral and octahedral complexes [29] [30]. The FTIR image is reported

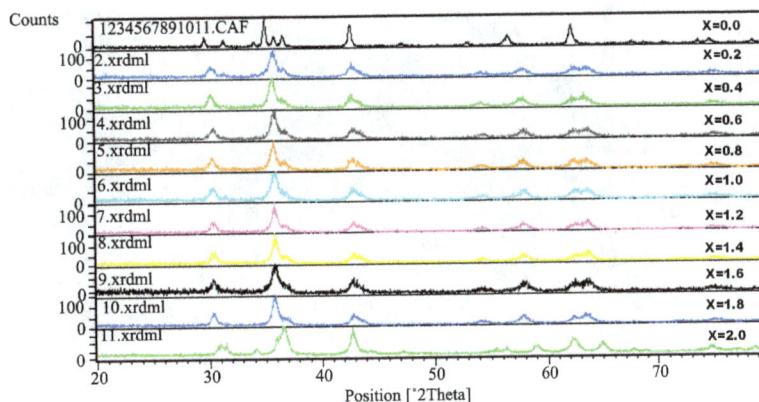

Figure 1. X-ray diffraction pattern of $Ni_{0.5}Zn_{0.5}Al_xFe_{2-x}O_4$ (x = 0.0, 0.2, 0.4, 0.6, 0.8, 1.0, 1.2, 1.4, 1.6, 1.8, 2.0) ferrite nano-particles.

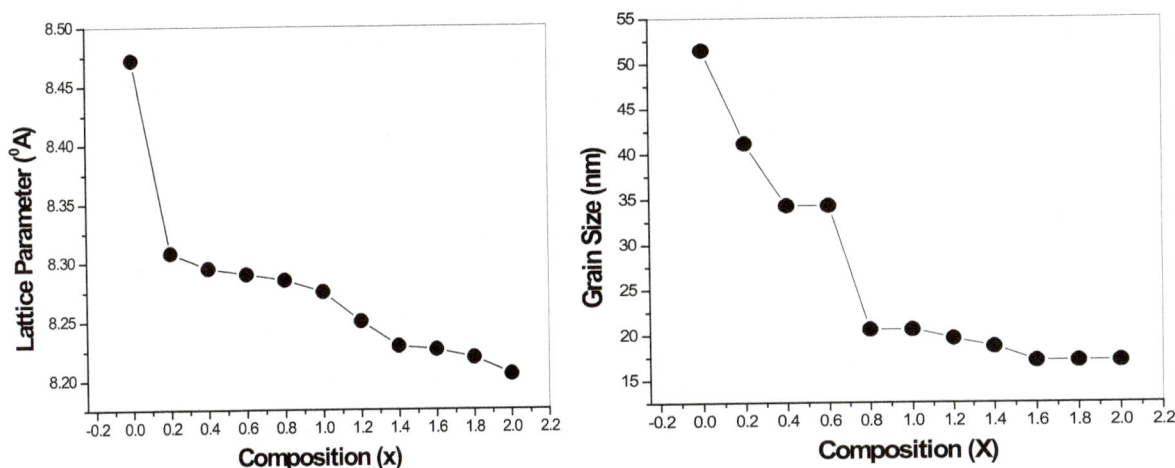

Figure 2. Lattice parameter, grain size with Al composition graphs of $Ni_{0.5}Zn_{0.5}Al_xFe_{2-x}O_4$ (x = 0.0, 0.2, 0.4, 0.6, 0.8, 1.0, 1.2, 1.4, 1.6, 1.8, 2.0) ferrite nano-particles.

the formation of spinel structure and the strong absorption bands around 600 cm^{-1} and 400 cm^{-1} which are characteristic of the tetrahedral and octahedral metal ions have been reported in the literature [31].

It is observed that the size of the particle is decreased with the Al content and for the composition X = 2 the particle size is found lowest. **Figure 6** shows the M-H curves of $Ni_{0.5}Zn_{0.5}Al_xFe_{2-x}O_4$ (x = 0.0, 0.2, 0.4, 0.6, 0.8, 1.0, 1.2, 1.4, 1.6, 1.8, 2.0) ferrite nanoparticles. The magnetization curves display narrow hysteresis for all the compositions except the composition X = 2.0 showed no hysteresis and both retentivity and coercivity parameters are almost found zero. The results reveal that the saturation magnetization of the nanoparticles decreased, while the coercivity kept at almost near zero value with increasing Al substitution. The saturation has not been reached even at its maximum applied field of 20 kOe. This phenomenon can be as signed to the fact that the magnitude of the magnetic field required to reach saturation magnetization depends on the size of the particles [32]. It is also clear that, all the compositions exhibited narrow loops, with a behaviour characteristic of soft magnetic materials (easy magnetization and demagnetization) [33]-[35]. All compositions exhibit low magnetization values and small coercive fields. The low values of magnetization can be explained on the basis of core-shell model which explains that the finite size effect of then a no particles leads to an on collinearity or canting of spins on their surface which there by reduces the magnetization. The decrease in the magnetization value may also be attributed to the presence of Al^{3+} paramagnetic ions with down-spin configuration instead of Fe^{3+} cations with up-spin configuration in the octahedral sites.

The hysteresis curves of $Ni_{0.5}Zn_{0.5}Al_xFe_{2-x}O_4$ (x = 2.0) ferrite nanoparticles are shown in **Figure 7**. The magnetization curves display no hysteresis and both retentivity and coercivity parameters are found almost zero and the ferrimagnetic hysteresis seems to have disappeared at room temperature. Verdes *et al.* [36] mentioned that

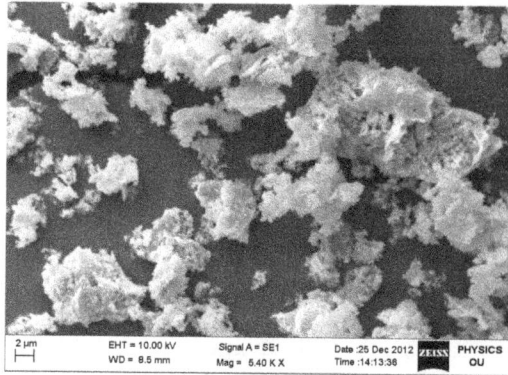

X=0.0

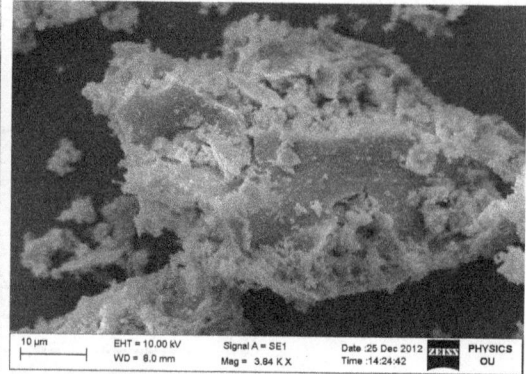

X=0.2

X=0.4

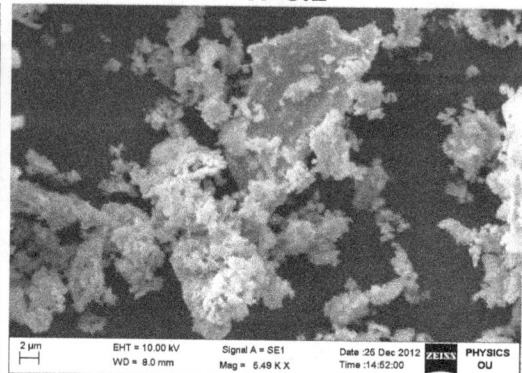

X=0.6

X=0.8

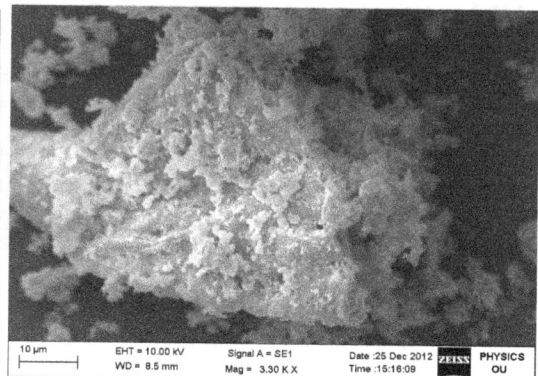

X=1.0

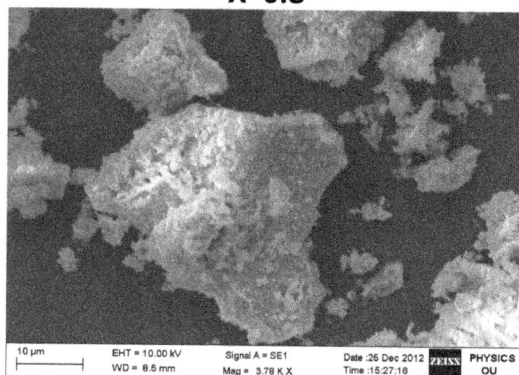

X=1.2

X=1.4

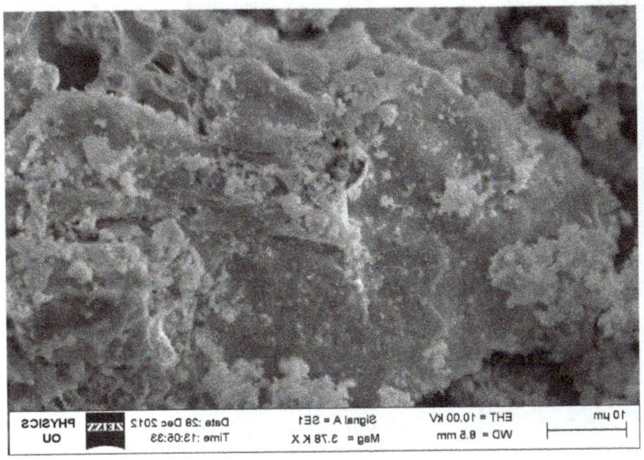

X=1.6

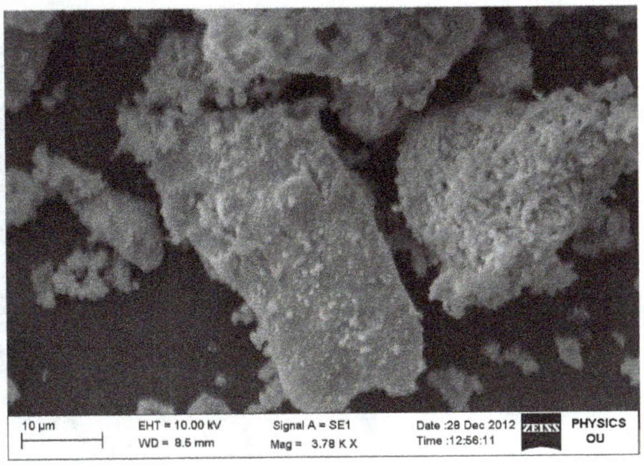

X=1.8

X=2.0

Figure 3. SEM images of $Ni_{0.5}Zn_{0.5}Al_xFe_{2-x}O_4$ (x = 0.0, 0.2, 0.4, 0.6, 0.8, 1.0, 1.2, 1.4, 1.6, 1.8, 2.0) ferrite nano-particles.

Table 1. Structural properties of $Ni_{0.5}Zn_{0.5}Al_xFe_{2-x}O_4$ (x = 0.0, 0.2, 0.4, 0.6, 0.8, 1.0, 1.2, 1.4, 1.6, 1.8, 2.0) ferrite nanoparticles.

Composition (x)	Lattice parameter (Å)	X-ray density (g/cm³)	Grain size (nm)	L_A (Å)	L_B (Å)
x = 0.0	8.472	5.373	51.47	3.668	2.994
x = 0.2	8.308	5.263	41.16	3.597	2.936
x = 0.4	8.295	5.192	34.29	3.591	2.932
x = 0.6	8.290	5.138	34.29	3.589	2.930
x = 0.8	8.285	5.013	20.58	3.587	2.928
x = 1.0	8.275	4.896	20.58	3.583	2.925
x = 1.2	8.250	4.804	19.60	3.572	2.916
x = 1.4	8.229	4.703	18.70	3.563	2.908
x = 1.6	8.226	4.570	17.15	3.561	2.907
x = 1.8	8.219	4.444	17.15	3.558	2.905
x = 2.0	8.205	4.328	17.15	3.552	2.900

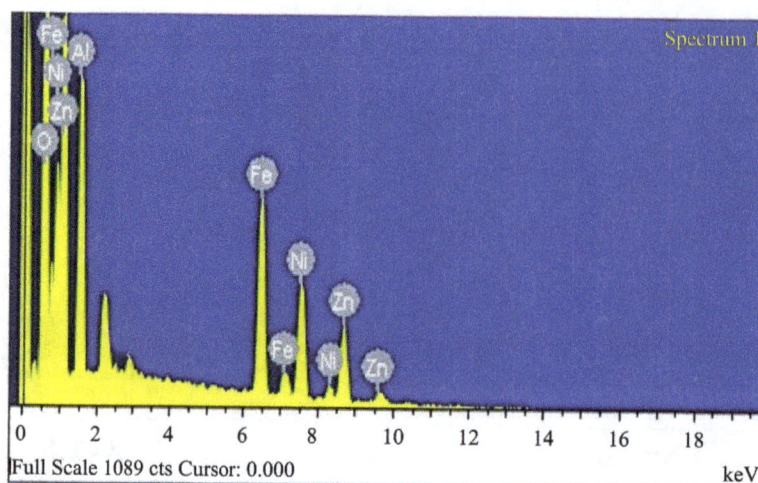

Figure 4. EDX pattern of $Ni_{0.5}Zn_{0.5}Al_xFe_{2-x}O_4$ (x = 1.0) ferrite nano-particles.

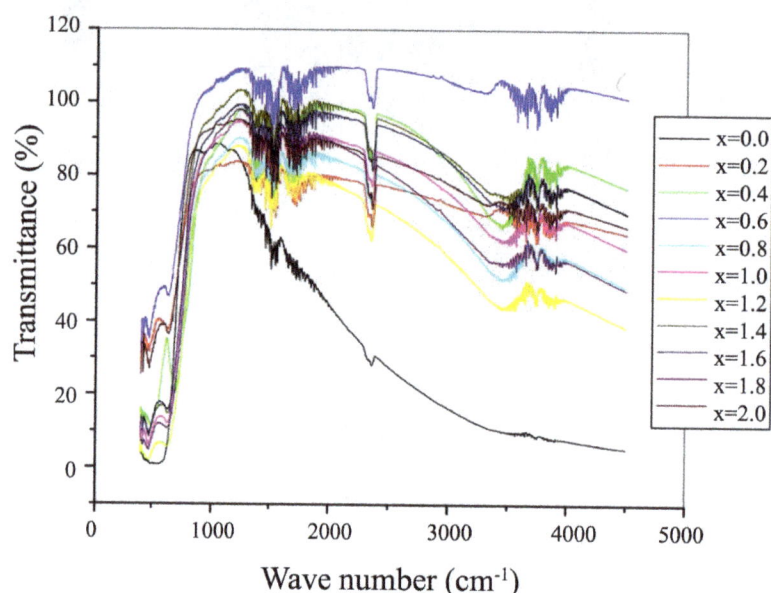

Figure 5. FTIR of $Ni_{0.5}Zn_{0.5}Al_xFe_{2-x}O_4$ (x = 0.0, 0.2, 0.4, 0.6, 0.8, 1.0, 1.2, 1.4, 1.6, 1.8, 2.0) ferrite nano-particles.

the low values of coercivity have been attributed to the particle-particle interactions among the nano-crystals owing to their extremely small size. The proximity of particles has a large effect on the hysteresis as they either become increasingly exchange coupled or show magnetostatic interactions with decreasing distance between the particles [37]. In other words, such magnetic behavior can be expressed in terms of superparamagnetism.

4. Conclusion

We successfully synthesized and characterized the $Ni_{0.5}Zn_{0.5}Al_xFe_{2-x}O_4$ (x = 0.0, 0.2, 0.4, 0.6, 0.8, 1.0, 1.2, 1.4, 1.6, 1.8, 2.0) ferrite nanoparticles using sol-gel auto combustion technique. XRD and the FTIR pattern showed that all the compositions were formed into single phase cubic spinel structure. The lattice parameters and the grain size were found decreasing with Al content. The SEM images showed the crystalline structure where as EDX patterns confirmed the compositional formation of the synthesized samples. It is observed that particle size is decreased with the Al content. For the composition $Ni_{0.5}Zn_{0.5}Al_xFe_{2-x}O_4$ (x = 2.0), the magnetization curves display no hysteresis and both retentivity and coercivity parameters are found almost zero and such magnetic behavior can be expressed in terms of superparamagnetism.

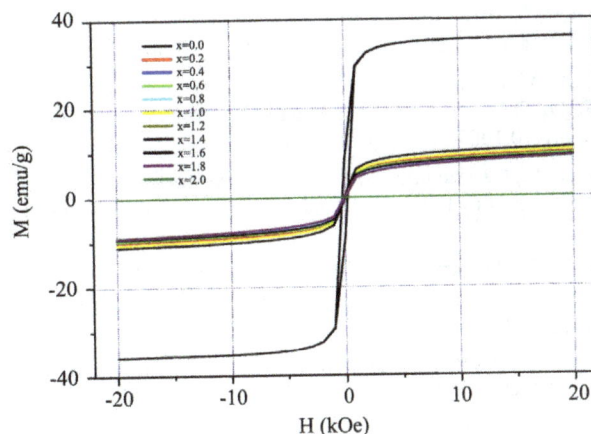

Figure 6. M-H loops of $Ni_{0.5}Zn_{0.5}Al_xFe_{2-x}O_4$ (x = 0.0, 0.2, 0.4, 0.6, 0.8, 1.0, 1.2, 1.4, 1.6, 1.8, 2.0) ferrite nano-particles.

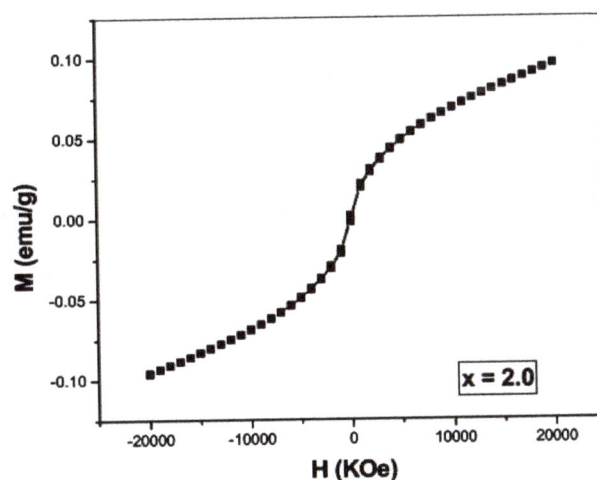

Figure 7. Composition (x) versus saturation magnetization of $Ni_{0.5}Zn_{0.5}Al_xFe_{2-x}O_4$ (x = 2.0) ferrite nano-particles.

Acknowledgements

The authors DP and PVR are grateful to the Principal, SNIST, Ghatkesar, Hyderabad for his support. The author KVK is thankful to Prof. M. Thirumala Chary, Principal, JNTUH CE, Nachupally (Kondagattu), Karimnagar-Dist., TS, India for his constant encouragement in bringing out this research work.

References

[1]   Rashad, M.M., Elsayed, E.M., Moharam, M.M., Abou-Shahba, R.M. and Saba, A.E. (2009) Structure and Magnetic Properties of $Ni_xZn_{1-x}Fe_2O_4$ Nanoparticles Prepared through Co-Precipitation Method. *Journal of Alloys and Compounds*, **486**, 759-767. http://dx.doi.org/10.1016/j.jallcom.2009.07.051

[2]   Krishna, K.R., Kumar, K.V., Ravindernathgupta, C. and Ravinder, D. (2012) Magnetic Properties of Ni-Zn Ferrites by Citrate Gel Method. *Advances in Materials Physics and Chemistry*, **2**, 149-154. http://dx.doi.org/10.4236/ampc.2012.23022

[3]   El-Sheikh, S.M., Rashad, M.M. and Harraz, F.A. (2013) Morphological Investigation and Magnetic Properties of Nickel Zinc Ferrite 1D Nanostructures Synthesized via Thermal Decomposition Method. *Journal of Nanoparticle Research*, **15**, 1-11. http://dx.doi.org/10.1007/s11051-013-1967-9

[4]   Rashad, M.M. and Nasr, M.I. (2012) Controlling the Microstructure and Magnetic Properties of Mn-Zn Ferrites Nanopowders Synthesized by Co-Precipitation Method. *Electronic Materials Letters*, **8**, 325-329. http://dx.doi.org/10.1007/s13391-012-1104-4

[5] Ravinder, D., Kumar, K.V. and Reddy, A.V.R. (2003) Preparation and Magnetic Properties of Ni-Zn Ferrite Thin Films. *Materials Letters*, **57**, 4162-4164. http://dx.doi.org/10.1016/S0167-577X(03)00091-0

[6] Saba, A.E., Elsayed, E.M., Moharam, M.M., Rashad, M.M. and Abou-Shahba, R.M. (2011) Structure and Magnetic Properties of $Ni_xZn_{1-x}Fe_2O_4$ Thin Films Prepared through Electrodeposition Method. *Journal of Materials Science*, **46**, 3574-3582. http://dx.doi.org/10.1007/s10853-011-5271-8

[7] Sorescu, M., Diamandescu, L., Swaminathan, R., McHenry, M.E. and Feder, M. (2005) Structural and Magnetic Properties of NiZn and Zn Ferrite Thin Films Obtained by Laser Ablation Deposition. *Journal of Applied Physics*, **97**, 10G105-1-10G105-3.

[8] Sorescu, M., Diamandescu, L., Peelamedu, R., Roy, R. and Yadoji, P. (2004) Structural and Magnetic Properties of NiZn Ferrites Prepared by Microwave Sintering. *Journal of Magnetism and Magnetic Materials*, **279**, 195-201. http://dx.doi.org/10.1016/j.jmmm.2004.01.079

[9] Toledo, J.A., Valenzuela, M.A., Bosch, P., Armendáriz, H., Montoya, A., Nava, N. and Vazquez, A. (2000) Effect of Al^{3+} Introduction into Hydrothermally Prepared $ZnFe_2O_4$. *Applied Catalysis A*, **198**, 235-245. http://dx.doi.org/10.1016/S0926-860X(99)00514-1

[10] Hashim, M., Alimuddin, Kumar, S., Shirsath, S.E., Kotnala, R.K., Shah, J. and Kumar, R. (2013) Synthesis and Characterizations of Ni^{2+} Substituted Cobalt Ferrite Nanoparticles. *Materials Chemistry and Physics*, **139**, 364-374. http://dx.doi.org/10.1016/j.matchemphys.2012.09.019

[11] Patange, S.M., Shirsath, S.E., Jadhav, S.P., Hogade, V.S., Kamble, S.R. and Jadhav, K.M. (2013) Elastic Properties of Nanocrystalline Aluminum Substituted Nickel Ferrites Prepared by Co-Precipitation Method. *Journal of Molecular Structure*, **1038**, 40-44. http://dx.doi.org/10.1016/j.molstruc.2012.12.053

[12] Zhao, L., Zhang, H., Xing, Y., Song, S., Yu, S., Shi, W., Guo, X., Yang, J., Lei, Y. and Cao, F. (2008) Studies on the Magnetism of Cobalt Ferrite Nanocrystals Synthesized by Hydrothermal Method. *Journal of Solid State Chemistry*, **181**, 245-252. http://dx.doi.org/10.1016/j.jssc.2007.10.034

[13] Mozaffari, M., Amighian, J. and Magn, J. (2003) Preparation of Al-Substituted Ni Ferrite Powders via Mechanochemical Processing. *Journal of Magnetism and Magnetic Materials*, **260**, 244-249.

[14] Mathew, D.S. and Juang, R.-S. (2007) An Overview of the Structure and Magnetism of Spinel Ferrite Nanoparticles and Their Synthesis in Microemulsions. *Chemical Engineering Journal*, **129**, 51-65. http://dx.doi.org/10.1016/j.cej.2006.11.001

[15] Fawzi, A., Sheikh, A.D. and Mathe, V.L. (2010) Structural, Dielectric Properties and AC Conductivity of $Ni_{(1-x)}Zn_xFe_2O_4$ Spinel Ferrites. *Journal of Alloys and Compounds*, **502**, 231-237. http://dx.doi.org/10.1016/j.jallcom.2010.04.152

[16] Shahane, G.S., Kumar, A., Arora, M., Pant, R.P., Lal, K. and Magn, J. (2010) Synthesis and Characterization of Ni-Zn Ferrite Nanoparticles. *Journal of Magnetism and Magnetic Materials*, **322**, 1015-1019.

[17] Kavas, H., Baykal, A., Toprak, M.S., Kseoglu, Y., Sertkol, M. and Aktas, B. (2009) Cation Distribution and Magnetic Properties of Zn Doped $NiFe_2O_4$ Nanoparticles Synthesized by PEG-Assisted Hydrothermal Route. *Journal of Alloys and Compounds*, **479**, 49-55. http://dx.doi.org/10.1016/j.jallcom.2009.01.014

[18] Yao, C., Zeng, Q., Goya, G.F., Torres, T., Liu, J., Wu, H., Ge, M., Zeng, Y., Wang, Y. and Jiang, J.Z. (2007) $ZnFe_2O_4$ Nanocrystals: Synthesis and Magnetic Properties. *The Journal of Physical Chemistry C*, **111**, 12274-12278. http://dx.doi.org/10.1021/jp0732763

[19] Hashim, M., Alimuddin, Kumar, S., Koo, B.H., Shirsath, S.E., Mohammed, E.M., Shah, J., Kotnala, R.K., Choi, H.K., Chung, H. and Kumar, R. (2012) Structural, Electrical and Magnetic Properties of Co-Cu Ferrite Nanoparticles. *Journal of Alloys and Compounds*, **518**, 11-18. http://dx.doi.org/10.1016/j.jallcom.2011.12.017

[20] Verma, V., Dar, M.A., Pandey, V., Singh, A., Annapoorni, S. and Kotnala, R.K. (2010) Magnetic Properties of Nano-Crystalline $Li_{0.35}Cd_{0.3}Fe_{2.35}O_4$ Ferrite Prepared by Modified Citrate Precursor Method. *Materials Chemistry and Physics*, **122**, 133-137. http://dx.doi.org/10.1016/j.matchemphys.2010.02.057

[21] Ghasemi, A., Ekhlasi, S. and Mousavinia, M. (2014) Effect of Cr and Al Substitution Cations on the Structural and Magnetic Properties of $Ni_{0.6}Zn_{0.4}Fe_{2-x}Cr_{x/2}Al_{x/2}O_4$ Nanoparticles Synthesized Using the Sol-Gel Auto-Combustion Method. *Journal of Magnetism and Magnetic Materials*, **354**, 136-145. http://dx.doi.org/10.1016/j.jmmm.2013.10.022

[22] Sridhar, R., Ravinder, D. and Kumar, K.V. (2012) Synthesis and Characterization of Copper Substituted Nickel Nano-Ferrites by Citrate-Gel Technique. *Advances in Materials Physics and Chemistry*, **2**, 192-199. http://dx.doi.org/10.4236/ampc.2012.23029

[23] Satar, A.A., El-Sayed, H.M., El-Shokrofy, K.M. and El-Tabey, M.M. (2005) Improvement of the Magnetic Properties of Mn-Ni-Zn Ferrite by the Non-magnetic Al^{3+}-Ion Substitution. *Journal of Applied Sciences*, **5**, 162-168. http://dx.doi.org/10.3923/jas.2005.162.168

[24] Chandra, P. and Mater, J. (1987) Effect of Aluminium Substitution on Electrical Conductivity and Physical Properties of Zinc Ferrite. *Journal of Materials Science Letters*, **6**, 651-652. http://dx.doi.org/10.1007/BF01770914

[25] Gunjal, R.P., *et al.* (2012) Dielectric Properties of Chromium Substituted Manganese Ferrites. *International Journal of Advanced Engineering Technology*, **3**, 16-17.

[26] Singh, H., Kumar, A. and Yadav, K.L. (2011) Structural, Dielectric, Magnetic, Magnetodielectric and Impedance Spectroscopic Studies of Multiferroic $BiFeO_3$-$BaTiO_3$ Ceramics. *Materials Science and Engineering: B*, **176**, 540-547.

[27] Mustafa, G., *et al.* (2015) Investigation of Structural and Magnetic Properties of Ce^{3+}-Substituted Nanosized Co-Cr Ferrites for a Variety of Applications. *Journal of Alloys and Compounds*, **618**, 428-436. http://dx.doi.org/10.1016/j.jallcom.2014.07.132

[28] Krishna, K.R., Ravinder, D., Kumar, K.V. and Lincon, C.A. (2012) Synthesis, XRD & SEM Studies of Zinc Substitution in Nickel Ferrites by Citrate Gel Technique. *World Journal of Condensed Matter Physics*, **2**, 153-159. http://dx.doi.org/10.4236/wjcmp.2012.23025

[29] Patil, S.A., Mahajan, V.C., Ghatage, A.K. and Lotke, S.D. (1998) Structure and Magnetic Properties of Cd and Ti/Si Substituted Cobalt Ferrites. *Materials Chemistry and Physics*, **57**, 86-91. http://dx.doi.org/10.1016/S0254-0584(98)00202-8

[30] Labde, B.K., Sable, M.C. and Shamkuwar, N.R. (2003) Structural and Infra-Red Studies of $Ni_{1+x}Pb_xFe_{2-2x}O_4$ System. *Materials Letters*, **57**, 1651-1655. http://dx.doi.org/10.1016/S0167-577X(02)01046-7

[31] Waldron, R.D. (1955) Infrared Spectra of Ferrites. *Physical Review*, **99**, 1727-1735. http://dx.doi.org/10.1103/PhysRev.99.1727

[32] Priyadharsini, P., Pradeep, A., Rao, P.S. and Chandrasekaran, G. (2009) Structural, Spectroscopic and Magnetic Study of Nanocrystalline Ni-Zn Ferrites. *Materials Chemistry and Physics*, **116**, 207-213. http://dx.doi.org/10.1016/j.matchemphys.2009.03.011

[33] Martinez, B., Obradors, X., Balcells, L., Rouanet, A. and Monty, C. (1998) Low Temperature Surface Spin-Glass Transition in γ-Fe_2O_3 Nanoparticles. *Physical Review Letters*, **80**, 181-184. http://dx.doi.org/10.1103/PhysRevLett.80.181

[34] Mozaffari, M., Manouchehri, S., Yousefi, M.H., Amighian, J. and Magn, J. (2010) The Effect of Solution Temperature on Crystallite Size and Magnetic Properties of Zn Substituted Co Ferrite Nanoparticles. *Journal of Magnetism and Magnetic Materials*, **322**, 383-388. http://dx.doi.org/10.1016/j.jmmm.2009.09.051

[35] Costaa, A.C.F.M., Silvaa, V.J., Ferreiraa, H.S., Costab, A.A., Cornejoc, D.R., Kiminamid, R.H.G.A. and Gamaa, L. (2009) Structural and Magnetic Properties of Chromium-Doped Ferrite Nanopowders. *Journal of Alloys and Compounds*, **483**, 655-657. http://dx.doi.org/10.1016/j.jallcom.2008.08.129

[36] Verdes, C., Thompson, S.M., Chantrell, R.W. and Stancu, A.L. (2002) Computational Model of Themagnetic and Transport Properties of Interacting Fine Particles. *Physical Review B*, **65**, 174417. http://dx.doi.org/10.1103/PhysRevB.65.174417

[37] Mathew, D.S. and Juang, R.S. (2007) An Overview of the Structure and Magnetism of Spinel Ferrite Nanoparticles and Their Synthesis in Microemulsions. *Chemical Engineering Journal*, **129**, 51-65. http://dx.doi.org/10.1016/j.cej.2006.11.001

4

Thermally Agitated Self Assembled Carbon Nanotubes and the Scenario of Extrinsic Defects

Chernet Amente[1], Keya Dharamvir[2]

[1]Physics Department, Addis Ababa Science and Technology University, Addis Ababa, Ethiopia
[2]Physics Department, Panjab University, Chandigarh, India
Email: chernet.geffe@ambou.edu.et

Abstract

Employing the arc discharge method we prepared carbon nanotubes, CNTs, in open air deionized water. Their morphology was studied varying the annealing temperature and characterizing by Raman Spectroscopy, Transmission Electron Microscopy (TEM), X-Ray Diffractogram (XRD) and Energy Dispersion X-Ray (EDX). According to the study, the CNTs are found self-assembled where the graphene sheets and/or defects are observed sort out themselves with enhancement of temperature.

Keywords

Arc Discharge, Carbon Nanotubes, Defects, Self Assembling, Thermal Agitation

1. Introduction

Since their discovery [1] carbon nanotubes are a front line research topic. These of needle-like configurations, observed during intensive research work on fullerene C60, were found differently fascinated and known to have single, double and multi walled [1] [2] structures. Properties of these systems have been studied theoretically and experimentally for over decades. The theoretical method involved various techniques including simulations by means of different algorithms [3]. The experimental method, however, required sample preparation and characterization phenomenon [1] [2] [4] [5]. In a sample preparation arc discharge, Chemical Vapor Deposition (CVD) and Laser Ablation [6] methods have been utilized at most. Characterizations and typical analysis have been done by Raman Spectroscopy, X-Ray Diffiractometer (XRD), Transmission Electron Microscopy (TEM),

Scanning Electron Microscopy (SEM), Atomic Force Microscopy (AFM), Energy Dispersion X-Ray (EDX), Scanning Tunneling Microscopy (STM), etc. depending on the type of the structure and morphology to be studied [7].

The quality and quantity of the nanotubes is understood as to depend on the type of the discharging method, annealing temperature and time, refluxing temperature and time, system geometry, the electric current and voltage applied, and type of acids used for the reflux [8]-[10]. In most cases samples are prepared from commercially available graphite rods, mounted on electrodes of the discharger and kept few millimeters of distances apart before driving one towards the other, in certain gas or liquid environment [1] [6] [9].

Following the realistic tight-binding band calculations by Hamada *et al.*, 1992 [11], experimental and theoretical works have been reporting that these tubes can be either metals [12] or semiconductors [13] [14]. Electronic band structure calculations have also predicted that the (n,m) indices determine the metallic or semiconducting behavior of CNTs [11] [15]. These nanotubes are understood as efficient sources of electron field emitters [16] [17] and enabled fabrication of remarkable varieties of field-effect transistors [18]-[20] for potential applications.

Studies indicate that, whether the product is single walled or multi walled depends on the amount of catalyst used [21]. Accordingly, the lesser the catalyst leads to multi walled carbon nanotubes production. This might result in the variation of sizes. Diameter of single walled carbon nanotubes (SWNTs) could vary up to 1.4 nm and millimeters of length [22], and that of multiwalled carbon nanotubes (MWNTs) up to hundreds of nanometers width [23] almost independent of the preparation temperature [24]. It has been reported that production of MWNTs by carbon arc discharge method does not require any catalyst. The remaining techniques, however, involved metal doping and resulted in producing lesser crystalline and many more defects [25].

In this research, we used the arc-discharge method and prepared carbon nanotubes in a deionized water. Structural analyses of the samples were done by spectra, diffraction and image pattern recording.

2. The Experiment

In sample preparation we used carbon graphite rods, of nearly 23 cm in length and 7.64 mm diameter, commercially available. These rods were cut into pieces and mounted on the electrodes of the arc used as anode and cathode, kept few millimeters apart in a chamber of deionized water cover, as shown in **Figure 1**, before the process.

A d.c. of 50 - 200 A driven by 40 V created a high temperature discharge between the two electrodes where high sparking and little smoke was observed. After frequent discharging the chamber is removed and the product soot along with water is transferred into a bigger beaker and kept covered with aluminum foil for about 6 hours until crude is formed at the bottom and then decanted. Subsequently, the crude is made open air dry at 100˚C for about 12 hours and the dried soot is collected for technical analysis.

3. Results and Discussion

The crystal purity and defect concentration of the graphite powder was tested by the mechanism of Raman spectra recording at room temperature, using RENISHAW-Raman equipment operating with Argon laser of one excitation and wavelength 514 nm.

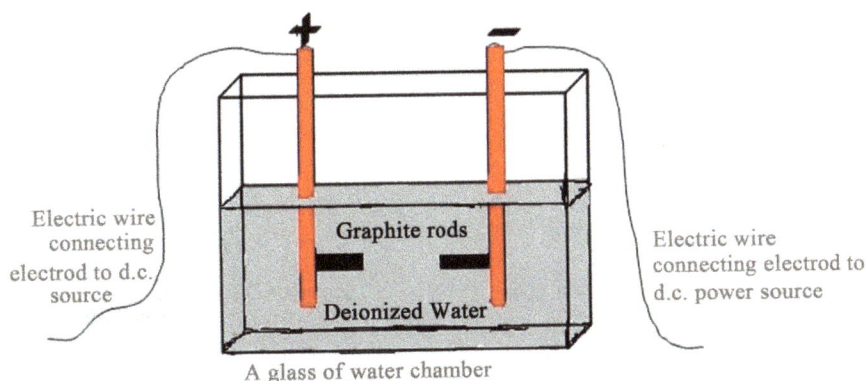

Figure 1. Schematic diagram of locally devised arcing equipment.

The I_D/I_G ratio indicated that the graphite used contains large amount of defects, as illustrated in **Figure 2(a)**. The collected data plots in **Figure 2(b)** and **Figure 2(c)** show that there is D-band (disorder induced phonon mode [26]) at 1348 cm^{-1} and strong peak G (graphite)-band at 1583 cm^{-1} Raman shift for the as prepared sample and the D-band at 1349 cm^{-1} and the G-band at 1582 cm^{-1} for the annealed sample, respectively. Further analysis indicates that there is a G-peak due to a high production of monolayer graphene [27] at 2710 cm^{-1}. The increase in I_D/I_G is, therefore, because of that annealing successfully altered the CNTs, perhaps, increasing the number of defects on their side walls [28]. The increase in D-band frequency might be due to the chemical charge transfer under different temperature treatment and/or amorphous carbon content most likely from the destruction of the CNTs [29]. The D-band (sp^3) is attributed as associated with vibrations of carbon atoms with dangling bonds in the terminal plane of disordered carbon/impurities whereas the G-band is related to the vibration of sp^2 bonded carbon atoms in a two-dimensional hexagonal lattice [30]. **Figure 2(c)** shows that there are D'-band peaks at 1614 cm^{-1} and 1629 cm^{-1} indicating the presence of randomly distributed impurities or surface charges in the graphene, resulting in splitting of the G-band in to G and D'-peaks. Further scrutiny shows that there is a G'-band (the second strongest after the G mode and the second overtone of the defect-induced D mode) at about 2708 cm^{-1} and attributed as useful in determining the number of graphene layers [31].

The X-Ray Diffractometer (XRD) D8 advanced, from Bruker A × S, of scan type locked coupled, scan angle 20° - 80° range, scan step 0.02°, scan speed 3°/min, max. power 40 kV/40mA, Cu tube, T/T horizontal, for which scan time is about 20 minutes for each sample and wavelength 1.5406 A° is used to identify the composition of the samples.

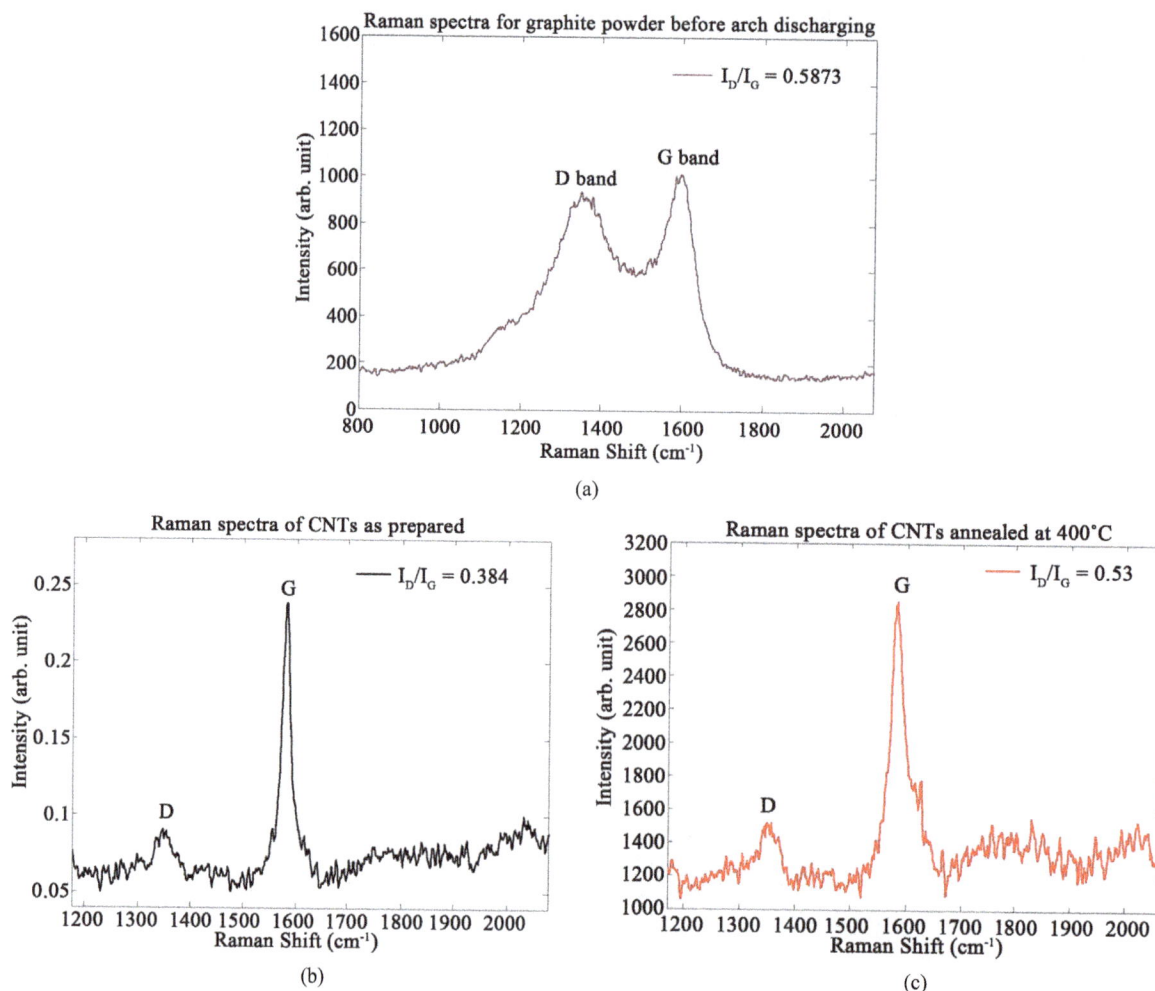

Figure 2. Raman spectra (a) of graphite powder before arc discharge (b) for the as prepared CNTs and (c) for CNTs annealed at 400°C for 30 minutes.

Accordingly, the peaks at about 26°, 44°, and 76° shown in **Figure 3** corresponds to indices of graphite C(002), C(101) and C(110) planes, respectively, indicating the presence of CNTs in the sample and the scattering of incident beam by (002) faces [32] which always exists, as far as X-ray diffraction pattern of CNT's are close to that of graphite [33], in agreement with experimental results obtained earlier and JCPDF #751621 indications [34].

The analysis was repeated, using another diffractometer, model name X'Pert PRO, company name PANalytical (formerly known as Philips) with scan type continues, scan angle 10° - 90°, step size 10,167/degree, time for step 20 sec, by CuKα_1 laser radiation, scan speed 0.1°/sec, λ = 1.540598 A° wavelength, generator power setting 40 kV, and current of 40 mA, for conformation.

The data plot of the result, shown in **Figure 4(b)**, also suggests that the reoccurrence of the peaks would come from the defects found in the raw material as far as reappeared in the unprocessed graphite powder, at about 28.2° for instance, and revealed by broadened line width at about 25.77° for graphite as well (see **Figure 4(a)**).

The EDX analysis indicates that there are Calcium (Ca) and Oxygen (O) constituents in the samples, perhaps, introduced during industrial preparation of graphite rod, as shown in **Figure 5**, and are known as introducing extra peaks at about 28°, 47°, and 56°, and could not be removed by annealing. Cu peaks are due to grip on which the sample was deposited for analysis.

Scherrer's formula [35] $t = k\lambda/\beta\cos\theta_B$ which is derived from Bragg's equation that has been utilized in determining crystallite inter planar spacing, is employed in estimating the nanotubes thickness t, where κ is the shape factor approximated to 0.9, $\beta = \Delta 2\theta \times \pi/180°$ is the line broadening at half the maximum intensity (at full width half of maximum intensity, FWHM), λ is the X-ray wave length and θ_B is the Bragg's angle.

One can understand, from **Table 1**, that the diffraction peaks at 2θ are slightly shifted as a result of further annealing (**Figure 3(c)**) perhaps due to formation of CNTs whose production has been overwhelmed by the defects. Moreover, the increase in thickness can be due to self-assembling of the CNTs.

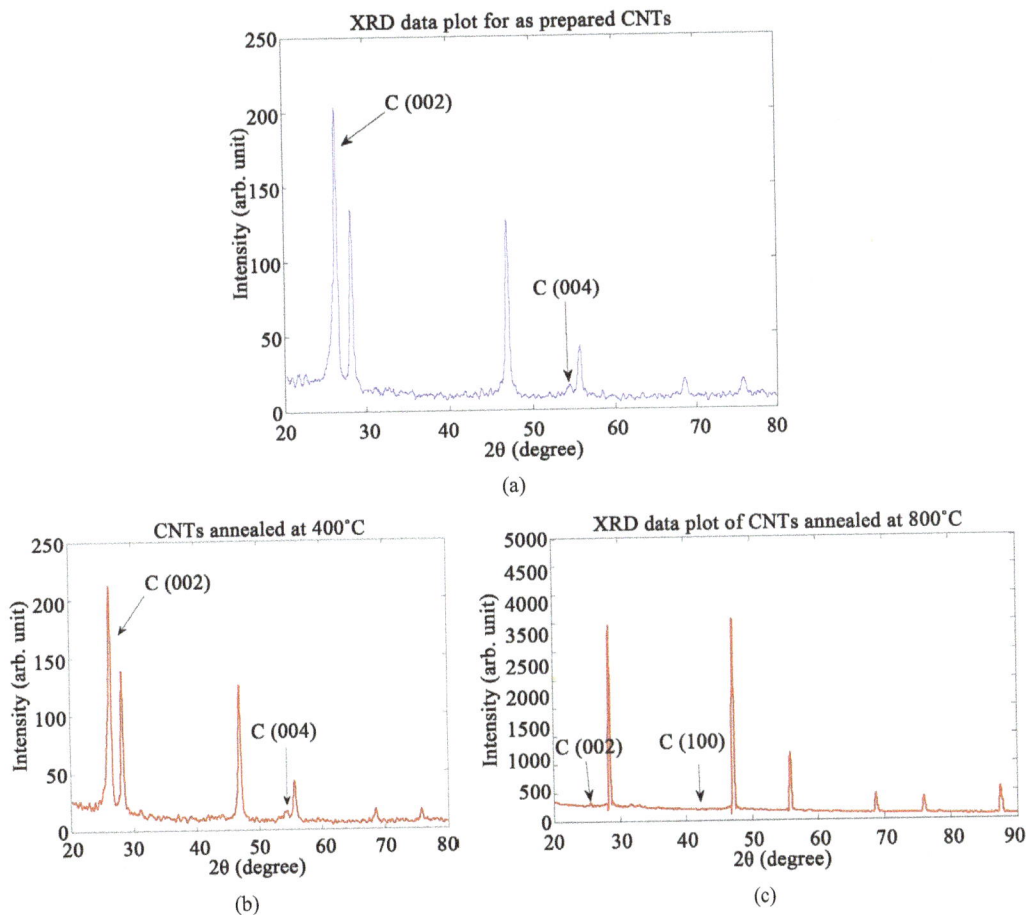

(a)

(b)

(c)

Figure 3. XRD profile of (a) the as prepared CNTs (b) CNTs annealed at 400°C and (c) CNTs annealed at 800°C.

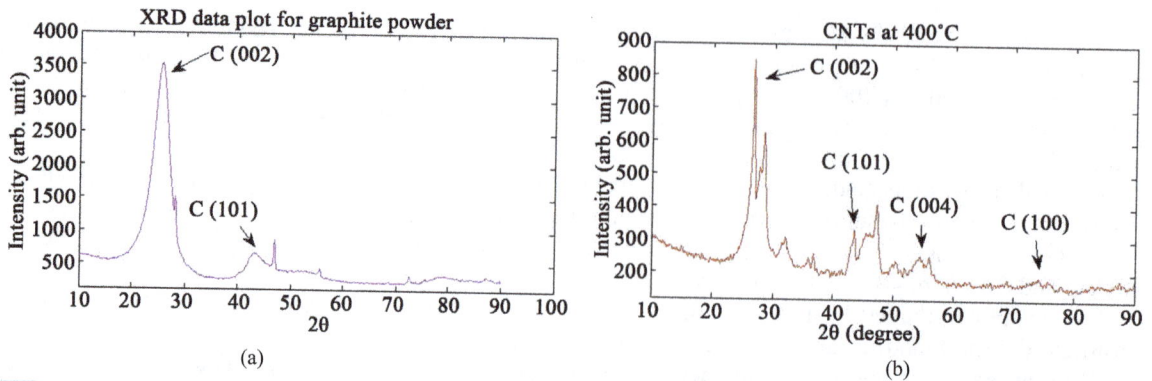

Figure 4. XRD profile of (a) Graphite powder (b) CNTs annealed at 400°C for comparison.

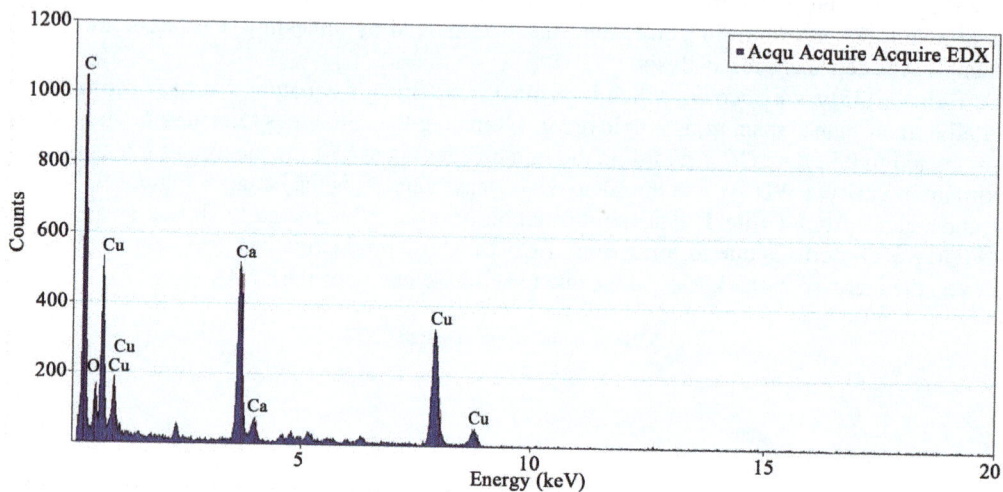

Figure 5. EDX pattern showing contents of the samples without metal doping.

Table 1. The XRD data of graphite powder and carbon nanotubes pre and post annealing referring to **Figure 3** and **Figure 4**.

	Graphite sample	Pristine CNTs	At 400°C	At 800°C
2θ (deg.)	25.63	26.28	26.79	25.58
$\Delta2\theta$ (deg.)	3.52733	0.95916	1.01797	0.19841
t ($A°$)	23.0881	85.0181	80.1906	410.4199

Table 2 shows that sizes of the attributed defects could decrease with increase in the Bragg's angle at 800°C annealing. However, there are no well defined circumstances in the case of the pristine and the 400°C annealed samples. It is worth mentioning that the size of those defects in the unprocessed graphite sample at about 28.2° is nearly 33.886 A°. This indicates that size of the defects has increased due to thermal agitation and perhaps burning of some of the CNTs which are known to self categorize as also shown in **Figure 8**.

Images of the CNTs were collected using transmission electron microscopy (TEM) model 7500, 2 keV HITACHI, maximum magnification 6×10^6 times and resolution 0.2 A°, after sonication of the nanopowder suspending in ethanol and exposing to ultrasonic waves for 3 hr. As shown in **Figure 6**, the as prepared sample has contained puffy colored impurities stacked on the surface of the tubes and also scattered elsewhere. After open air annealing at 400°C the density of these impurities and/or amorphous carbon [10] is reduced and the clarity of the CNTs is improved, as in **Figure 7**. This shows that further annealing to certain temperature limit may give better and more purified CNT products. The measured internal diameter of these tubes is known to vary nearly from 1.0 nm - 7.0 nm and the external diameter ranges between 6 nm - 26 nm, where their length extends to about 0.4 μm.

We have also further annealed the CNTs at 800°C for 30 minutes and cooled back to room temperature. The color feature of the sample powder was found changed to grey and reduced in quantity, in agreement with pre-

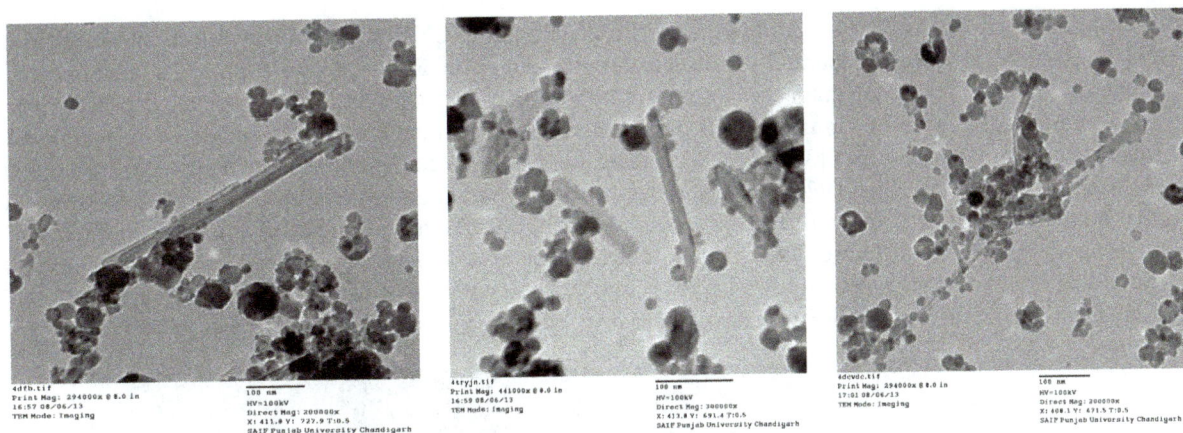

Figure 6. TEM images of the as prepared CNTS from different points of focus.

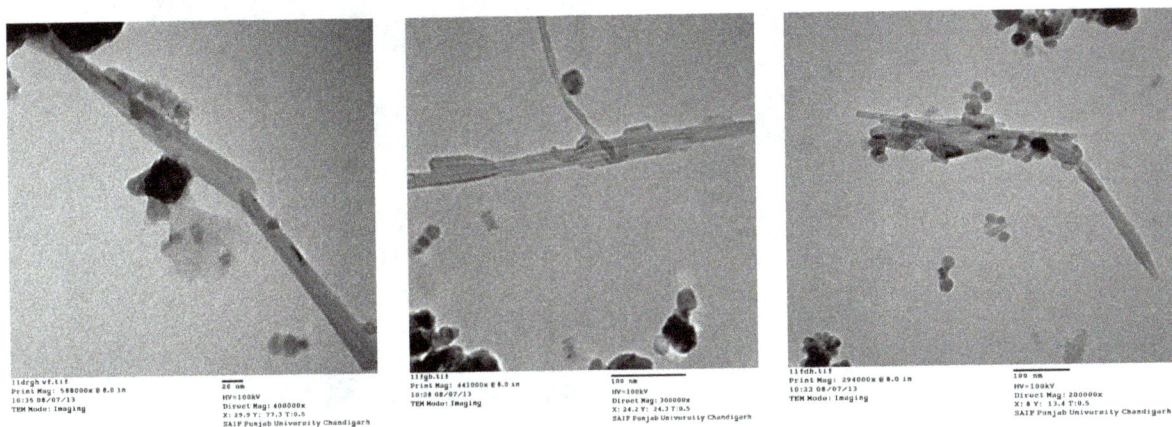

Figure 7. TEM images of CNTs post annealing at 400°C for 30 minutes, collected from different points of focus.

Table 2. Calculated thickness (particle size, t) of impurities from XRD data pre and post annealing, referring to **Figure 3**, and applying the Scherrer's formula.

For the as Prepared CNTs					
2θ (deg.)	28.136	46.936	55.617	68.198	76.128
$\Delta 2\theta$ (deg.)	1.46519	0.55271	2.34239	7.04736	4.95001
t (A°)	55.8742	156.6317	38.3281	13.6078	20.3761
For CNTs annealed at 400°C					
2θ (deg.)	28.16245	46.93296	55.66761	68.48652	75.96305
$\Delta 2\theta$ (deg.)	1.18242	0.55498	1.38371	4.72368	4.95001
t (A°)	69.2402	155.9893	64.8982	20.3364	20.3761
For CNTs annealed at 800°C					
2θ (deg.)	28.36802	47.09762	55.68	68.76009	75.96692
$\Delta 2\theta$ (deg.)	0.16689	0.19188	0.33996	0.72157	2.51296
t (A°)	490.7903	151.4543	264.3844	133.3474	40.0926

viously obtained results [25]. According to the TEM images the CNTs are extremely clean and condensed in comparison to the lower temperature annealed samples shown in **Figure 6** and **Figure 7**. There were large amounts of heavy sheets of graphene layers, amorphous carbon and/or other impurities and perhaps broken CNTs produced, as seen in **Figures 8(b)-8(d)**. These implies that some of the CNTs might have burned and are unstable at higher temperatures [36], suggesting reduction of the annealing temperature and/or time could resolve the scenario of obtaining high yield. Moreover, the amount of impurities scattered around the CNTs shown in **Figure 6** are known to decrease with further annealing and the CNTs are progressively freed from those surface bonded agents. These CNTs are found in bundle rather than scattered as the graphene layers and/or amorphous carbon does as well. An interesting feature is, therefore, the separation of the system into different category. The CNTs bundled as in **Figure 8(a)**; graphene sheets of edge thickness about 20 nm forced to be arranged in

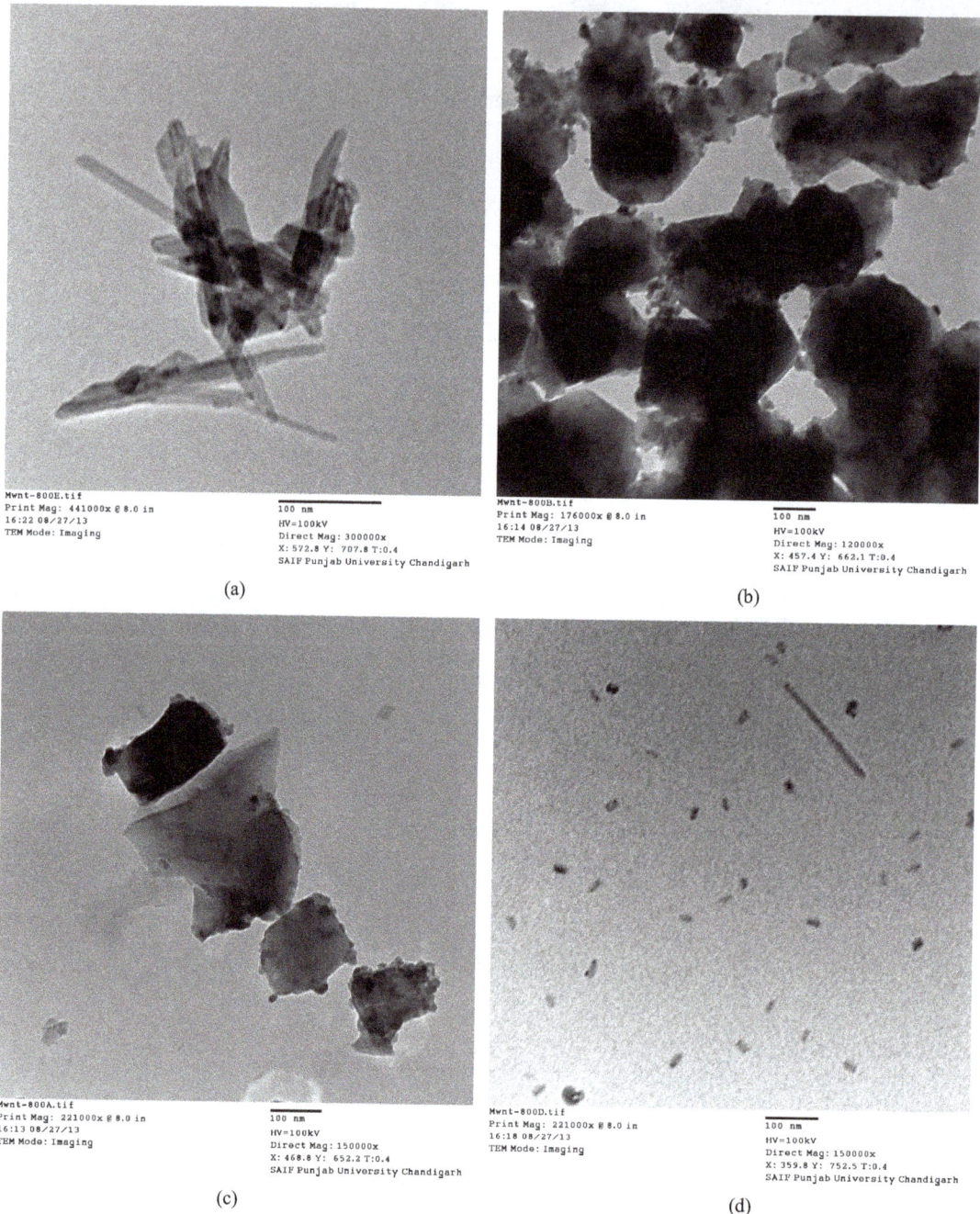

(a)

(b)

(c)

(d)

Figure 8. TEM images CNTs post annealing at 800˚C for 30 minutes.

order, as in **Figure 8(b)**; micron sized amorphous system gathered and attached to one another as in **Figure 8(c)**; and black spotted objects, thought as broken CNTs, of size 5 nm - 30 nm in different category as in **Figure 8(d)**. These indicate that during further annealing structures of the same category get sorted and move to one side assisting enhancement of the purification process.

4. Conclusion

The prepared and purified CNTs in a deionized water environment are understood to have the same feature as those prepared in N_2, H_2, He or their mixture. The challenging scenario was finding the source of extra peaks observed during XRD analysis, which could have been resolved by EDX analysis that clearly shown that Ca, O and/or their compound existed in the sample. According to the TEM images, the CNTs are identified as to self-assemble further being thermally agitated with increase in temperature. The increase in I_D/I_G ratio can be attributed as indicator for structural defects due to annealing the CNTs and presence of impurities. It is also understood that elemental analysis should be done for the contents of the sample source (graphene) prior to any course of action in order to easily manage experimental procedures and fix treatment technique for the removal of defects. Finally, the water environment preparation of CNTs is understood as the most cost effective mechanism relative to those has been in use. Thermal agitation is also known to self-categorizing constituents in the sample; signifying complete removal of defects requires additional treatment technique.

Acknowledgements

We acknowledge the financial support from the C. V. Raman fellow ship for African researchers and Mr. Charanjit Singh, department of chemistry Panjab University, for his invaluable assistance.

References

[1] Iijima, S. (1991) Helical Microtubules of Graphitic Carbon. *Nature*, **354**, 56-58. http://dx.doi.org/10.1038/354056a0

[2] Iijima, S. and Ichihashi, T. (1993) Single-Shell Nanotubes of 1-nm Diameter. *Nature*, **363**, 603-605. http://dx.doi.org/10.1038/363603a0

[3] Feng, D. (2005) Theoretical Study of the Stability of Defects in Single-Walled Carbon Nanotubes as a Function of Their Distance from the Nanotube End. *Physical Review B*, **72**, 1-7.

[4] Bethune, D.S., Kiang, C.H., de Vries, M.S., Gorman, G., Savoy, R., Vasques, J. and Beyers, R. (1993) Cobalt-Catalyzed Growth of Carbon Nanotubes with Single-Atomic-Layer Walls. *Nature*, **363**, 605-607. http://dx.doi.org/10.1038/363605a0

[5] Yuhuang, W., Myung, J.K., Hongwei, S., Carter, K., Hua, F., Lars, M.E., Wen-Fang, H., Sivaram, A., Robert, H.H. and Richard, E.S. (2005) Continued Growth of Single-Walled Carbon Nanotubes. *Nano Letters*, **5**, 997-1002. http://dx.doi.org/10.1021/nl047851f

[6] Journet, C. and Bernier, P. (1998) Production of Carbon Nanotubes. *Applied Physics A*, **67**, 1-9. http://dx.doi.org/10.1007/s003390050731

[7] Bellucci, S., Gaggiotti, G., Marchetti, M., Micciulla, F., Mucciato, R. and Regi, M. (2007) Atomic Force Microscopy Characterization of Carbon Nanotubes. *Journal of Physics: Conference Series*, **61**, 99-104.

[8] Yoshinori, A., Xinluo, Z., Sakae, I. and Iijimaa, S. (2002) Mass Production of Multiwalled Carbon Nanotubes by Hydrogen Arc Discharge. *Journal of Crystal Growth*, **237-239**, 1926-1930. http://dx.doi.org/10.1016/S0022-0248(01)02248-5

[9] Shi, Z., Lian, Y., Zhou, X., Gu, Z., Zhang, Y., Iijima, S., Zhou, L., Yue, T.K. and Zhang, S. (1999) Mass Production of Single-Wall Carbon Nanotubes by Arc Discharge Method. *Carbon*, **37**, 1449-1453. http://dx.doi.org/10.1016/S0008-6223(99)00007-X

[10] Stancu, M., Ruxanda, G., Ciuparu, D. and Dinescu, A. (2011) Purification of Multiwall Carbon Nanotubes Obtained by AC Arc Discharge Method. *Optoelectronics and Advanced Materials*, **R5**, 846-850.

[11] Hamada, N., Sawada, S. and Oshiyama, A. (1992) New One-Dimensional Conductors: Graphitic Microtubules. *Physical Review Letters*, **68**, 1579-1581. http://dx.doi.org/10.1103/PhysRevLett.68.1579

[12] Tans, S.J., Devoret, M.H., Dai, H., Thess, A., Smalley, R.E., Georliga, L.J. and Dekker, C. (1997) Individual Single-Wall Carbon Nanotubes as Quantum Wires. *Nature*, **386**, 474-477. http://dx.doi.org/10.1038/386474a0

[13] Tans, S.J., Verschueren, R.M. and Dekker, C. (1998) Room Temperature Transistor Based on a Single Carbon Nanotube. *Nature*, **393**, 49-52. http://dx.doi.org/10.1038/29954

[14] McEuen, P.L., Fuhrer, M.S. and Park, H. (2002) Single-Walled Carbon Nanotube Electronics. *IEEE Transitions on Nanotechnology*, **1**, 78-85. http://dx.doi.org/10.1109/TNANO.2002.1005429

[15] Garau, C., Frontera, A., Quinonero, D., Costa, A., Ballester, P. and Dey, P.M. (2003) Lithium Diffusion in Single-Walled Carbon Nanotubes: A Theoretical Study. *Chemical Physics Letters*, **374**, 548-555. http://dx.doi.org/10.1016/S0009-2614(03)00748-6

[16] de Heer, W.A., Chatelain, A. and Ugarte, D. (1995) A Carbon Nanotube Field-Emission Electron Source. *Science*, **270**, 1179-1180. http://dx.doi.org/10.1126/science.270.5239.1179

[17] Jensen, A., Hauptmann, J.R., Nygård, J., Sadowski, J. and Lindelof, P.E. (2004) Hybrid Devices from Single Wall Carbon Nanotubes Epitaxially Grown into a Semiconductor Heterostructure. *Nano Letters*, **4**, 349-352. http://dx.doi.org/10.1021/nl0350027

[18] Martel, R., Schmidt, T., Shea, H.R., Hertel, T. and Avouris, P. (1998) Single- and Multi-Wall Carbon Nanotube Field-Effect Transistors. *Applied Physics Letters*, **73**, 2447-2449. http://dx.doi.org/10.1063/1.122477

[19] Tans, S.J., Verschueren, A.R.M. and Dekker, C. (1998) Room-Temperature Transistor Based on a Single Carbon Nanotube. *Nature*, **393**, 49-52. http://dx.doi.org/10.1038/29954

[20] Alexander, A.K., Sergey, B. Lee, M.Z., Baughman, R.H. and Zakhidov, A.A. (2010) Electron Field Emission from Transparent Multiwalled Carbon Nanotube Sheets for Inverted Field Emission Displays. *Carbon*, **48**, 41-46. http://dx.doi.org/10.1016/j.carbon.2009.08.009

[21] Chai, S.P., Zein, S.H.S. and Mohamed, A.R. (2004) A Review on Carbon Nanotubes Production via Catalytic Methane Decomposition. 1st National Postgraduate Colloquium School of Chemical Engineering USM NAPCOl, 60-69.

[22] Huang, S., Cai, X. and Liu, J. (2003) Growth of Millimeter-Long and Horizontally Aligned Single-Walled Carbon Nanotubes on Flat Substrates. *Journal of the American Chemical Society*, **125**, 5636-5637. http://dx.doi.org/10.1021/ja034475c

[23] Zhang, H., Fu, X., Yin, J., Zhou, C., Chen, Y., Li, M. and Wei, A. (2005) The Effects of MWNTs with Different Diameters on the Electrochemical Hydrogen Storage Capability. *Physics Letters A*, **339**, 370-377. http://dx.doi.org/10.1016/j.physleta.2005.03.013

[24] Mahanandia, P., Schneider, J.J., Engel, M., Stühn, B., Subramanyam, S.V. and Nanda, K.K. (2011) Studies towards Synthesis, Evolution and Alignment Characteristics of Dense, Millimeter Long Multiwalled Carbon Nanotube Arrays, Beilstein. *Journal of Nanotechnology*, **2**, 293-301. http://dx.doi.org/10.3762/bjnano.2.34

[25] Grobert, N. (2007) Carbon Nanotubes Becoming Clean. *Materials Today*, **10**, 28-35. http://dx.doi.org/10.1016/S1369-7021(06)71789-8

[26] Zhao, X. and Ando, Y. (1998) Raman Spectra and X-Ray Diffraction Patterns of Carbon Nanotubes Prepared by Hydrogen Arc Discharge. *Japanese Journal of Applied Physics*, **37**, 4846-4849.

[27] Iqbal, M.W., Singh, A.K., Iqbal, M.Z. and Eom, J. (2012) Raman Fingerprint of Doping Due to Metal Adsorbates on Graphene. *Journal of Physics Condensed Matter*, **24**, Article ID: 335301.

[28] Jeong, Y., Kim, J. and Lee, G.W. (2010) Optimizing Functionalization of Multiwalled Carbon Nanotubes Using Sodium Lignosulfonate. *Colloid and Polymer Science*, **288**, 1-6. http://dx.doi.org/10.1007/s00396-009-2127-8

[29] Dresselhaus, M.S., Rao, A.M. and Dresselhaus, G. (2004) Raman Spectroscopy in Carbon Nanotubes. *Encyclopedia of Nanoscience and Nanotechnology*, **9**, 307-338.

[30] Li, H., He, X., Kang, Z., Huang, H., Liu, Y., Liu, J., Lian, S., Tsang, C.H.A., Yang, X. and Lee, S.-T. (2010) Water-Soluble Fluorescent Carbon Quantum Dots and Photocatalyst Design. *Angewandte Chemie International Edition*, **49**, 4430-4434.

[31] Akhavan, O. (2011) Photocatalytic Reduction of Graphene Oxides Hybridized by ZnO Nanoparticles in Ethanol. *Carbon*, **49**, 11-18. http://dx.doi.org/10.1016/j.carbon.2010.08.030

[32] Caoa, A., Xua, C., Lianga, J., Wu, D. and Wei, B. (2001) X-Ray Diffraction Characterization on the Alignment Degree of Carbon Nanotubes. *Chemical Physics Letters*, **344**, 13-17. http://dx.doi.org/10.1016/S0009-2614(01)00671-6

[33] Khani, H. and Moradi, O. (2013) Influence of Surface Oxidation on the Morphological and Crystallographic Structure of Multi-Walled Carbon Nanotubes via Different Oxidants. *Journal of Nanostructure in Chemistry*, **3**, 73.

[34] Wang, Z., Ba, D., Liu, F., Cao, P., Yang, T., Gu, Y. and Gao, H. (2005) Synthesis and Characterization of Large Area Well-Aligned Carbon Nanotubes by ECR-CVD without Substrate Bias. *Vacuum*, **77**, 139-144. http://dx.doi.org/10.1016/j.vacuum.2004.08.012

[35] Scherrer, P. (1918) Bestimmung der Größe und der innerenStruktur von Kolloidteilchen Mittels Röntgenstrahlen. P, Nachrichten von der Gesellschaft der Wissenschaften, Gttingen. *Mathematisch-Physikalische Klasse*, **2**, 98-100.

[36] Ajayan, P.M. (1999) Nanotubes from Carbon. *Chemical Reviews*, **99**, 1787-1799. http://dx.doi.org/10.1021/cr970102g

5

Silica-Based Nanocoating Doped by Layered Double Hydroxides to Enhance the Paperboard Barrier Properties

Vânia M. Dias[1], Alena Kuznetsova[2], João Tedim[2], Aleksey A. Yaremchenko[2],
Mikhail L. Zheludkevich[2], Inês Portugal[1], Dmitry V. Evtuguin[1*]

[1]CICECO/Department of Chemistry, University of Aveiro, Aveiro, Portugal
[2]CICECO/Department of Materials and Ceramic Engineering, University of Aveiro, Aveiro, Portugal
Email: *dmitrye@ua.pt

Abstract

Paperboard is an environment-friendly multi-layer material widely used for packaging applications. However, for food packaging paperboard lacks essential barrier properties towards oxygen and water vapor. Conventional solutions to enhance these barrier properties (e.g. paperboard film coating with synthetic polymers) require special manufacturing facilities and difficult the end-of-life disposal and recycling of the paperboard. Paperboard coating with silica-based formulations is an eco-friendly alternative hereby disclosed. Silica-nanocoatings were prepared by sol-gel synthesis, with or without the addition of Zn(2)-Al-NO$_3$ layered double hydroxides (LDHs), and applied on the surface (ca 2 g/m^2) of industrial paperboard samples by a roll-to-roll technique. The physicochemical features of silica-nanocoatings were studied by FTIR-ATR, SEM/EDS, XRD analysis and surface energy measurements. The barrier properties of uncoated and silica-coated paperboard were accessed by water vapor transmission rate (WVTR) and oxygen permeability (J_{o2}) measurements. The best barrier results were obtained for paperboard coated with a mixture of tetraethoxysilane (TEOS) and 3-aminopropyltriethoxysilane (APTES), with and without the incorporation of LDHs.

Keywords

Paperboard, Silica-Based Formulations, Layered Double Hydroxides, Sol-Gel Synthesis, Barrier Properties

---

*Corresponding author.

1. Introduction

Paperboard is a thick paper-based multi-layer material with numerous applications due to its structural stability, lightweight, printable surface, attractive shelf-display and ease of recovery for recycling or end-of-life disposal. However, in comparison with traditional glass and plastic containers paperboard is unsuitable for some food-packaging applications due to poorer barrier protection against grease, water vapor, oxygen, odors and other substances. Conventional solutions to improve these barrier properties include paperboard wax impregnation or surface coating with petroleum-derived polymer films [1]-[4]. Since both strategies reduce the recyclability and biodegradability features of paperboard in recent years several biopolymers (e.g. starch, cellulose, chitosan, pectin, xylan, isolated soy protein, pullulan among others) have been screened as alternative sustainable coatings for packaging applications [5]-[11]. In addition, nano-sized fillers such as silicate, titanium dioxide and clays, can be added to reinforce the mechanical, thermal and barrier properties of the biopolymers [6] [11]. Alternatively, inorganic nanoparticles can be synthesized *in situ* in the matrix of the biopolymer, thus forming organic-inorganic hybrid coatings [10]-[12].

The sol-gel process is widely used to produce high purity organic-inorganic composite materials (derived from metals, glass, and ceramics, among others) with excellent mechanical strength and thermal/chemical stability [13]-[15]. The method involves a series of hydrolysis and polycondensation reactions of organometallic precursors (metal alkoxides) that generate nanoparticles randomly dispersed in a polymeric matrix [16]-[18]. Sol-gel techniques have been used in paper-science for conservation of historic documents, to enhance paper printing quality, and to improve the hydrophobicity of paper [2]. In fact, it has been revealed that silica nanoparticles dispersed on a paper surface form an impermeable interface between a three-dimensional silica network and cellulose fibres [16]-[19]. Moreover, the hydrophobicity of silica-modified surfaces can be enhanced by the incorporation of long-chain aliphatic groupsin the silica network, for example by using a main silica precursor (e.g. tetraethoxysilane) mixed with a secondary organically modified alkoxysilane [19]-[23]. In particular, the use of alkoxysilane bearing amine groups enables the anchoring of specific substances (e.g. catalytic species, sensors or scavengers) that extend the range of application of cellulose-based materials [24] [25].

Layered double hydroxides (LDHs) are plate-like materials possessing high aspect-ratio and ion-exchange capacity [26]-[28]. The incorporation of LDHs in polymeric matrices has high potential to enhance the polymer's barrier properties (the high aspect-ratio increases the length and tortuosity of the diffusion path) without sacrificing the inherent processability and mechanical properties of the pure polymer [10] [26] [27] [29]. The structure of LDHs is based on the brucite structure ($Mg(OH)_2$), where some Mg^{2+} cations are partially substituted by trivalent cations inducing a net positive charge in the hydroxide sheets, balanced by the presence of exchangeable anions and water molecules in the interlayer galleries [26] [29]. The anion-exchange capacity of LDHs, allied to the high aspect ratio and good interfacial interactions with polymer matrices are very appealing for diverse applications such as corrosion protection coatings [27] [30] and polymeric-nanocomposites for food packaging materials [28] [31]. Nevertheless the benefits of LDHs materials for the modification of paperboard packaging materials have not been reported before.

The main goal of this work was to study the impact of silica-based nanocoating, with and without the incorporation of LDHs, on the water vapor and oxygen barrier properties of paperboard.

2. Materials and Methods

2.1. Materials

Industrial samples of paperboard (210 g/m^2 and 0.29 - 0.30 mm thickness) with a conventional surface treatment applied on the front side (30 - 35 g/m^2 of a slurry of $CaCO_3$ and Al_2O_3 particles mixed in a synthetic resin emulsion) were supplied by Prado Karton S.A. (Tomar-Portugal). Hereafter this material will be designated as "uncoated" paperboard.

Silica precursors namely tetraethoxysilane (TEOS), diethoxydimethylsilane (DEDMS), 3-aminopropyltriethoxysilane (APTES), and octyltriethoxysilane (OTES) all with 96% - 98% purity, nitric acid (HNO$_3$, p.a. 65% grade) and a low-molecular-weight grade polyethylene glycol (PEG-400) were supplied by Sigma-Aldrich Chem. Co. Precipitated calcium carbonate (PCC) was supplied by the industrial group Portucel-Soporcel (Portugal). Zinc nitrate hexahydrate (Zn(NO$_3$)$_2$·6H$_2$O, ≥99%), aluminum nitrate nonahydrate (Al(NO$_3$)$_3$·9H$_2$O, ≥98.5%), sodium hydroxide (NaOH, ≥98%) and sodium nitrate (NaNO$_3$, ≥99.5%) were supplied by Sigma-Aldrich Chem. Co. and used as received for the synthesis of LDHs.

2.2. Preparation of Silica-Based Formulations and Paperboard Coating

Silica-based formulations were prepared by sol-gel synthesis [17] [18] using a mineral acid catalyst (HNO_3) for the hydrolysis of TEOS at pH 1.2—Equation (1). Typically, a mixture of TEOS (0.5 - 1.0 mol), water (0.5 - 2.0 mol), HNO_3 (50 - 300 µl) and PEG-400 (250 - 500 µl) was allowed to react at room temperature (~20˚C) under mechanical stirring (250 rpm) during 40 min. Afterwards, PCC (30 - 50 mg) was added to raise the pH above the isoelectric point of silica (pH 3 - 4) in order to promote the condensation reactions—Equation (2). Some formulations were prepared in a similar way but using TEOS (95% v/v) and a secondary silica precursor (5% v/v) such as DEDMS, APTES or OTES in the hydrolysis step.

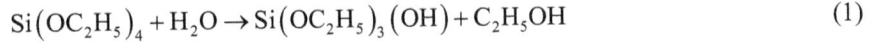

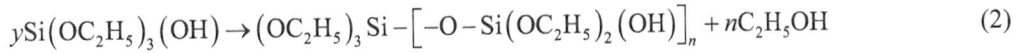

$$Si(OC_2H_5)_4 + H_2O \rightarrow Si(OC_2H_5)_3(OH) + C_2H_5OH \tag{1}$$

$$ySi(OC_2H_5)_3(OH) \rightarrow (OC_2H_5)_3 Si-\left[-O-Si(OC_2H_5)_2(OH)\right]_n + nC_2H_5OH \tag{2}$$

Silica formulations were applied on the front side (treated surface) of the "uncoated" paperboard surface by roll-to-roll technique using a Mathis LAB reverse roll coater type RRC-BW 350 mm pilot size-press, at fixed coating speed (20 m/min). The distance between the cylinders was adjusted to load a silica formulation on the paperboard surface within the range 2 - 3 g/m^2. Coatings were cured by infrared-heating (ca. 105˚C, 5 min).

2.3. Preparation of Silica-Based Formulations Containing LDHs

Zn-Al layered double hydroxides intercalated with nitrates(Zn(2)-Al-NO_3 LDHs) were synthesized by the co-precipitation method [30] using decarbonized (boiled and cooled under a flow of nitrogen gas) distilled water in the preparation of solutions.A mixed solution of zinc/aluminum nitrate (Zn^{2+} 0.05 M, Al^{3+} 0.025 M) was added-dropwise to an aqueous solution of $NaNO_3$ (1.5 M) under vigorous stirring and nitrogen atmosphere, at pH 9 - 10 (controlled by the addition of an aqueous solution of NaOH (2.0 M)). At the end of the reaction the resulting suspension was treated hydrothermally (100˚C during 4 hours) to promote crystal growth. After cooling down to room temperature, LDHs crystals were recovered by centrifugation (10000 r.p.m., 5 min), washed several times with decarbonized distilled water and stored in a closed container in the form of a slurry.

When required, LDHs were added to the silica-based formulations immediately after the condensation reaction and applied on the paperboard surface as previously described (2.2).

2.4. Analyses of Paperboard Surfaces

Contact angles (θ) of uncoated and coated paperboard surfaces were measured at room temperature (23˚C - 25˚C) on a DataPhysics Instrument OCA20, using the sessile drop method (drop volume ca 2.0 µl) and three different probe-liquids namely diiodomethane (Aldrich, 99% purity GC), formamide (Sigma, 99% purity GC) and distilled water (MilliQ grade).

The Owens-Wendt-Rable-Kaeble (OWRK) model [32] was used to assess the total surface energy (γ_s) of paperboard surfaces, and the corresponding polar (γ_s^p) and dispersive (γ_s^d) components, using measured contact angles (θ), liquid-probe surface tension data (**Table 1**) and Equations (3) and (4) [18] [32] [33].

$$\gamma_s = \gamma_s^d + \gamma_s^p \tag{3}$$

$$\left((1+\cos\theta)/2\right) \times \left(\gamma_l / (\gamma_l^d)^{1/2}\right) = (\gamma_s^p)^{1/2} \times (\gamma_l^p/\gamma_l^d)^{1/2} + (\gamma_s^d)^{1/2} \tag{4}$$

where, γ_l is the total surface tension of the liquid, and γ_l^p and γ_l^d are the polar and dispersive components of the surface tension, respectively. Plotting the right-hand-side of Equation (4) as a function of $(\gamma_l^p/\gamma_l^d)^{1/2}$ enables the calculation of γ_s^p and γ_s^d from the parameters of the linear regression (Equation (4)). The wetting envelopes of the paperboard surfaces were obtained by representing γ_s^p versus γ_s^d at a fixed contact angle (θ = 90˚) [34].

Silica-coated paperboard samples were characterized by Fourier Transform Infra-red spectroscopy with Attenuated Total Reflectance (FTIR_ATR) analysis with the spectra being recorded in absorbance mode in the range 4000 - 400 cm^{-1}, co-adding 200 scans at 8 cm^{-1} resolution (Mattson FT-IR spectrometer Model 7000 equipped

Table 1. Surface tension and corresponding polar and dispersive components of the standard liquids used for surface energy calculations [32] [33].

Liquid	Surface tension, γ_l (mJ/m^2)	Polar component, γ_l^p (mJ/m^2)	Dispersive component, γ_l^d (mJ/m^2)
Water	72.8	51.0	21.8
Formamide	58.0	20.4	37.6
Diiodomethane	50.8	2.3	48.5

with Golden Gate diamond ATR cell). Penetration of the silica formulations was evaluated by Scanning Electron Microscopy/Energy Dispersive Spectroscopy (SEM/EDS) analysis (SEM Hitachi SU-70 microscope operating at 15 kV; EDS Brucker Quantax 400 detector) of transversal cuts of paperboard.

The structure of LDHs in the silica formulations was analyzed by X-ray diffraction (XRD) using a Philips X'Pert MPD diffractometer (Bragg-Brentano geometry, Cu Kα radiation, and the exposition corresponded to 5 s per step of 0.02° over the angular range $4 < 2\theta < 65°$).

Physical and mechanical properties of paperboard samples were evaluated by standardized tests: tensile strength (N/m) was determined according to NP EN ISO 1924-2 and used to evaluate the Tensile Index (*i.e.* tensile strength divided by grammage (g/m^2)); burst strength (kPa) was determined by NP EN ISO 2758; Bendtsen roughness (cm^3/min) and porosity were both determined following NP EN ISO 8791; air permeation was measured by the Gurley method (NP 795 ISO 5636-5).

2.5. Paperboard Barrier Properties

Water vapor barrier properties of paperboard surfaces were measured in triplicate, for each sample, using standard procedures (ASTM E96-95) and the "desiccant method" [35]. Accordingly, circular paperboard samples were sealed to the open mouth of test-cups (area 19.6 cm^2; internal depth 2.0 cm) containing a desiccant, namely anhydrous calcium chloride (0% relative humidity (RH)) previously dried at 200°C (2 h), and weighed. Afterwards, the sealed test-cups were placed in a test chamber with forced air circulation, at 26°C ± 1°C and 52% RH (obtained with a saturated aqueous solution of magnesium nitrate hexahydrate [35]), and weighed periodically over a three day period. The amount of water transferred through the paperboard sample, *i.e.* the test-cup weight gain ($\Delta W = W_t - W_{t=0}$), along time (*t*) was used to evaluate the water vapor transmission rate (WVTR, g·m^{-2}·day^{-1}) from the linear representation of the experimental data, as expressed by Equation (5)

$$\Delta W = A \times \text{WVTR} \times t \tag{5}$$

where A is the exposed area of the paperboard sample (19.6 cm^2).

Oxygen permeation through uncoated and silica-coated paperboard samples was measured by a proven technique used for the characterization of glass-ceramic and ceramic dense membranes [36]-[38]. The experimental setup comprised one sample holder (alumina tube) and two electrochemical yttrium-stabilized zirconia (YSZ) oxygen sensors. The paperboard sample was hermetically sealed on the top-opening of the alumina tube with cyanoacrylate adhesive. The outside of the paperboard sample was exposed to air while the inner surface was swept by a nitrogen flow (50 cm^3·min^{-1}, with oxygen partial pressure ca. 2 Pa). Oxygen partial pressure in the sweep gas flow at the inlet ($P(O_2)_{in}$) and outlet ($P(O_2)_{out}$) of the sample holder under steady-state conditions was determined using YSZ sensors. In the course of experiment, the paperboard sample was at room temperature and under zero total pressure gradients. Specific oxygen permeation fluxes (J_{O_2} , m^3·m^{-2}·day^{-1}) were calculated using Equation (6)

$$J_{O_2} = \left(F_{N_2}/S\right) \times \left[P\left(O_2\right)_{out} - P\left(O_2\right)_{in}\right]/P \tag{6}$$

where F_{N_2} is the inlet nitrogen flow rate (50 cm^3·min^{-1}), P is the total pressure (Pa) and S is the sample's exposed surface area (0.785 cm^2).

3. Results and Discussion

3.1. Paperboard Coating with Silica-Based Formulations

Uncoated paperboard was supplied with a conventional surface treatment applied on the front side to improve surface smoothness and opacity (the $CaCO_3$ and Al_2O_3 white particles mixed in a synthetic resin mask the darker internal layers of paperboards). It is presumed that silica based formulations applied in small amounts (2 - 3 g/m^2) do not change the general appearance of the paperboard surface, but hopefully improve their barrier properties. This is confirmed by the oxygen and water vapor barrier properties results presented (qualitatively) in **Table 2** for paperboard coated with silica formulations prepared using various precursors (alkoxysilanes) with and without the incorporation of LDHs. Moreover, these results indicate that the barrier properties are influenced by the nature of the functional groups introduced in the silica network. By means of example, the TEOS_APTES silica network without (III) and with (IIIa) LDHs is schematically represented in **Figure 1**.

The SEM-EDS images of uncoated and silica-coated paperboard are presented in **Figure 2** to analyze the surface morphology and the penetration of silica (represented in red colour). Uncoated paperboard presents a smooth surface (**Figure 2**, front side) whereas silica-coated paperboard exhibit some cracks more pronounced for the TEOS_OTES coating (IV)than for the TEOS_DEDMS (II) and TEOS_APTES coating (III) (**Figure 2**, front side and cross-section) that apparently influence the barrier properties (to be discussed further on). Besides, distribution of silica appears to be more homogeneous for coating (IV) than for coating (II) and (III). In all cases, silica penetrates deep inside the paperboard layers (**Figure 2**, cross-section) due to the relatively low viscosity of the formulations, even though most part of the silica is retained on the treated surface. The incorporation of

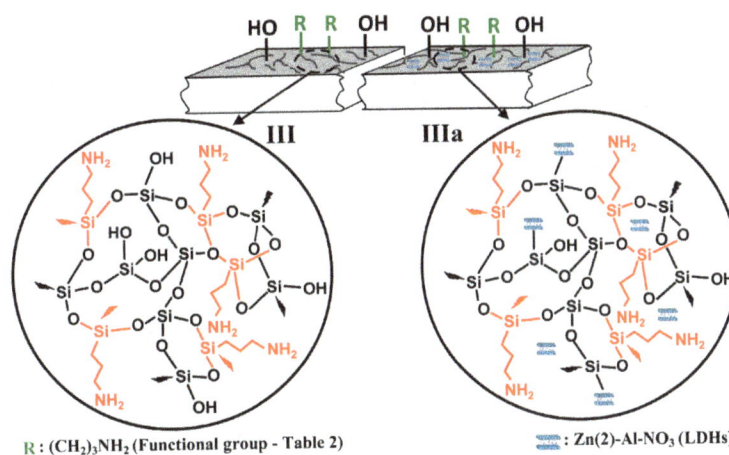

R : $(CH_2)_3NH_2$ (Functional group - Table 2)   : Zn(2)-Al-NO_3 (LDHs)

Figure 1. Schematic representation of the silica-network formed on the paperboard surface upon coating with TEOS_APTES formulation with (IIIa) and without (III) incorporation of LDHs.

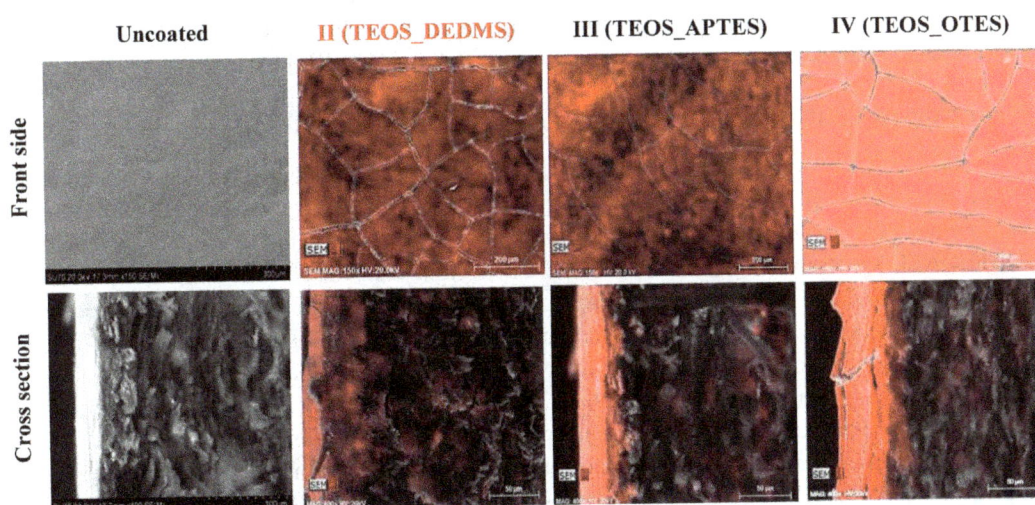

| Uncoated | II (TEOS_DEDMS) | III (TEOS_APTES) | IV (TEOS_OTES) |

Figure 2. SEM-EDS images (front side and cross section) of paperboard before (uncoated) and after coating with TEOS_DEDMS (II),TEOS_APTES (III) or TEOS_OTES (IV) formulations applied on the front side of the paperboardsurface (located on the left side of the images).

Table 2. Barrier properties (qualitative) for uncoated and silica coated paperboard.

				Barrier property	
	Functional group (R)	Cracks ∿		H_2O	O_2
	Uncoated	------------	--------	+	+++
I	TEOS	-OH	+++	++	+++++
Ia	TEOS_LDHs			+++	+++++
II	TEOS_DEMDS	-CH₃	++	++++	++++++
IIa	TEOS_DEDMS_LDHs			+++++	+++++
III	TEOS_APTES	-(CH₂)₃NH₂	+	++++	++++++
IIIa	TEOS_APTES_LDHs			++++++	+++++
IV	TEOS_OTES	-(C₇H₁₄)CH₃	++++	++++	++
IVa	TEOS_OTES_LDHs			++++++	+

LDHs has no significant effect on silica distribution (SEM images are similar with and without LDHs) neither on cracks formed on the paperboard surface (**Table 2**).

The chemical composition and uniformity of the silica-based coatings was further accessed by FTIR_ATR analysis (**Figure 3**). In fact, uncoated paperboard presents a sharp band at 1385 - 1400 cm^{-1} (assigned to the presence of nitrate anions [39] introduced in the industrial paperboard treatment process) almost absent in the spectra of coated paperboards. Since this technique examines only the surface of the sample one may conclude that paperboard was completely covered by the silica coatings that turn nitrate anions inaccessible towards FTIR-ATR analysis.

The most significant peaks in the spectra of silica-coated paperboards (**Figure 3(a)**) have been assigned to silanol groups (Si-OH stretching at 950 cm^{-1}) and silica oxide moieties (e.g. Si-O-Si bending at 450 cm^{-1}; asymmetric and symmetric Si-O-Si stretching at 1050 cm^{-1} and 800 cm^{-1}, respectively) thus confirming the occurrence of hydrolysis and condensation reactions of the alkoxysilanes [17] [40]. In addition, the use of silica co-precursors led to the appearance of characteristic small peaks (almost unnoticeable) in the FTIR-ATR spectra. For instance, the small peak at 1265 cm^{-1} in the spectrum of TEOS_DEDMS (II) is attributed to the presence of CH_3 groups [22]; the absorption band at 2930 cm^{-1} in the TEOS_OTES spectrum (IV) corresponds to the asymmetric stretching vibrations of C–H bonds [21]; in the TEOS_APTES spectrum (III) the band at 1548 cm^{-1} is attributed to N–H vibrations of amino groups and the bands at 2929 cm^{-1} and 1461 cm^{-1} are assigned respectively to stretching and bending vibrations of aminopropyl CH_2 groups [23] [41].

3.2. Paperboard Coating with Silica-Based Formulations and LDHs

Zn(2)-Al-NO_3 type LDHs are expected to act as barrier promoters when added to silica-coated paperboard due to the high anion-exchange capacity of these nanomaterials [27]. The incorporation of LDHs in the silica coating without their degradation or loss of structural integrity was examined by FTIR-ATR and XRD analysis.

The FTIR-ATR spectra of silica-coated paperboard without and with LDHs (**Figure 3(a)** and **Figure 3(b)**, respectively) are quite similar except for the peak in the region 1350 - 1385 cm^{-1}, superposed with a shoulder near 1400 cm^{-1}. The presence of these bands, assigned to symmetric and asymmetric stretching modes of nitrate anions [30], undoubtedly confirm the incorporation of LDHs in the nanocoating.

The XRD patterns presented in **Figure 4** indicate that Zn(2)-Al-NO_3 LDHs particles (obtained by co-precipitation) are single-phase materials displaying characteristic reflection peaks at low 2θ angles $\approx$ 10°, 20° and 30° (ascribed to (00l) reflections) and a basal plane spacing (d, calculated from the position of (00l) reflections) in the range 0.89 - 0.90 nm. These results are in agreement with literature data for Zn(2)-Al LDHs intercalated with NO_3^- [30] [42]. Furthermore, the fundamental reflections associated with LDHs are still visible when LDHs are incorporated into silica formulations, thereby confirming the successful preparation of silica nanocoating with LDHs particles.

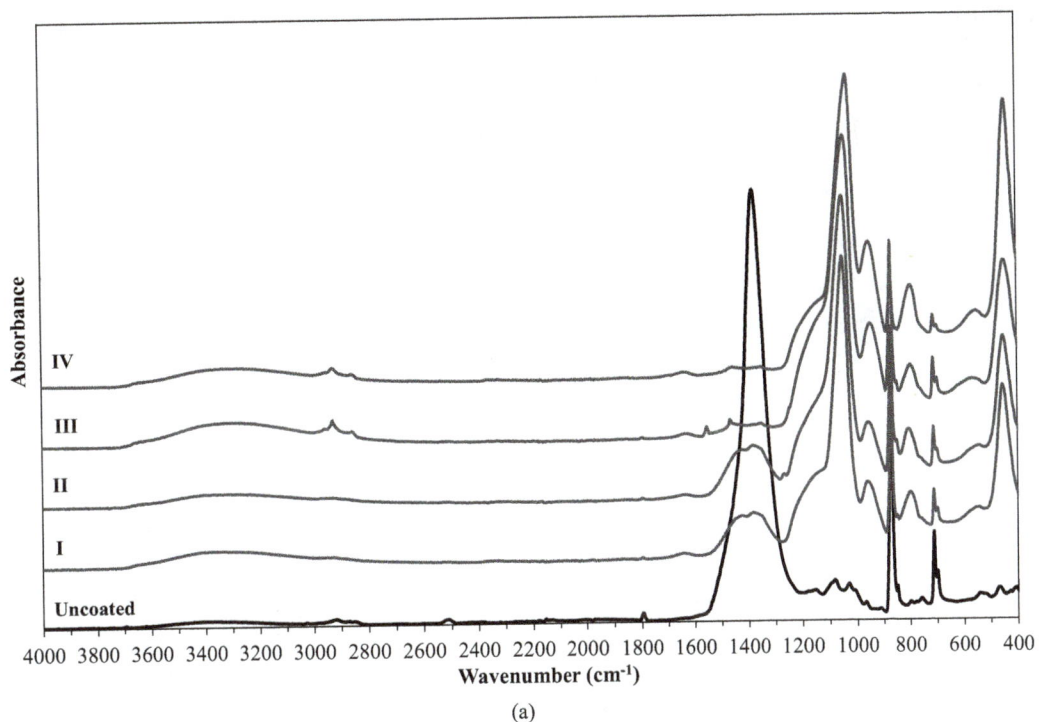

(a)

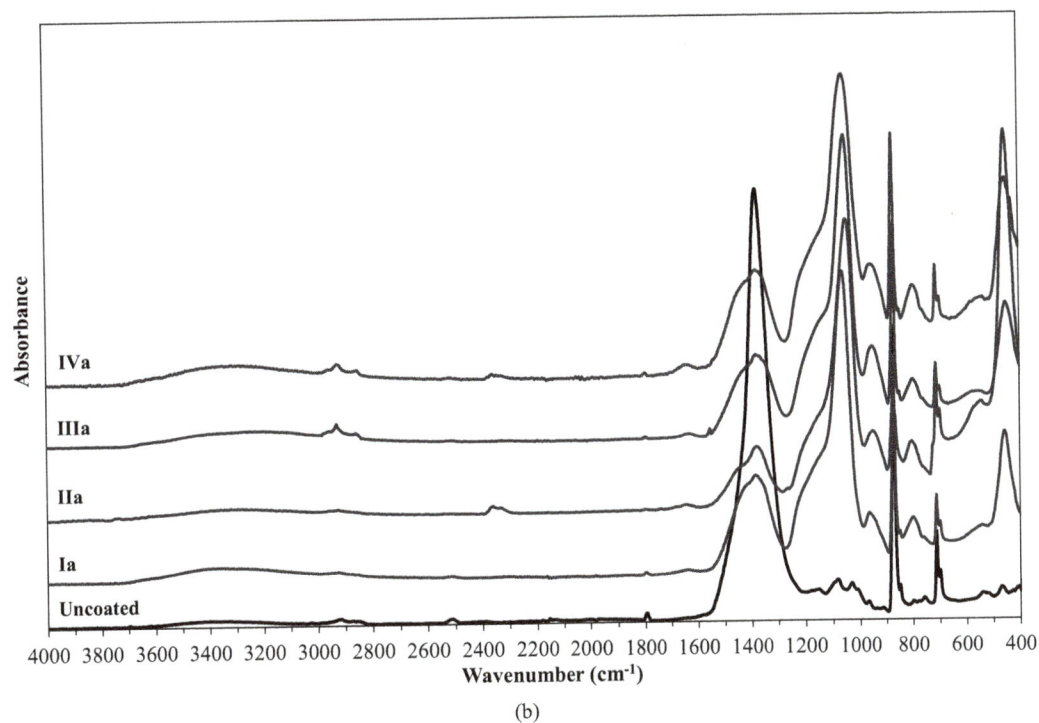

(b)

Figure 3. FTIR-ATR spectra of paperboard before (uncoated) and after coating with several silica based formulations without (a) and with (b) incorportation of LDHs.

3.3. Physical and Mechanical Properties of Silica-Coated Paperboard

The physical and mechanical properties of industrial paperboard uncoated and coated with various silica-based formulations (without (I to IV) and with (Ia to IVa) LDHs) are presented in **Table 3**. The results reveal that silica coating (I to IV) consistently improves paperboard properties such as tensile index (indicative of fibre

Table 3. Physical and mechanical properties and barrier properties of industrial paperboard before (uncoated) and after coating with various silica based formulations with (Ia to IVa) and without (I to IV) incorporation of LDHs.

	Load (g/m²)	Tensile index (Nm/g)	Bendtsen roughness[a] (ml/min)	Burst strength (kPa)	WVTR (g·m⁻²·day⁻¹)	J_{O_2} (m³·m⁻²·day⁻¹)
Uncoated	-----	65	88	535	549 ± 21	1.51
I	2.09	66	129	590	300 ± 5	0.68
Ia	2.03	67	118	593	270 ± 3	1.04
II	2.42	67	145	572	231 ± 3	0.56
IIa	2.14	67	138	606	199 ± 3	0.94
III	2.05	66	147	582	222 ± 4	0.49
IIIa	2.15	68	138	593	132 ± 3	0.78
IV	2.07	66	204	569	226 ± 4	2.58
IVa	2.10	68	216	613	131 ± 3	1.92

(a) (*i.e.* the airflow leaking between the tested surface and the head of the equipment).

strength) and burst strength (indicative of pressure tolerance before rupture). The addition of LDHs (Ia to IVa) causes a further increase in burst strength and tensile index (except for TEOS_DEDMS_LDHs coating (IIa)). Overall, the basic mechanical properties of paperboard are maintained or improved by silica nanocoating, especially when LDHs are incorporated in the silica formulations.

Surface roughness (Bendtsen roughness *i.e.* the airflow (ml/min) leaking between the tested surface and the head of the equipment) increases with silica coating and is reduced by the addition of LDHs (except for TEOS_OTES_LDHs coating (IVa)). Moreover, the nanocoating lowers the porosity of paperboard (measured as Bendtsen porosity *i.e.* airflow required for air to leak across the paperboard) from 12 ml/min (uncoated) to 10 ml/min or lower (silica-coated).

3.4. Surface Wettability of Silica-Coated Paperboard

The amount of gas or vapor permeating a solid membrane is affected by the diffusivity and solubility of the permeant in the membrane material. Hence the affinity of water to paperboard surface plays an important role. Uncoated and silica-coated paperboard surfaces were characterized by contact angle measurements to evaluate water affinity by means of the polar and dispersive components of the total surface energy (calculated with the OWRK model [32]). The results presented in **Figure 5** reveal that silica coating, the nature of the functional groups incorporated in the silica network and the addition of LDHs have a significant impact on the surface energy parameters, especially on its polar component.

The surface energy of uncoated paperboard (~32 mN/m) is dominated by its dispersive component. Coating with TEOS-based formulations introduces free hydroxyl groups on the paperboard surface that increase the total surface energy (~44 mN/m) and its polar component (~ 19 mN/m) and thus confer an undesirable hydrophilic character. To control hydrophilicity alternative formulations were prepared with TEOS ($Si(OC_2H_5)_4$) partially replaced by a secondary silica precursor (5% w/w) bearing different functionalities, as presented in **Table 2** (formulations II, III and IV): DEDMS has two methyl groups ($(CH_3)_2Si(OC_2H_5)_2$), APTES has one aminopropyl group ($H_2N(CH_2)_3Si(OC_2H_5)_3$) and OTES has one octyl group ($(C_8H_{17})Si(OC_2H_5)_3$). As expected, the paperboard's total surface energy and especially its polar component decreased significantly in comparison with the results obtained for paperboard coated only with TEOS (formulation I) (**Figure 5**). The total surface energy follows the order I > III > IV > II whereas the polar component follows the order I > III > II > IV. This can be better understood bearing in mind that the replacement of TEOS with OTES (formulation IV) introduces long chain hydrophobic –Si–octyl groups that increase the surface hydrophobicity (lower polar component); the replacement of TEOS with DEDMS (formulation II) reduces the number of free hydroxyl and alkoxy groups on the silica network; the replacement of TEOS with APTES (formulation III) introduces amine groups, which can form hydrogen bonds with water molecules conferring a hydrophilic character to the surface [43].

In general, the polar component of the surface energy is slightly higher for the formulation with LDHs (Ia to

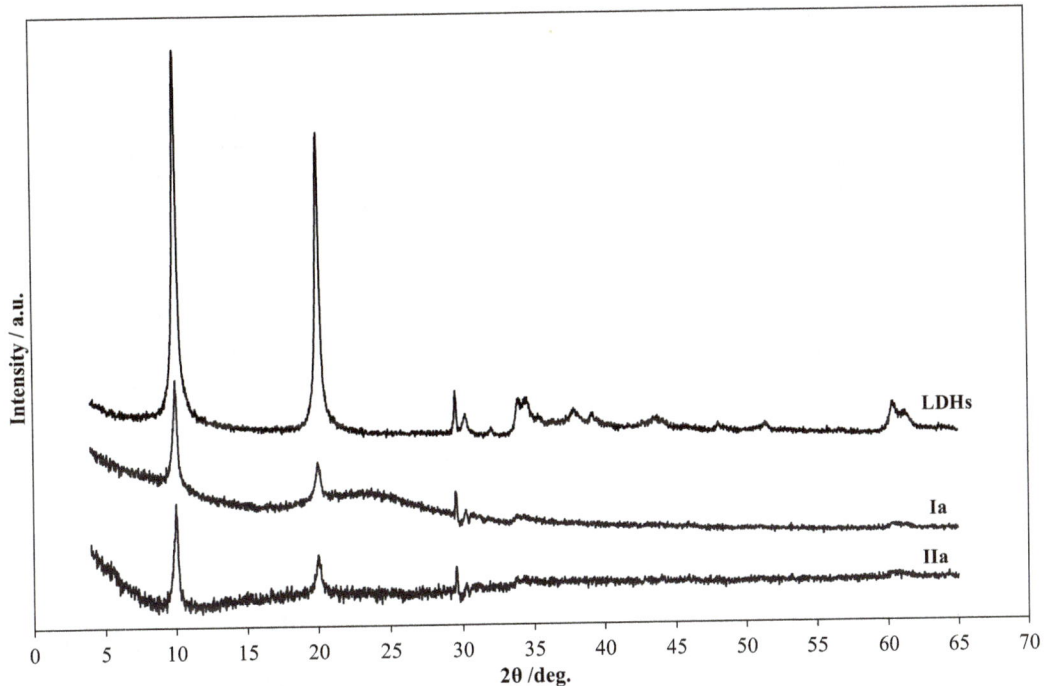

Figure 4. X-ray diffractograms of LDHs and silica formulations (Ia and IIa) containing LDHs (TEOS_LDHs (Ia) and TEOS_DEDMS_LDHs (IIa)).

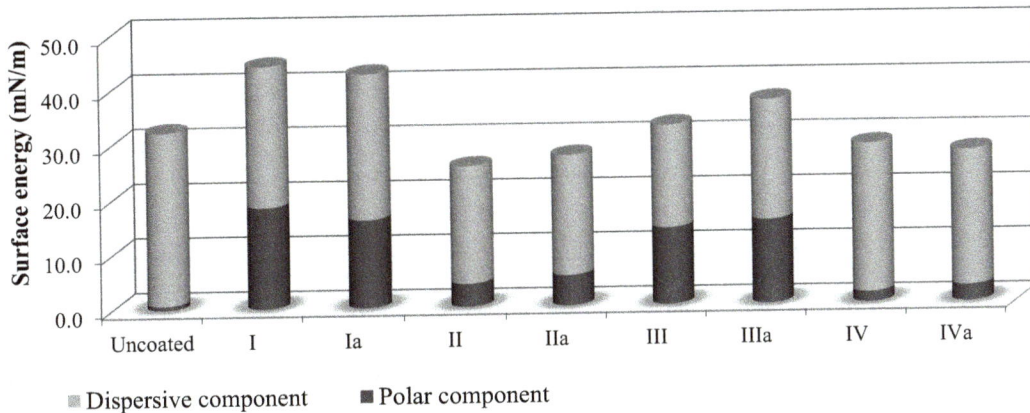

Figure 5. Polar and dispersive components of the total surface energy of paperboard before (uncoated) and after coating with several silica based formulations, without (I to IV) and with (Ia to IVa) incorporation of LDHs.

IVa) in comparison to the formulation without LDHs (I to IV) (**Figure 5**). This can be explained by the hydrophilic character of LDH-NO$_3$, which displays capacity to retain water internally, between the layers, and externally, in the hydroxide layers [27].

The polar and dispersive components of the surface energy were used to draw the surface wetting envelopes (γ_l^p *versus* γ_s^d at a fixed contact angle), a very useful tool to predict the contact angle of a particular liquid with the solid surface [34]. The contact angle selected ($\theta = 90°$) describes the frontier situation for wettable ($\theta < 90°$) and non-wettable ($\theta > 90°$) surfaces based on the general contact angle theory of Young [44] [45], *i.e.* a solid surface is defined as hydrophobic or hydrophilic when the contact angle of a water droplet is larger or smaller than 90°, respectively [44]-[46]. For better visualization of surface wettability the water coordinates are represented by a dark point in **Figure 6**: when the point is located outside the wetting envelope curve the water contact angle with the surface is greater than 90°, *i.e.* the surface is non-wettable. Accordingly, the following conclusions

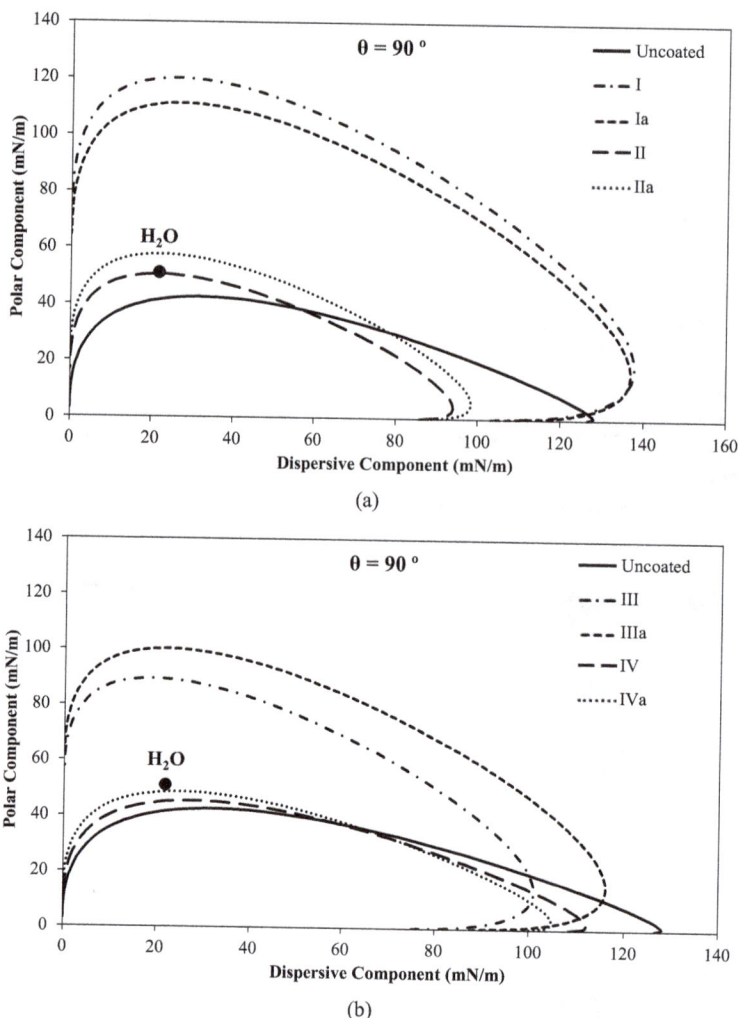

Figure 6. Wetting envelope at a fixed contact angle ($\theta = 90°$) for paperboard before (uncoated) and after coating with several silica based formulations without (I to IV) and with (Ia to IVa) incorporation of LDHs.

might be drawn from the analysis of the curves presented in **Figure 6**: wettability of uncoated paperboard is low; paperboard coated only with TEOS (formulation I—**Figure 6(a)**) presents high wettability (the water point is located far from the 90° wetting envelope); the partial replacement of TEOS with DEDMS (formulation II—**Figure 6(a)**) or OTES (formulation IV—**Figure 6(b)**) reduces wettability due to the lower number of free hydroxyl groups remaining on the surface and because of the long chain –Si–octyl groups introduced by OTES (formulation IV) [20] [21]; the partial replacement of TEOS with APTES (formulation III) creates a wettable surface due to the hydrophilic character conferred by the amine groups [43], as previously discussed. Finally, the incorporation of LDHs imparts an insignificant change in the wettability of paperboard coated with TEOS (formulation Ia—**Figure 6(a)**) even though wettability increases slightly for the other formulations (IIa—**Figure 6(a)**, IIIa and IVa—**Figure 6(b)**).

3.5. Water Vapor and Oxygen Barrier Properties

Water vapor transmission rate (WVTR) and oxygen permeability (J_{O2}) were measured to evaluate the barrier properties of uncoated and silica coated paperboards with and without LDHs. The results are presented qualitatively in **Table 2** and quantitatively in **Table 3**.

WVTR for uncoated paperboard (550 g·m^{-2}·day^{-1}) was considerably reduced after coating with TEOS (formulation I) (~45%) or TEOS mixed with a secondary precursor (formulations II, III or IV) (~60%). A further reduction (10% to 40%) is attained by the incorporation of LDHs (formulations Ia to IVa). These results demonstrate the potential of silica nanocoatings to expand the applicability of paperboards in food packaging. However, WVTR values are still below the levels furnished by synthetic polymers in coated paperboards used for food containers (10 $g \cdot m^{-2} \cdot day^{-1}$) [5].

Paperboard coated by non-modified silica (formulation I, TEOS)presents better water vapor barrier properties than uncoated paperboard in spite of the lower water repellence *i.e.* higher polarity (**Figure 5**) and wettability (**Figure 6(b)**) of the coated surface. This apparent controversy can be explained by the formation of an impermeable surface between the three-dimensional silica network and cellulose fibersin the paperboard via strong hydrogen bonding involving the hydroxyl groups of silica and cellulose [16]-[18]. The use of alkoxysilane co-precursors does not disturb the silica network and lowers the concentration of free hydroxyl groups on the surface, thus conferring hydrophobicity to the silica-coated paperboard (formulation (II)—**Figure 7**). This is corroborated by the surface energy data (**Figure 5** and **Figure 6**) and explains the lower WVTR values of paperboard coated with mixed formulations II, III or IV (**Table 3**). Incorporation of LDHs in the silica formulations (formulation (IIa)—**Figure 7**) provides an additional reduction of the WVTR values (formulations Ia to IVa—**Table 3**) not only because of their ability to retain water molecules in the interlayer region as water scavengers [27] [30] but also because of their high aspect-ratio (flat structure) that increases the length and tortuosity of the diffusion path [29] [47]. A final undesirable aspect that ought to be resolved is the formation of cracks in the paperboard coated surface (SEM images **Figure 2** and **Table 2**) that provide pathways for water vapor diffusion, therefore diminishing the barrier properties.

Silica-based nanocoatings prepared with TEOS alone (formulation (I)—**Figure 7**) or mixed with a co-precursor such as DEDMS (formulation (II)—**Figure 7**) or APTES (formulation (III)—**Figure 1**) enhanced the oxygen barrier properties of paperboard (**Table 2** and **Table 3**). In fact, oxygen permeability of uncoated paperboard (1.51 $m^3 \cdot m^{-2} \cdot day^{-1}$) was reduced by 55%, 63% and 68% after coating with formulations I, II and III, respectively. This behaviorwas previously discussed and ascribed to the dense and impermeable silica network formed on the paperboard surface (**Figure 7**) [16]-[18] [48]. Apparently, the nature of the co-precursor has a small effect on oxygen permeability except for formulation IV (TEOS_OTES). Most likely the presence of bulky organic radicals (–Si–octyl groups) in the silica formulation affects adversely the density and physical integrity of the inorganic network as revealed by the SEM analysis. In fact, the front and cross-section images of silica-coated paperboard (**Figure 2**) show more and deeper cracks for formulation IV (TEOS_OTES) than for formulation II (TEOS_DEDMS) and formulation III (TEOS_APTES) thus explaining the particularly high oxygen permeability rate observed for the first formulation (2.578 $m^3 \ m^{-2} \cdot day^{-1}$) [21] [23] [48]. The incorporation of LDHs has a similar effect decreasing the barrier properties towards oxygen (J_{O2}increases about 50% - 70% when compared with the formulation without LDHs) and once more this may be assigned to modifications in the paperboard's silica network (formulation (II)—**Figure 7**).

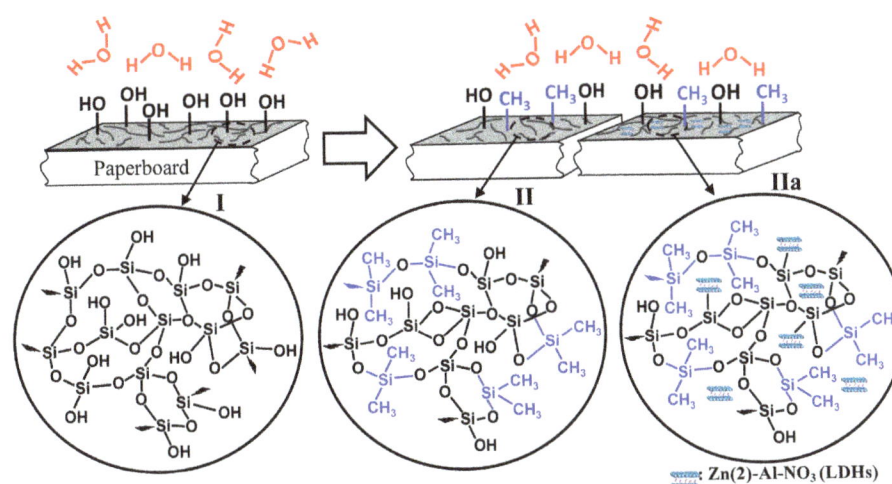

Figure 7. Schematic representation of the interaction of water molecules with the silica-network formed on the paperboard surface (TEOS formulation (I), TEOS_DEDMS formulation (II), TEOS_ DEDMS formulation with LDHs (IIa)).

4. Conclusion

The results of this study reveal that low grammage (2 - 3 g/m^2) coating of industrial paperboard with silica formulations prepared with TEOS alone or TEOS (95% v/v) mixed with a co-precursor (5% v/v of DEDMS, APTES or OTES) is a very promising strategy to impart water and oxygen barrier properties suitable for packaging applications. The addition of high aspect-ratio (flat structure) particles of Zn(2)-Al-NO$_3$ (LDHs) to the silica formulation lowers the water vapor transmission rate (WVTR) due to the water scavenging capacity of LDHs and also due to a physical barrier that decreases the vapor diffusivity. The silica modification using co-precursors containing aliphatic moieties decreased both the polar and dispersive components of the surface energy of deposited silica domains conveying water repulsion effects to the paperboard surface. The compilation of these two factors (water repelling functionalities and physical barrier) in specific formulations (TEOS_APTES_LDHs and TEOS_OTES_LDHs) lowered WVTR by nearly 80% when compared to the "uncoated" paperboard. However, the implementation of LDHs and bulky aliphatic functionalities in silica formulations deteriorate the oxygen barrier properties of silica coatings. The compromise solution, *i.e.* TEOS_APTES silica formulation without LDHs, exhibited much lower WVTR and oxygen permeability values than for uncoated paperboard, precisely 60% and 70% lower, respectively. The critical point for gas barrier in silica-based coatings was suggested to be the violation of the coating physical integrity *i.e.* formation of microcrackson the paperboard induced by shrinkage upon the curing process. Further work is required to obtainsilica formulations that provide higher dimensional stability to the silica network.

Acknowledgements

Authors acknowledge the financial support from NANOBARRIER EU FP7 project (FP7-NMP-2011-LARGE-5, ref. N° 280759) and CICECO-Aveiro Institute of Materials grant financed by FCT (PEst-C/CTM/LA0011/2013). The financial support from the project CICECO-Aveiro Institute of Materials (Ref. FCT UID/CTM/50011/ 2013) is greatly acknowledged. Authors also thank the technical assistance of paperboard mill Prado Karton SA. Dr. João Tedim also thanks FCT for researcher grant IF/00347/2013.

References

[1]   Hirvikorpi, T., Vähä-Nissi, M., Harlin, A. and Karppinen, M. (2010) Comparison of Some Coating Techniques to Fabricate Barrier Layers on Packaging Materials. *Thin Solid Films*, **518**, 5463-5466. http://dx.doi.org/10.1016/j.tsf.2010.04.018

[2]   Samyn, P. (2013) Wetting and Hydrophobic Modification of Cellulose Surfaces for Paper Applications. *Journal of Materials Science*, **48**, 6455-6498. http://dx.doi.org/10.1007/s10853-013-7519-y

[3]   Rahman, M.S. (2007) Handbook of Food Preservation. 2nd Edition, CRC Press, 479-480. http://dx.doi.org/10.1201/9781420017373

[4]   Siracusa, V. (2012) Food Packaging Permeability Behaviour: A Report. *International Journal of Polymer Science*, **2012**, 1-11. http://dx.doi.org/10.1155/2012/302029

[5]   Han, J., Salmieri, S., Le Tien, C. and Lacroix, M. (2010) Improvement of Water Barrier Property of Paperboard by Coating Application with Biodegradable Polymers. *Journal of agricultural and food chemistry*, **58**, 3125-3131. http://dx.doi.org/10.1021/jf904443n

[6]   Stepien, M., *et al.* (2012) Surface Chemical Characterization of Nanoparticle Coated Paperboard. *Applied Surface Science*, **258**, 3119-3125. http://dx.doi.org/10.1016/j.apsusc.2011.11.048

[7]   Khwaldia, K., Arab-Tehrany, E. and Desobry, S. (2010) Biopolymer Coatings on Paper Packaging Materials. *Comprehensive Reviews in Food Science and Food Safety*, **9**, 82-91. http://dx.doi.org/10.1111/j.1541-4337.2009.00095.x

[8]   Farris, S., Unalan, I.U., Introzzi, L., Fuentes-Alventosa, J.M. and Cozzolino, C.A. (2014) Pullulan-Based Films and Coatings for Food Packaging: Present Applications, Emerging Opportunities, and Future Challenges. *Journal of Applied Polymer Science*, **131**, 1-12. http://dx.doi.org/10.1002/app.40539

[9]   Schmid, M., *et al.* (2014) Water Repellence and Oxygen and Water Vapor Barrier of PVOH-Coated Substrates before and after Surface Esterification. *Polymers*, **6**, 2764-2783. http://dx.doi.org/10.3390/polym6112764

[10]   Vartiainen, J., Vähä-nissi, M. and Harlin, A. (2014) Biopolymer Films and Coatings in Packaging Applications—A Review of Recent Developments. *Materials Sciences and Applications*, **5**, 708-718. http://dx.doi.org/10.4236/msa.2014.510072

[11]   Othman, S.H. (2014) Bio-Nanocomposite Materials for Food Packaging Applications: Types of Biopolymer and Na-

Silica-Based Nanocoating Doped by Layered Double Hydroxides to Enhance the Paperboard Barrier...

49

no-Sized Filler. *Agriculture and Agricultural Science Procedia*, **2**, 296-303.
http://dx.doi.org/10.1016/j.aaspro.2014.11.042

[12] Bang, G. and Kim, S.W. (2012) Biodegradable Poly(lactic acid)-Based Hybrid Coating Materials for Food Packaging Films with Gas Barrier Properties. *Journal of Industrial and Engineering Chemistry*, **18**, 1063-1068.
http://dx.doi.org/10.1016/j.jiec.2011.12.004

[13] Rahman, I.A. and Padavettan, V. (2012) Synthesis of Silica Nanoparticles by Sol-Gel: Size-Dependent Properties, Surface Modification, and Applications in Silica-Polymer Nanocomposites—A Review. *Journal of Nanomaterials*, **2012**, 1-15. http://dx.doi.org/10.1155/2012/132424

[14] Pandey, S. and Mishra, S.B. (2011) Sol-Gel Derived Organic-Inorganic Hybrid Materials: Synthesis, Characterizations and Applications. *Journal of Sol-Gel Science and Technology*, **59**, 73-94. http://dx.doi.org/10.1007/s10971-011-2465-0

[15] Wang, S.X., Mahlberg, R., Nikkola, J., *et al.* (2011) Surface Characteristics and Wetting Properties of Sol-Gel Coated Base Paper. *Surface and Interface Analysis*, **44**, 539-547. http://dx.doi.org/10.1002/sia.3841

[16] Wang, S.X., Jämsä, S., Mahlberg, R., *et al.* (2014) Treatments of Paper Surfaces with Sol-Gel Coatings for Laminated Plywood. *Applied Surface Science*, **288**, 295-303. http://dx.doi.org/10.1016/j.apsusc.2013.10.024

[17] Sequeira, S., Evtuguin, D.V., Portugal, I. and Esculcas, A.P. (2007) Synthesis and Characterisation of Cellulose/Silica Hybrids Obtained by Heteropoly Acid Catalysed Sol-Gel Process. *Materials Science and Engineering: C*, **27**, 172-179.
http://dx.doi.org/10.1016/j.msec.2006.04.007

[18] Portugal, I., Dias, V.M., Duarte, R.F. and Evtuguin, D.V. (2010) Hydration of Cellulose/Silica Hybrids Assessed by Sorption Isotherms. *Journal of Physical Chemistry B*, **114**, 4047-4055. http://dx.doi.org/10.1021/jp911270y

[19] Latthe, S.S., Imai, H., Ganesan, V., Kappenstein, C. and Venkateswara Rao, A. (2009) Optically Transparent Superhydrophobic TEOS-Derived Silica Films by Surface Silylation Method. *Journal of Sol-Gel Science and Technology*, **53**, 208-215. http://dx.doi.org/10.1007/s10971-009-2079-y

[20] Purcar, V., Stamatin, I., Cinteza, O., *et al.* (2012) Fabrication of Hydrophobic and Antireflective Coatings Based on Hybrid Silica Films by Sol-Gel Process. *Surface and Coatings Technology*, **206**, 4449-4454.
http://dx.doi.org/10.1016/j.surfcoat.2012.04.094

[21] Parale, V.G., Mahadik, D.B., Mahadik, S.A., *et al.* (2013) OTES Modified Transparent Dip Coated Silica Coatings. *Ceramics International*, **39**, 835-840. http://dx.doi.org/10.1016/j.ceramint.2012.05.079

[22] Mah, S.K. and Chung, I.J. (1995) Effects of Dimethyldiethoxysilane Addition on Tetraethylorthosilicate Sol-Gel Process. *Journal of Non-Crystalline Solids*, **183**, 252-259. http://dx.doi.org/10.1016/0022-3093(94)00631-8

[23] Zhang, X., Wu, W., Wang, J. and Tian, X. (2008) Direct Synthesis and Characterization of Highly Ordered Functional Mesoporous Silica Thin Films with High Amino-Groups Content. *Applied Surface Science*, **254**, 2893-2899.
http://dx.doi.org/10.1016/j.apsusc.2007.10.022

[24] Gamelas, J.A.F., Evtyugina, M.G., Portugal, I. and Evtuguin, D.V. (2012) New Polyoxometalate-Functionalized Cellulosic Fibre/Silica Hybrids for Environmental Applications. *RSC Advances*, **2**, 831-839.
http://dx.doi.org/10.1039/C1RA00371B

[25] Gamelas, J.A.F., Evtuguin, D.V. and Esculcas, A.P. (2007) Transition Metal Substituted Polyoxometalates Supported on Amine-Functionalized Silica. *Transition Metal Chemistry*, **32**, 1061-1067.
http://dx.doi.org/10.1007/s11243-007-0277-4

[26] Newman, S.P. and Jones, W. (1998) Synthesis, Characterization and Applications of Layered Double Hydroxides Containing Organic Guests. *New Journal of Chemistry*, **22**, 105-115. http://dx.doi.org/10.1039/a708319j

[27] Tedim, J., Kuznetsova, A., Salak, A.N., *et al.* (2012) Zn-Al Layered Double Hydroxides as Chloride Nanotraps in Active Protective Coatings. *Corrosion Science*, **55**, 1-4. http://dx.doi.org/10.1016/j.corsci.2011.10.003

[28] Uysal, U.I., Cerri, G., Marcuzzo, E., Cozzolino, C.A. and Farris, S. (2014) Nanocomposite Films and Coatings Using Inorganic Nanobuilding Blocks (NBB): Current Applications and Future Opportunities in the Food Packaging Sector. *RSC Advances*, **4**, 29393-29428. http://dx.doi.org/10.1039/C4RA01778A

[29] Azeredo, H., Mattoso, L. and McHugh, T. (2011) Nanocomposites in Food Packaging—A Review Advances. In: Reddy, B., Ed., *Diverse Industrial Applications of Nanocomposites*, InTech, Rijeka, 57-78.

[30] Poznyak, S.K., Tedim, J., Rodrigues, L.M., *et al.* (2009) Novel Inorganic Host Layered Double Hydroxides Intercalated with Guest Organic Inhibitors for Anticorrosion Applications. *ACS Applied Materials & Interfaces*, **1**, 2353-2362.
http://dx.doi.org/10.1021/am900495r

[31] Coiai, S., Scatto, M., Conzatti, L., *et al.* (2011) Optimization of Organo-Layered Double Hydroxide Dispersion in LDPE-Based Nanocomposites. *Polymers for Advanced Technologies*, **22**, 2285-2294.
http://dx.doi.org/10.1002/pat.1759

[32] Owens, D.K. and Wendt, R.C. (1969) Estimation of the Surface Free Energy of Polymers. *Journal of Applied Polymer*

Science, **13**, 1741-1747. http://dx.doi.org/10.1002/app.1969.070130815

[33] Figueiredo, A.B., Evtuguin, D.V., Monteiro, J., *et al.* (2011) Structure-Surface Property Relationships of Kraft Papers: Implication on Impregnation with Phenol-Formaldehyde Resin. *Industrial Engineering Chemistry Research*, **50**, 2883-2890. http://dx.doi.org/10.1021/ie101912h

[34] Janssen, D., De Palma, R., Verlaak, S., Heremans, P. and Dehaen, W. (2006) Static Solvent Contact Angle Measurements, Surface Free Energy and Wettability Determination of Various Self-Assembled Monolayers on Silicon Dioxide. *Thin Solid Films*, **515**, 1433-1438. http://dx.doi.org/10.1016/j.tsf.2006.04.006

[35] ASTM E 95-96 (1995) Standard Test Methods for Water Vapor Transmission of Materials (E96-E95). *Annual Books of ASTM Standards*, **552**, 785-792.

[36] Kovalevsky, A.V., Yaremchenko, A.A., Kolotygin, V.A., *et al.* (2011) Processing and Oxygen Permeation Studies of Asymmetric Multilayer $Ba_{0.5}Sr_{0.5}Co_{0.8}Fe_{0.2}O_{3-\delta}$ Membranes. *Journal of Membrane Science*, **380**, 68-80. http://dx.doi.org/10.1016/j.memsci.2011.06.034

[37] Goel, A., Tulyaganov, D.U., Kharton, V.V., *et al.* (2007) Effect of BaO Addition on Crystallization, Microstructure, and Properties of Diopside-Ca-Tschermak Clinopyroxene-Based Glass-Ceramics. *Journal of the American Ceramic Society*, **90**, 2236-2244. http://dx.doi.org/10.1111/j.1551-2916.2007.01743.x

[38] Yaremchenko, A.A., Kharton, V.V., Avdeev, M., Shaula, A.L. and Marques, F.M.B. (2007) Oxygen Permeability, Thermal Expansion and Stability of $SrCo_{0.8}Fe_{0.2}O_{3-\delta}$-$SrAl_2O_4$ Composites. *Solid State Ionics*, **178**, 1205-1217. http://dx.doi.org/10.1016/j.ssi.2007.05.016

[39] Miller, F.A. and Wilkins, C.H. (1952) Infrared Spectra and Characteristic Frequencies of Inorganic Ions. *Analytical Chemistry*, **24**, 1253-1294. http://dx.doi.org/10.1021/ac60068a007

[40] Nalwa, H.S. (2003) Handbook of Organic-Inorganic Hybrid Materials and Nanocomposites: Hybrid Materials. American Scientific Publishers, Valencia, 280-285.

[41] Hsiao, V.K.S., Waldeisen, J.R., Zheng, Y.B., *et al.* (2007) Aminopropyltriethoxysilane (APTES)-Functionalized Nanoporous Polymeric Gratings: Fabrication and Application in Biosensing. *Journal of Materials Chemistry*, **17**, 4896-4901. http://dx.doi.org/10.1039/b711200a

[42] Tedim, J., Poznyak, S.K., Kuznetsova, A., *et al.* (2010) Enhancement of Active Corrosion Protection via Combination of Inhibitor-Loaded Nanocontainers. *ACS Applied Materials & Interfaces*, **2**, 1528-1535. http://dx.doi.org/10.1021/am100174t

[43] Wu, Z., Xiang, H., Kim, T., Chun, M.S. and Lee, K. (2006) Surface Properties of Submicrometer Silica Spheres Modified with Aminopropyltriethoxysilane and Phenyltriethoxysilane. *Journal of Colloid and Interface Science*, **304**, 119-124. http://dx.doi.org/10.1016/j.jcis.2006.08.055

[44] Butt, H., Graf, K. and Kappl, M. (2003) Physics and Chemistry of Interfaces. Wiley-VCH, Weinheim.

[45] Yuan, Y. and Lee, T.R. (2013) Contact Angle and Wetting Properties. In: Bracco, G. and Holst, B., Eds., *Surface Science Techniques Springer Series, Surface Sciences*, Volume 51, Springer Berlin Heidelberg, Berlin and Heidelberg, 3-34. http://dx.doi.org/10.1007/978-3-642-34243-1

[46] Verplanck, N., Coffinier, Y., Thomy, V. and Boukherroub, R. (2007) Wettability Switching Techniques on Superhydrophobic Surfaces. *Nanoscale Research Letters*, **2**, 577-596. http://dx.doi.org/10.1007/s11671-007-9102-4.

[47] Chen, H., Zhang, F., Fu, S. and Duan, X. (2006) *In Situ* Microstructure Control of Oriented Layered Double Hydroxide Monolayer Films with Curved Hexagonal Crystals as Superhydrophobic Materials. *Advanced Materials*, **18**, 3089-3093. http://dx.doi.org/10.1002/adma.200600615

[48] Lee, S., Oh, K.K., Park, S., Kim, J.-S. and Kim, H. (2010) Scratch Resistance and Oxygen Barrier Properties of Acrylate-Based Hybrid Coatings on Polycarbonate Substrate. *Korean Journal of Chemical Engineering*, **26**, 1550-1555. http://dx.doi.org/10.1007/s11814-009-0263-y

6

Development of Nanostructure Formation of $Fe_{73.5}Cu_1Nb_3Si_{13.5}B_9$ Alloy from Amorphous State on Heat Treatment

Md. Khalid Hossain[1]*, Jannatul Ferdous[2], Md. Manjurul Haque[2], A. K. M. Abdul Hakim[3]

[1]Institute of Electronics, Atomic Energy Research Establishment, Savar, Dhaka, Bangladesh
[2]Department of Applied Physics, Electronics & Communication Engineering, Islamic University, Kushtia, Bangladesh
[3]Department of Glass and Ceramic Engineering, Bangladesh University of Engineering and Technology, Dhaka, Bangladesh
Email: *khalid.baec@yahoo.com

Abstract

Iron-based amorphous alloys have attracted technological and scientific interests due to their excellent soft magnetic properties. The typical nanocrystalline alloy with the composition of $Fe_{73.5}Cu_1Nb_3Si_{13.5}B_9$ known as FINEMENT has been studied for structural properties analysis. Recently, it is found that after proper annealing the amorphous alloy like $Fe_{73.5}Cu_1Nb_3Si_{13.5}B_9$ has a transition to the nanocrystalline state, thus exhibiting good magnetic properties. The alloy in the form of ribbon of 10 mm width and 25mm thickness with the composition of $Fe_{73.5}Cu_1Nb_3Si_{13.5}B_9$ was prepared by rapid quenching method. The prepared ribbon sample has been annealed for 30 min in a controlled way in the temperature range 490°C - 680°C. By analyzing X-ray diffraction (XRD) patterns, various structural parameters such as lattice parameters, grain size and silicon content of the nanocrystalline Fe(Si) grains, crystallization behavior and nanocrystalline phase formation have been investigated. In the nanocrystalline state, Cu helps the nucleation of α-Fe(Si) grains while Nb controls their growth, Si and B has been used as glass forming materials. Thus on the residual amorphous, the nanometric Fe(Si) grains develops. From broadening of fundamental peaks, the optimum grain size has been determined in the range of 7 - 23 nm.

Keywords

Fe Based Alloy, Rapid Solidification, Crystallization Behavior, Nanocrystalline Phase Formation

---

*Corresponding author.

1. Introduction

A new class of iron based alloys was introduced in by Yoshizawa, Oguma and Yamauchi [1]. This novel material with the composition of $Fe_{73.5}Cu_1Nb_3Si_{13.5}B_9$ has a trade name FINEMENT. The particular about the new material is its ultrafine microstructure of bcc Fe-Si with the grain sizes of 10 - 15 nm that are due to the presence of Cu and Nb [1] [2] from which their soft magnetic properties derive lastly. The magnetic properties of $Fe_{73.5}Cu_1Nb_3Si_{13.5}B_9$ and related alloys were reported by many publications; here only some of the references are given [2]-[7].

Nanocrystalline materials constitute a new class of condensed matter having interesting properties, which are mostly microstructure dependent. These materials are first formed into amorphous ribbons and then annealed above the crystallization temperature to form the nanocrystalline microstructure that consists of bcc Fe(Si) nanograins embedded on amorphous matrix. The crystallization of Fe-Si-B amorphous ribbons contains 1 at.% Cu and 3 at.% Nb. The crystallization of bcc Fe(Si) solid solution from amorphous state takes place according to the basic scheme characteristic to the hypo-eutectic glasses [8]: am1→Fe(Si) + am2, where am1 and am2 are the initial amorphous precursor and the remainder amorphous phase respectively. There are mainly two phases in the alloys: a bcc Fe-Si solid solution and some residual amorphous phase. The average grain size of the bcc Fe(Si) phase is about 10 nm [2]. The addition of Cu and Nb results in the formation of an ultrafine grain structure [1] [2].

Recently, a generalization of the random anisotropy model, taking into account the two phase character of nanocrystalline materials, has been developed [9] and it explains the previously mentioned hardening as well as other features which cannot be understood without the generalization. Since the unique properties of nanostructured materials are dictated by the dimensions of the crystallites, it is very advantageous to control the size of the particles by controlling the annealing temperature of the specimens.

The aim of the present research work is to synthesize Fe-based alloys of $Fe_{73.5}Cu_1Nb_3Si_{13.5}B_9$ composition in the amorphous state by using rapid solidification technique and study their structural properties by varying the annealing condition. The annealing of Fe-based soft nanocomposite magnetic materials has been performed in air. X-ray diffraction technique has been used for the characterization of nanostructured phases.

2. Experimental

Rapid solidification technique was used to prepare the ribbons with composition $Fe_{73.5}Cu_1Nb_3Si_{13.5}B_9$. The amorphous ribbons ware prepared at a temperature 1500°C in an argon atmosphere (0.2 to 0.3 atoms). The dimension of the ribbon was 10 mm width and 25 mm thickness. The purity of the material is Fe (99.98%), Cu (99+%), Nb (99.8%), Si (99.9%) and B (99.5%) as obtained from Johnson Mathey (Alfa Aesar Inc.). **Figure 1** shows the schematic diagram of melt spin system.

A PHILIPS PW3040 X′ Pert PRO X-ray diffractometer was used to study the crystalline phases of the prepared samples in the Materials Science Division, Atomic Energy Centre, Dhaka-1000, Bangladesh. The powder specimens were exposed to CuK_α radiation with a primary beam of 40 kV and 30 mA with a sampling pitch of 0.02° and time for each step data collection was 1.0 sec. A 2θ scan was taken from 10° to 90° to set possible fundamental peaks where Ni filter was used to reduce CuK_α radiation. All the data of the samples were analyzed using computer software "X′ PERT HIGHSCORE". X-ray diffraction patterns were carried out to confirm the crystal structure. Instrumental broadening of the system was determined from θ - 2θ scan of standard Si. **Figure 2** and **Figure 3** show the X-ray diffraction technique and block diagram of the PHILIPS PW 3040 X′ Pert PRO XRD system respectively.

3. Results

To determine the crystalline phases, X-ray diffraction studies have been performed for samples annealed at different temperatures. **Figure 4** and **Figure 5** represent the X-ray diffraction spectra of quenched alloy and the alloy annealed at different temperatures for 30 min. In the figures, the parenthesis represents the indices of the reflecting planes of the phases. The bcc α-Fe(Si) phases are found, identified by using standard software. All the results of lattice parameter, grain size and silicon content at % for different annealing temperatures of composition are listed in **Table 1**.

In **Figure 6**, the lattice parameter of various annealed samples in the temperature range 490°C - 680°C have

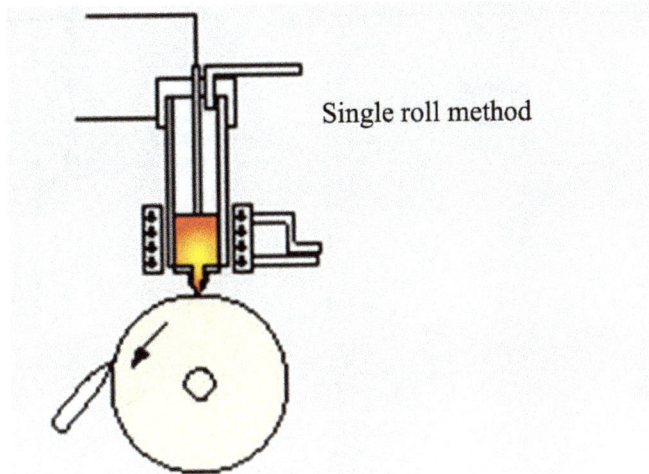

Figure 1. Schematic diagram of melt spin system.

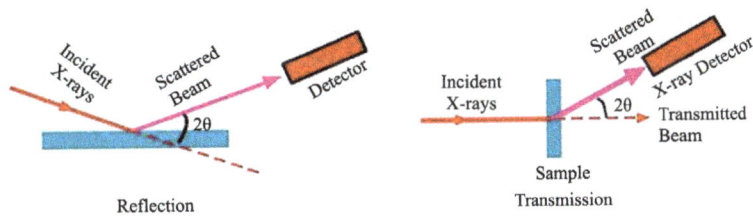

Figure 2. X-ray diffraction technique.

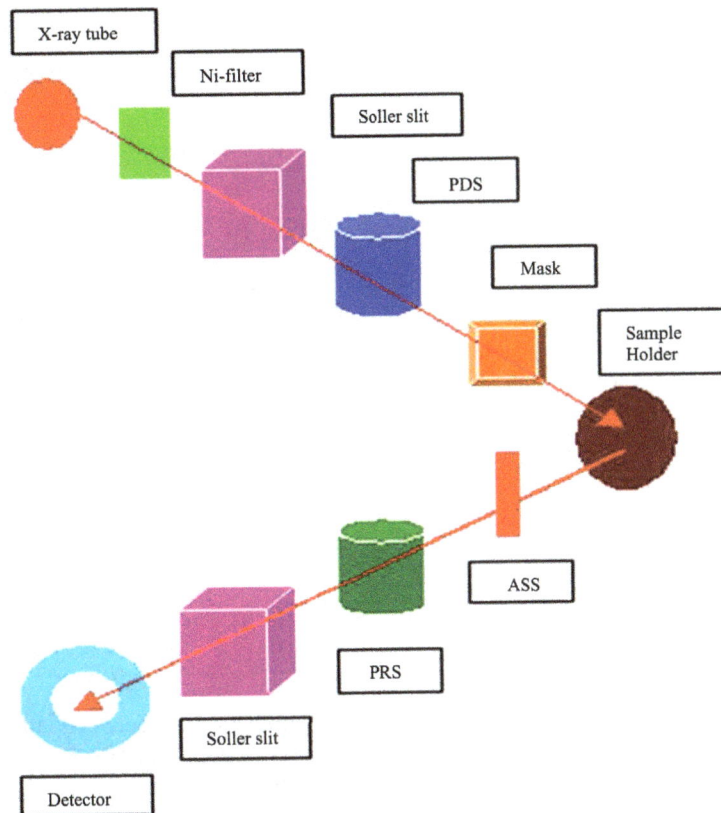

Figure 3. Block diagram of the PHILIPS PW 3040 X' Pert PRO XRD system.

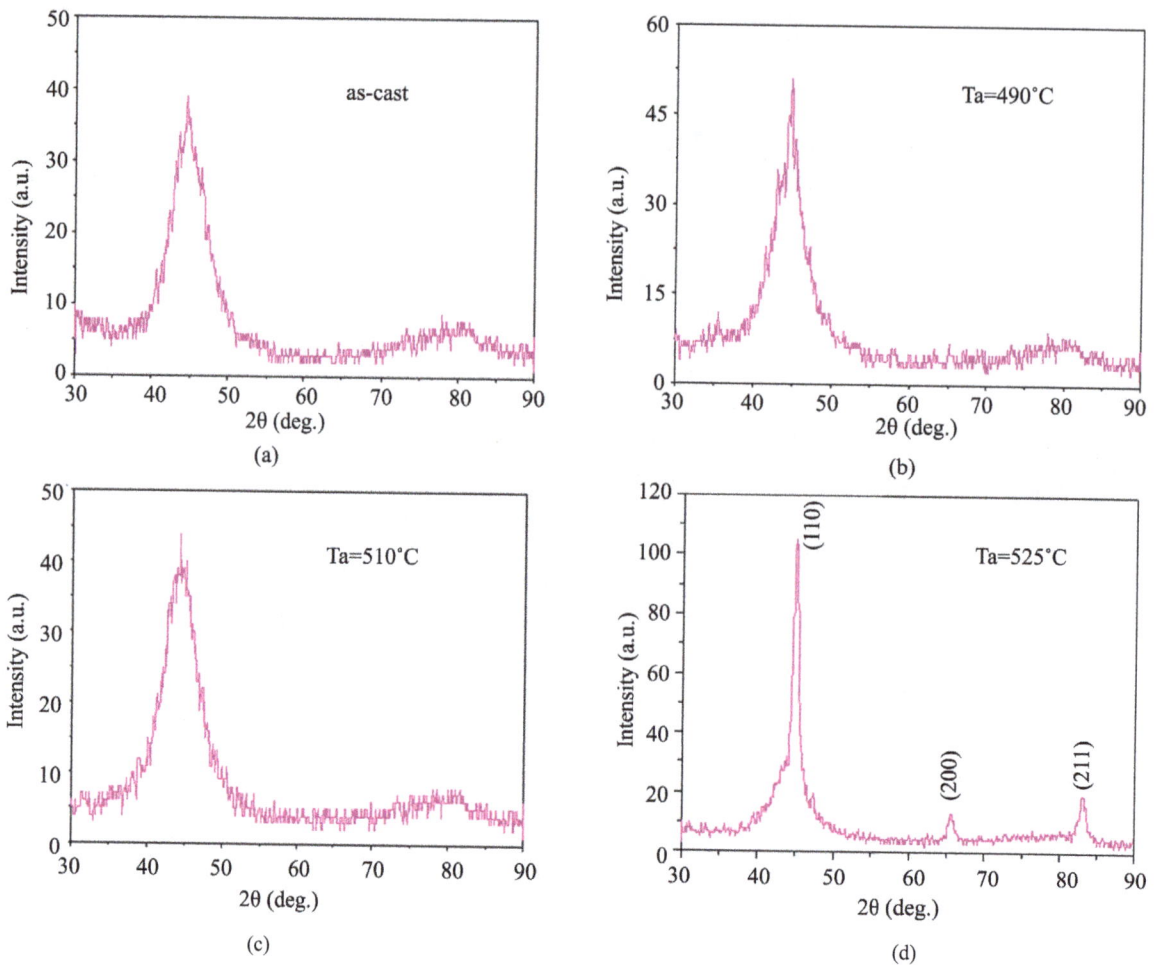

Figure 4. XRD patterns of $Fe_{73.5}Cu_1Nb_3Si_{13.5}B_9$ ribbon sample (a) as-cast and annealed samples at (b) 490°C (c) 510°C (d) 525°C for 30 min.

Table 1. Experimental XRD data of nano-crystalline $Fe_{73.5}Cu_1Nb_3Si_{13.5}B_9$ amorphous ribbon at different annealing temperatures.

Annealing temperature, Ta (°C)	Lattice parameter a_o (Å)	Grain size Dg (nm)	Silicon content (% Si)
As-cast	…	…	…
490	2.8615	7	6.45
510	2.8420	8	15.6
525	2.8373	10	17.8
545	2.8304	13	21.1
560	2.8307	14	20.8
600	2.8351	18	18.8
680	2.8482	23	12.7

been presented. In **Figure 6**, the established quantitative relationship [10] between lattice parameter was used to determine the silicon content of α-Fe(Si) nanograins. Nanocrystalline grain of α-Fe(Si) is formed from amorphous precursor, when the sample is annealed above the crystallization temperature. By using the Scherrer's formula, the mean grain size of the nanograins was determined from the X-ray fundamental line (110) has been presented in **Figure 8**.

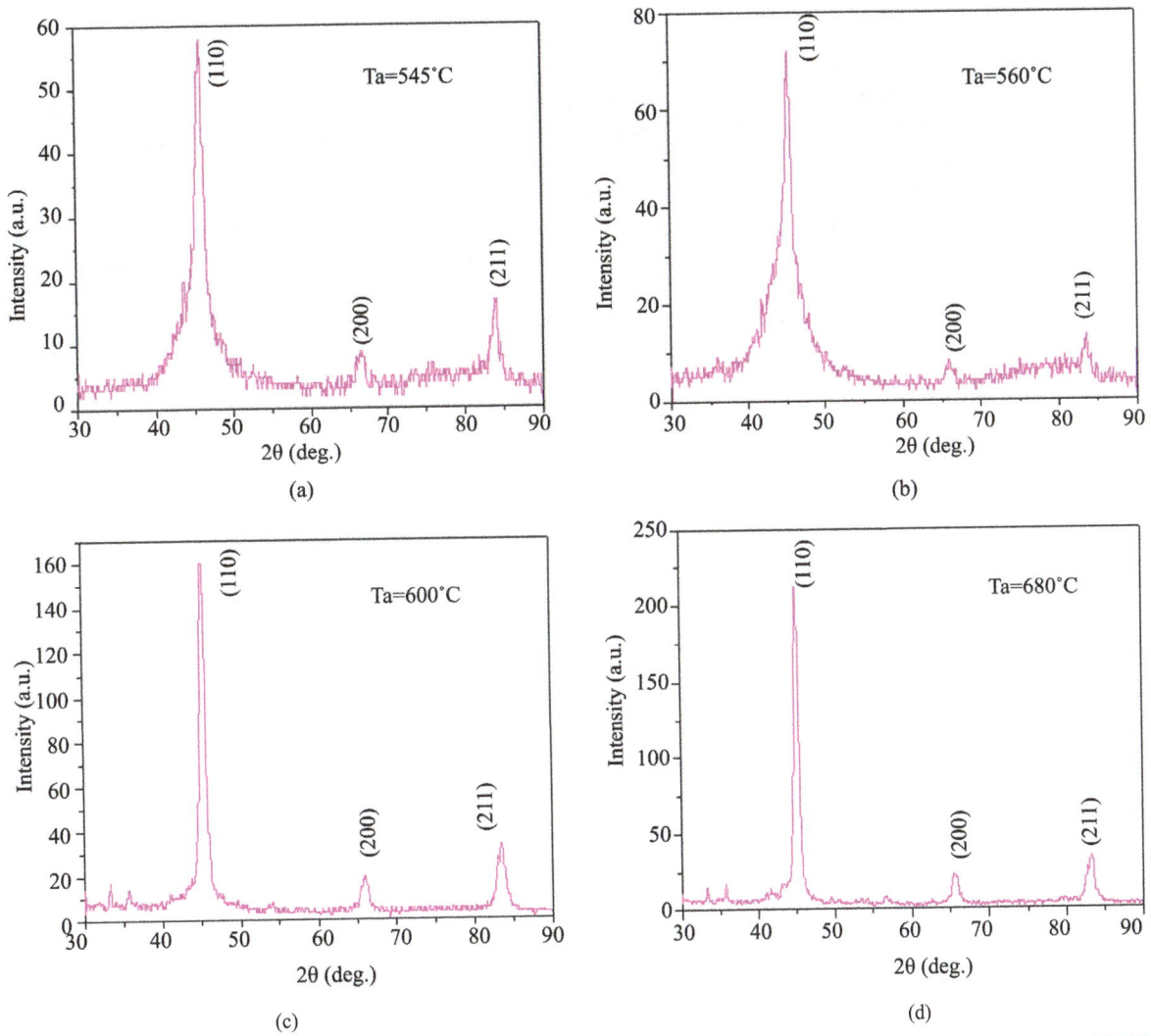

Figure 5. XRD patterns of $Fe_{73.5}Cu_1Nb_3Si_{13.5}B_9$ ribbon samples annealed at (a) 545°C (b) 560°C (c) 600°C and (d) 680°C for 30 min.

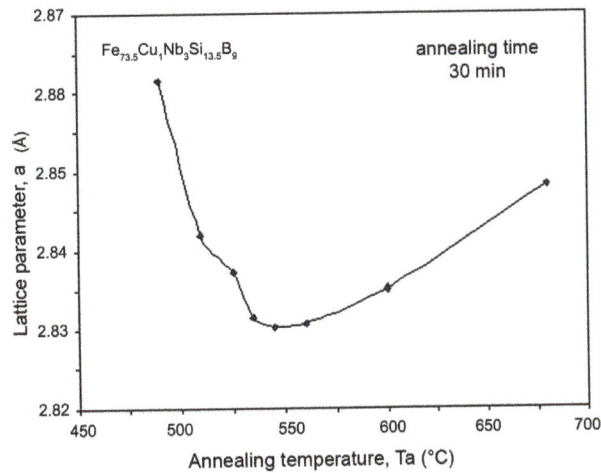

Figure 6. Variation of lattice constant with annealing temperature.

4. Discussion

Crystalline phases that developed through heat treatment have been studied by X-ray diffraction and presented in **Figure 4** and **Figure 5**. In the as cast condition the sample is in the amorphous state having no sharp peaks of crystalline phase since no crystalline phase have formed due to rapid quenching. When the sample is annealed above the crystallization temperature, nanocrystalline grain of α-Fe(Si) was formed in the amorphous precursor. X-ray diffraction results indicate that no α-Fe phases are present in the alloys annealed below 510°C. The appearance of broader diffused pattern is also the characteristic of amorphous material was observed bellow the annealing temperature 510°C. Crystalline phase was developed on the amorphous ribbon, when the alloys were annealed at or above 525°C, which have been identified as a bcc α-Fe(Si) solid solution produced in the amorphous matrix. The peak of the bcc Fe phase in the alloys is increased with increasing of the annealing temperature of the crystalline nanograins of bcc Fe-Si phase was observed from the intensity of the diffraction lines in patterns obtained under the same conditions.

Figure 6 shows the lattice parameter of the bcc Fe-Si phase versus annealing temperature for the alloys. The lattice parameter of the bcc Fe-Si phase decreases with increasing annealing temperature up to 545°C and then the value of lattice parameter increases up to annealing temperature 680°C. It is observed that the lattice parameter slightly decreases with annealing temperature. It means that the silicon content in α-Fe(Si) alloy increases with annealing temperature since it is well known that the lattice parameter of bcc α-Fe(Si) alloys decreases with the increases of silicon content [11] [12]. Since the lattice parameter of α-Fe(Si) phases are always smaller than that of pure α-Fe, the value of which is 2.866 Å. Thus it can be assumed that the decrease of lattice parameter is due to the contraction of α-Fe lattice as a result of diffusion of the silicon with smaller atomic size into the iron lattice with larger atomic size forming a substitutional solid solution during the crystallization process to form α-Fe(Si). But the metalloid element B is practically insoluble in α-Fe (<< 0.01 at.%) and the solubilities [13] of Cu and Nb are low (< 0.2 at.% Cu, <0.1 at.% Nb only above 550°C). Hence, the nanocrystalline phase consists essentially of Fe and Si.

In **Figure 7**, silicon content of α-Fe (Si) nanograins have been presented. A gradual increase of Si content in nanocrystalline phase with increasing annealing temperature is observed. This can be explained by the fact that the element Si from the amorphous phase diffuses into α-Fe space lattice by diffusion during the crystallization process to form α-Fe(Si) nanograins. This means that the crystallization behaviour of this material is a diffusion controlled process with temperature as controlling parameter. So the longer annealing temperature results in more diffusion of Si enriching the Fe(Si) nanograins. **Figure 8** shows the grain size increases gradually and attains a limiting value of 7 to 14 nm until 560°C, which is very suitable for the exchange coupling through residual amorphous matrix [14]. A sudden increase of grain size above 650°C is observed, achieving a value of 18 and 23 nm at 600°C and 680°C, respectively. Because annealing at higher temperature above 680°C leads to the precipitation of Fe-borides [15]. The nanocrystalline material obtained in this way displays improved mechanical and magnetic properties in comparison to the ones achieved by conventional annealing [6] [16] [17].

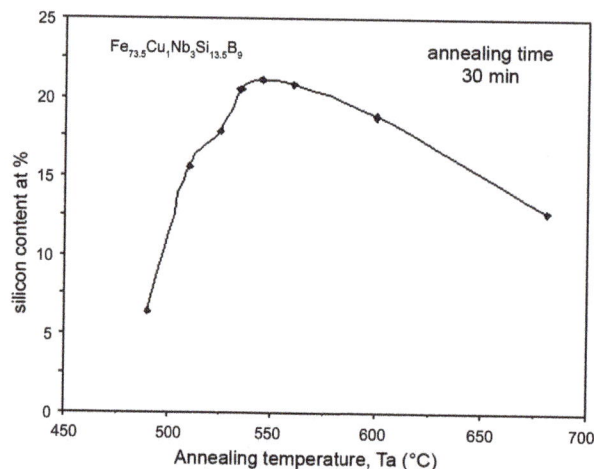

Figure 7. Variation of silicon content at.% with annealing temperature.

Development of Nanostructure Formation of Fe73.5Cu1Nb3Si13.5B9 Alloy from Amorphous...

57

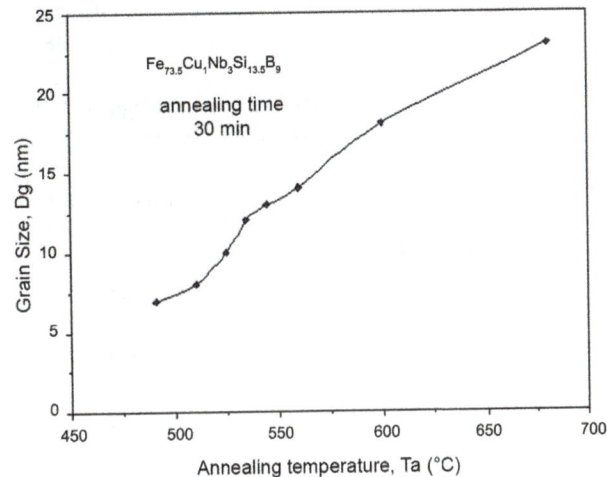

Figure 8. Average grain size as a function of annealing temperature.

5. Conclusion

X-ray diffraction results show that the grain size has been obtained in the range of 7 nm to 21 nm at different stage of annealing and Si content has been reached up to 21.1 at.%. The annealing of magnetic ribbon $Fe_{73.5}Cu_1Nb_3Si_{13.5}B_9$ at 545°C for 30 min leads to the favorable nanocrystalline structure of bcc Fe(Si) mixed with a remaining amorphous fraction. The crystallization behavior of FINEMENT $Fe_{73.5}Cu_1Nb_3Si_{13.5}B_9$ is still an interesting subject of further research.

Acknowledgements

The authors are thankful to Materials Science Division, Atomic Energy Centre, Dhaka-1000, Bangladesh for providing the amorphous $Fe_{73.5}Cu_1Nb_3Si_{13.5}B_9$ ribbon materials and also for extending experimental facilities. We also thank Dr. Sheikh Manjura Hoque, Materials Science Division, Atomic Energy Centre, Dhaka-1000, for her valuable suggestions for the research work.

References

[1] Yoshizawa, Y., Oguma, S. and Yamauchi, K. (1998) New Fe-Based Soft Magnetic Alloys Composed of Ultrafine Grain Structure. *Journal of Applied Physics*, **64**, 6044. http://dx.doi.org/10.1063/1.342149

[2] Yoshizawa, Y. and Yamauchi, K. (1990) Fe-Based Soft Magnetic Alloys Composed of Ultrafine Grain Structure. *Materials Transactions*, **31**, 307. https://www.jim.or.jp/journal/e/31/04/307.html

[3] Yoshizawa, Y., Yamauchi, K., Yamane, T. and Sugihar, H. (1988) Common Mode Choke Cores Using the New Fe-Based Alloys Composed of Ultrafine Grain Structure. *Journal of Applied Physics*, **64**, 6047. http://dx.doi.org/10.1063/1.342150

[4] Yoshizawa, Y. and Yamauchi, K. (1989) Effects of Magnetic Field Annealing on Magnetic Properties in Ultrafine Crystalline Fe-Cu-Nb-Si-B Alloys. *IEEE Transaction on Magnetics*, **25**, 3324-3326. http://dx.doi.org/10.1109/20.42291

[5] Köster, U., Schünemann, U., Blank-Bewersdorff, M., Brauer, S., Sutton, M. and Stephenson, G.B. (1991) Nanocrystalline Materials by Crystallization of Metal-Metalloid Glasses. *Materials Science and Engineering: A*, **133**, 611-615. http://dx.doi.org/10.1016/0921-5093(91)90146-E

[6] Herzer, G. (1989) Grain Structure and Magnetism of Nanocrystalline Ferromagnets. *IEEE Transaction on Magnetics*, **25**, 3327. http://dx.doi.org/10.1109/20.42292

[7] Hilzinger, H.R. (1990) Recent Advances in Rapidly Solidified Soft Magnetic Materials. *Journal of Magnetism and Magnetic Materials*, **83**, 370-374. http://dx.doi.org/10.1016/0304-8853(90)90552-2

[8] Lovas, A., Kiss, L.F., Varga, B., Kamasa, P., Balogh, I. and Bakonyi, I. (1998) Survey of Magnetic Properties during and after Amorphous-Nanocrystalline Transformation. *Journal de Physique IV*, **8**, 291-298. http://dx.doi.org/10.1051/jp4:1998268

[9] Hernando, A., Vázquez, M., Kulik, T., and Prados, C. (1995) Analysis of the Dependence of Spin-Spin Correlations on the Thermal Treatment of Nanocrystalline Materials. *Physical Review B*, **51**, 1381. http://dx.doi.org/10.1103/PhysRevB.51.3581

[10] Franco, V., Conde, C.F., Conde, A. and Kiss L.F. (2000) Superparamagnetic Behaviour in an $Fe_{76}Cu_1Nb_3Si_{10.5}B_{9.5}$ Alloy. *Journal of Magnetism and Magnetic Materials*, **215-216**, 400-403. http://dx.doi.org/10.1016/S0304-8853(00)00170-0

[11] Zhi, J., He, K.-Y., Cheng, L.-Z. and Fu, Y.-J. (1996) Influence of the Elements Si/B on the Structure and Magnetic Properties of Nanocrystalline $(Fe,Cu,Nb)_{77.5}Si_xB_{22.5-x}$ Alloys. *Journal of Magnetism and Magnetic Materials*, **153**, 315-319. http://dx.doi.org/10.1016/0304-8853(95)00548-X

[12] Yoshizawa, Y., Oguma, S. and Yamauchi, K. (1958) A Handbook of the Lattice Spacing and Structure of Metals and Alloys. Pergamon, London, 6044.

[13] Kubaschewski, O. (1982) Iron-Binary Phase Diagrams. Springer Verlag, Berlin, Heidelberg. http://link.springer.com/book/10.1007%2F978-3-662-08024-5

[14] Herzer, G. (1997) Nanocrystalline Soft Magnetic Alloys. In: Buchow, K.H.J., Ed., *Handbook of Magnetic Material*, Vol. 10, Elsevier B.V., Amsterdam.

[15] Allia, P., Baricco, M., Knobel, M., Tiherto, P. and vanai, F. (1994) Soft Nanocrystalline Ferromagnetic Alloys with Improved Ductility Obtained through dc Joule Heating of Amorphous Ribbons. *Journal of Magnetism and Magnetic Materials*, **133**, 243-247. http://dx.doi.org/10.1016/0304-8853(94)90536-3

[16] Dahlgren, M., Grössingera, R., Hernandob, A., Holzera, D., Knobelc, M. and Tibertod, P. (1996) Magnetic Properties of $Fe_{86}Zr_7Cu_1B_6$ at Elevated Temperatures. *Journal of Magnetism and Magnetic Materials*, **160**, 247-248. http://dx.doi.org/10.1016/0304-8853%2896%2900180-1

[17] Herzer, G. (1990) Grain Size Dependence of Coercivity and Permeability in Nanocrystalline Ferromagnets. *IEEE Transaction on Magnetics*, **26**, 1397-1402. http://dx.doi.org/10.1109/20.104389

7

Unusual Oxidation Behaviour of Mesoporous Silicates towards Lignin Model Phenolic Monomer

Rojalin Sadual[1], Sushanta K. Badamali[1*], Sudhir E. Dapurkar[2], Rajesh K. Singh[3]

[1]Department of Chemistry, Utkal University, Bhubaneswar, India
[2]Tata Chemicals Limited, Pune, India
[3]Department of Chemistry, North Orissa University, Baripada, India
Email: [*]skbadamali@gmail.com

Abstract

Oxidation of the lignin model monomer *apocynol*, 1-(4-hydroxy-3-methoxyphenoxy)-ethanol catalysed by mesoporous silica catalysts *i.e.* MCM-41, MCM-48, SBA-15 using H_2O_2 as an oxidant has been studied. Selectively, 2-methoxybenzoquinone was obtained along with acetovanillone. Such unprecedented oxidation behaviour of these metal free siliceous catalysts is attributed to the polar internal surface, high surface area as well as the pore architecture. On the other hand, the studied reaction was found to be non-selective when a commercial grade mesoporous silica *i.e.* Silica-5 was used as catalyst for comparison. Among the various silica catalysts studied, MCMs gave highest conversion and selectivity towards 2-methoxybenzoquinone under very mild reaction conditions.

Keywords

Lignin, Apocynol, Methoxybenzoquinone, Acetovanillone, Mesoporous Silica

1. Introduction

The discovery of ordered mesoporous silicate materials by the Mobil researchers in 1992 [1] with high surface area (700 - 1500 $m^2 \cdot g^{-1}$), tuneable pore size (2 - 20 nm), and large pore volume (0.4 - 1.0 $cm^3 \cdot g^{-1}$) has attracted great interest for their potential applications in the fields of catalysis, chromatography, sensing, shape selective catalysis and support material [2] [3]. Since their discovery, two of the stable members of this family, MCM-41

---

[*]Corresponding author.

having unidimensional pore structure and MCM-48 with three-dimensional pore system have been exploited extensively in the field of catalysis. The scope of application of mesoporous materials is further broadened with the advent of SBA-n molecular sieve and SBA-15 in particular is being widely studied [4]. However, studies on the use of metal free siliceous materials in catalysis are rare and limited. Particularly, purely siliceous framework possesses substantial amounts (~30% - 40%) of silanol (defect sites) groups [5]. And various active functionalities have often been implanted *via* these hydroxyl groups to enable these silicates to be catalytically active. Notably few interesting reports are available on the oxidation behaviour of silica based materials. Morasas and Harrington have reported the oxidative and hydroxylative conversions of polar aromatic compounds over quartz surface [6]. And the peroxidative activity of silica is chiefly attributed to the polar nature of the silica surface [7]. Similarly, the photocatalytic decarboxylation of organic compounds catalyzed in the presence of mesoporous silicas has studied by Itoh *et al.* [8]. Recently, we have reported the oxidative ability of mesoporous silica materials towards lignin model compound under microwave irradiation [9] [10]. It has been demonstrated that α-methylvanillyl alcohol (*apocynol*) can selectively be transformed to either acetovanillone or 2-methoxybenzoquinone, using mesoporous silica and H_2O_2 as oxidant under microwave activation. These motivating results further stimulated us to continue the study on the reaction under conventional heating rather than microwave activation, as the former was less expensive and industrially important.

Among the lignin model phenolic monomer, Apocynol, 1-(4-hydroxy-3-methoxyphenoxy)-ethanol is commonly studied over different catalysts in order to understand the reaction pathway, which can possibly be extended to lignin based biomass. Recently, much attention is focussed on selective transformation of lignin derived fragments to fine chemicals through catalysis [11]. Wozniak *et al.* [12] have studied the liquid phase oxidation of apocynol over potassium nitrosodisulfonate (Fremy's salt) and corresponding benzoquinones were selectively (~80%) obtained over a reaction period of 2 h. Similarly, 2-methoxybenzoquinone was obtained as the common product with maximum selectivity of 50% by using Co(salen) catalysts [13] [14]. Realising the importance of benzoquinones as precursors in various organic reactions, currently the possible generation of quinones from lignin based guaiacyl units is being investigated [15]. In the present study, we report on the unusual oxidation behaviour of the siliceous mesoporous materials towards apocynol using H_2O_2 as an oxidant under mild reaction conditions with highest selectivity towards 2-methoxybenzoquinone. For a comparison commercial grade mesoporous silica (Silica-5, Grace, Germany) was also used as catalyst as received.

2. Experimental

2.1. Preparation of Mesoporous Molecular Sieve

Mesoporous silica catalysts; MCM-48, MCM-41 and SBA-15 were synthesized hydrothermally as per the procedure previously reported [1] [4] [16].

For the synthesis of MCM-48, 0.67 g of sodium hydroxide was dissolved in 17 ml of distilled water and stirred. In another beaker, 7.11 g of hexadecyltrimethylammonium bromide was dissolved in 28.4 ml of distilled water and stirred for 30 min. Then, 7.1 ml of tetraethyl orthosilicate was added to the sodium hydroxide solution with continuous stirring. To this mixture, the surfactant solution was added slowly under stirring conditions and was left out for 30 min so that a uniform solution is formed. The gel was then transferred to a teflon bottle and aged at 373 K for 72 h. The solid product obtained was recovered by repeated washing followed by drying in air oven at 352 K for 6 h. The template was removed by calcining at 823 K for 3 h in flow of nitrogen followed by 3 h in air.

Mesoporous silica SBA-15 was synthesized in an acidic medium with poly(alkylene oxide) triblock copolymers, such as poly(ethylene oxide)-poly(propylene oxide)-poly(ethylene oxide) (PEO-PPO-PEO) as the surfactant, tetraethyl orthosilicate as a source of silica. A total of 4.0 g of Pluronic123 was first added to 120 mL of 2 M HCl along with 30 mL of deionised water under constant stirring until a clear solution was obtained. To this 8.5 g of tetraethyl orthosilicate was added followed by continuous stirring at 313 K for 24 h. The gel was transferred to a teflon bottle and aged at 373 K for 24 h. The solid product was recovered by filtration and repeated washing with deionised water, followed by drying at 343 K for overnight. The template was removed by calcining at 823 K for 6 h in air.

The synthesis of MCM-41 was carried out, in a basic medium with hexadecyltrimethylammonium bromide (CTAB) as surfactant and sodium silicate as the silica source. In a typical synthesis procedure, 40 g of water, 18.7 g of sodium silicate, 1.2 g of sulfuric acid were combined under continuous stirring. The resulting mixture was then allowed for stirring for 15 min. To this, the surfactant solution, prepared by mixing 17.0 g CTAB with 50 ml water, was added and stirred for 30 min. Another 20 g of water was then added to this gel, after this, the resulting gel was crystallized at 373 K for 6 d. The solid was then recovered by washing repeatedly followed by filtration. The recovered products were then dried at 353 K for 12 h. The as-synthesized product was then calcined at 823 K for 3 h in flowing nitrogen, followed by 3 h in flowing air.

2.2. Preparation of Apocynol

Apocynol was synthesized as per the procedure outlined elsewhere [10] with some minor modifications and characterized by melting point, ^1H, ^{13}C NMR (Varian), GC (Varian) and GC-MS (Agilent) studies.

2.3. Characterization of the Catalysts

The synthesized materials were characterized by powder X-ray diffraction (XRD) and nitrogen physisorption measurements. XRD patterns were recorded using Brucker D8 diffractometer with Cu Kα radiation. BET-surface area and pore size were measured by using ASAP 2020, Micromeritics analytical system.

2.4. Catalytic Activity Studies

In a typical catalytic reaction, apocynol (1.0 mmol, 168 mg), acetonitrile (5 mL), the catalyst (100 mg) and 35% aqueous H_2O_2 (3.0 mmol, 0.34 mL) were placed in a round bottomed flask and allowed to react with varying reaction temperature and duration. The reaction mixture was analysed by GC (Varian-450 instrument), with VF-1ms column (15 m × 0.25 mm × 0.39 mm film thickness) and flame ionization detector (FID). Substrate conversion and product selectivity were determined using an external standard method, in a similar procedure described earlier [10].

3. Results and Discussion

The XRD patterns of the calcined mesoporous silica catalysts are shown in **Figure 1**. XRD patterns of calcined, MCM-41 (**Figure 1(a)**) and MCM-48 (**Figure 1(b)**) show a strong reflection observed in between $2\theta = 2.2°$ - 3° in addition to several weak reflections in the range of $2\theta = 3°$ - 7° indicated the formation of mesoporous structure. Similarly, in case of SBA-15 (**Figure 1(c)**), a strong reflection at $2\theta = 1.06°$ and several weak reflections between $2\theta = 1.5°$ - 3.0° confirms its structure. The BET surface area and pore size were determined to be 978 $m^2 \cdot g^{-1}$ and 2.6 nm for MCM-48 while that for MCM-41 and SBA-15 was found to be 850 $m^2 \cdot g^{-1}$, 2.5 nm and 940 $m^2 \cdot g^{-1}$, 6.1 nm respectively [17].

Figure 2 shows the FT-IR spectra of calcined MCM-41 (a) MCM-48 (b) and SBA-15 (c) respectively. All the calcined spectra, show absence of the characteristic band for surfactant template molecule at 2928, 2854 and 1490 cm^{-1} leaving behind other characteristic peaks of mesoporous material. In addition to above features, absorption bands noticed at around 1072, 808 and 460 cm^{-1} are originating from the asymmetric, symmetric stretching and bending vibration for Si-O-Si bonds [18] respectively. A strong band at 966 cm^{-1} seen clearly, in all the spectra, is attributed to terminal Si-O stretching of silanol groups. The occurrence of broad bands between 3000 - 3800 cm^{-1} in case of all the samples indicates the presence of surface hydroxyl groups, characteristic of mesoporous structure [19]. It is worth mentioning that both MCM-41 and MCM-48 show intense band in this range compared to SBA-15. This is attributed to the abundantly presence of incompletely condensed silica as evidenced from ^{29}Si MAS-NMR data [20]. The predominant presences of silanol ($\equiv$SiOH) groups impart hydrophilic nature to these MCM-n type materials.

4. Catalytic Experiments

We have undertaken a detail investigation on the oxidation of apocynol, **1** (**Scheme 1**), catalyzed by different catalysts under varying reaction conditions such as temperature, substrate: oxidant, duration and the results are summarized in **Table 1**.

Over all the silica catalysts used, it was noticed that within a very short period (15 min) major substrate was reacted yielding both **2** and **3**. Acetovanillone is the early oxidation product [21], and 2-methoxybenzoquinone is believed to be formed by the generation of phenoxy radical and subsequent cleavage of side chain [22].

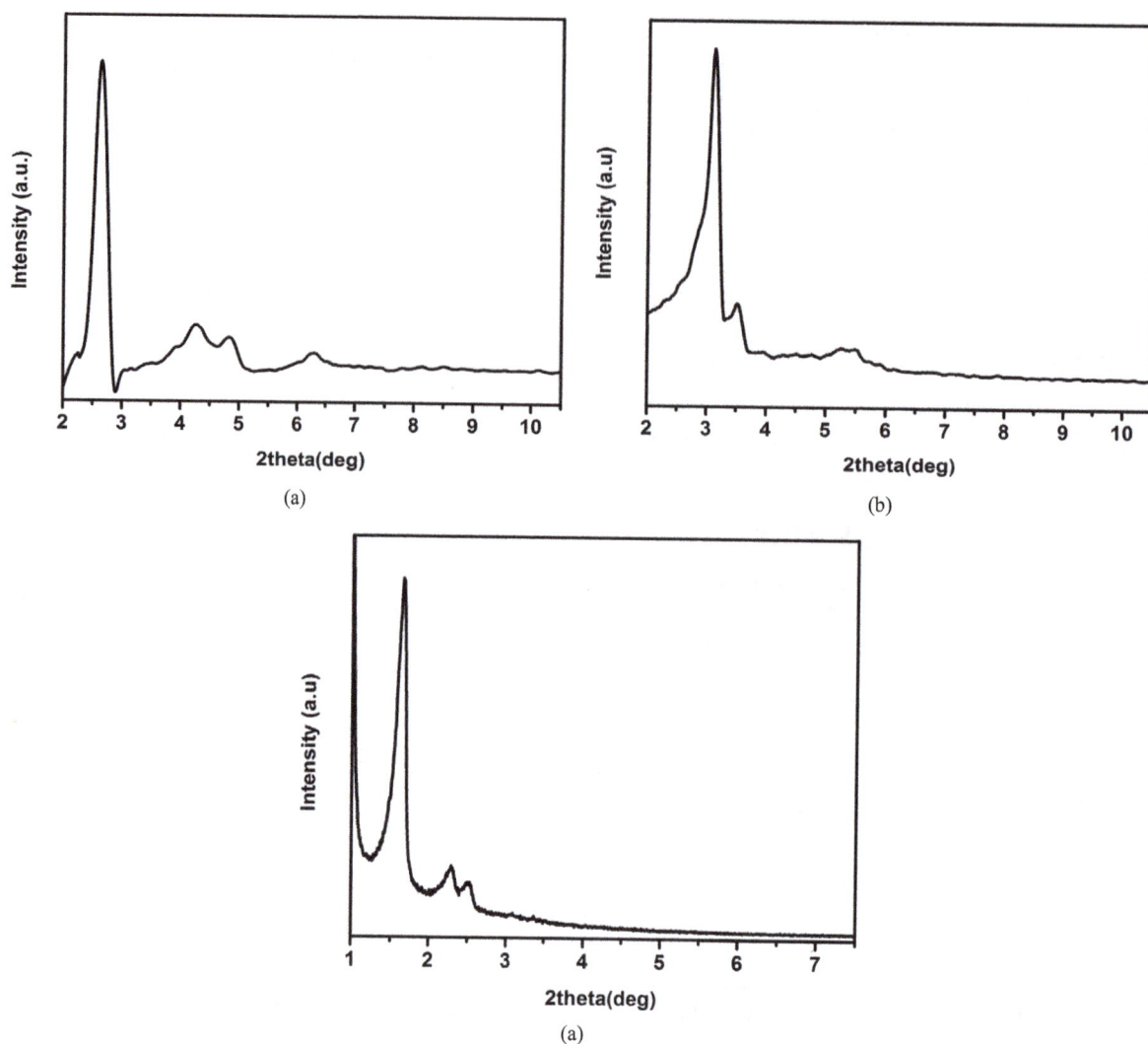

Figure 1. XRD patterns of calcined (a) MCM-41; (b) MCM-48 and (c) SBA-15.

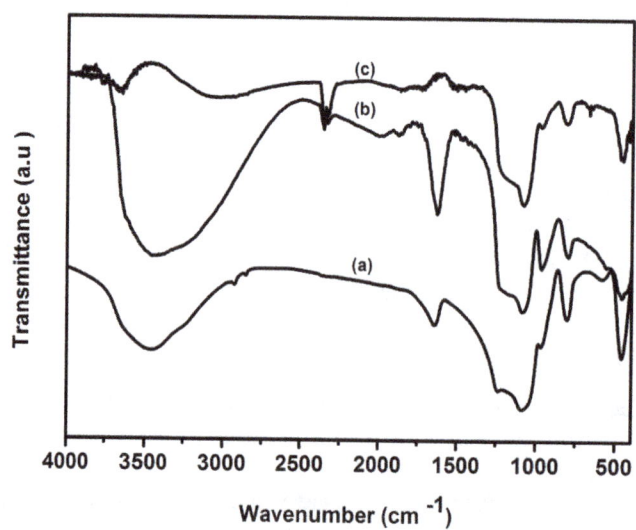

Figure 2. FT-IR spectra of calcined (a) MCM-41; (b) MCM-48 and (c) SBA-15.

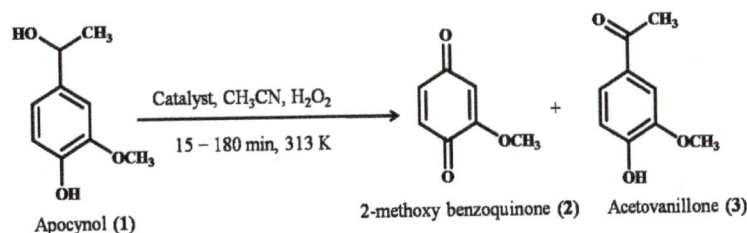

Scheme 1. Reaction profile of oxidation of apocynol (1).

Table 1. Reaction profile of oxidation of apocynol (1).

Entry	Catalyst	Duration (min)	Conversion (%)	Product distribution (%)	
			(apocynol)	(2-methoxy benzoquinone)	(acetovanillone)
1	MCM-41	15	84	65	19
		30	96	77	4
		180	98	84	
2	MCM-48	15	95	65	16
		30	94	67	-
		180	>99	40	-
3	SBA-15	15	83	49	22
		30	89	40	21
		180	93	20	-
4	Silica-5	15	76	31	28
5	H_2O_2 only	15	80	22	15
6	Blank	180	-	-	-

Reaction conditions: 1.0 mmol apocynol, 3 mmol H_2O_2, 5 mL acetonitrile, 100 mg catalyst at 313 K, Duration 15 - 180 min.

Interestingly, about 98% of substrate conversion was achieved over MCM-41 with best quinone selectivity (~84%) among the catalysts screened. In general an optimum conversion and selectivity was obtained within 30 min of reaction. Continuing the reactions beyond 30 min resulted in an increase of substrate conversion with fall in selectivity, when MCM-48 and SBA-15 were used as catalysts. However, MCM-41 maintained a steady increase in conversion and selectivity.

It is well evident that apocynol is quite stable irrespective of 3 h of heating in the solvent (entry-6). And in presence of H_2O_2 (entry-5), 1 showed high conversion resulting in 2-methoxybenzoquinone (2), acetovanillone (3) along with high molecular weight (unidentified) products. Additional products detected in the higher retention time, are attributed to the possible polymerisation of quinone to form high molecular weight compounds. In order to understand the influence of nature of silica and porous structure, Silica-5 having high surface area with irregular porous structure was employed as catalyst, which gave a lower substrate conversion with nearly equal selectivity towards 2-methoxybenzoquinone and acetovanillone (entry-4). These results indicated the fact that an ordered porous silicate surface as well as the pore architecture governs both the conversion and selectivity.

Another significant feature of this reaction has been, with increase in reaction duration the apocynol conversion remained almost same, while significant fall in acetovanillone yield. In order to understand this aspect, we further investigated the reaction profile by varying substrate: oxidant amount. Representative results of

MCM-48 catalyst are presented in **Table 2**. It is well evident that with increase in the oxidant amount the selectivity of the 2-methoxybenzoquinone was enhanced with drop in acetovanillone yield. These observations point toward that 2-methoxybenzoquinone is the stable product which is formed at the expense of acetovanillone and similar trend was observed over other catalysts studied. Based on the results obtained, the product formation is likely to occur through a plausible pathway (**Scheme 2**). The polar surface along with high surface area favours the adsorption of both apocynol and hydrogen peroxide (H_2O_2). Subsequently, binding of H_2O_2 onto the -OH group leads to activation of the oxidant resulting in possible formation of hydroxyl radicals. The hydroxyl radical formed either abstracts a phenolic hydrogen atom from the phenolic substrate, **1** (path-1) and produces a reactive phenoxy radical, **4** or it abstracts a proton from the benzylic carbon (path-2), forming the reactive radical, **6**. By the path-1, **4**, transformed to intermediate, **5**, through addition of another hydroxyl radical. Intermediate **5**, undergoes rearrangement to form 2-methoxybenzoquinone, **2**. On the other hand, through path-2, intermediate **6** undergoes deprotonation via another hydroxyl radical leading to the formation of acetovanillone, **3**.

The unlikely oxidation behaviour of the mesoporous silicas is attributed to the abundance of surface hydroxyl groups, polar internal surface, high surface area and pore geometry. At this juncture the facts suggest that,

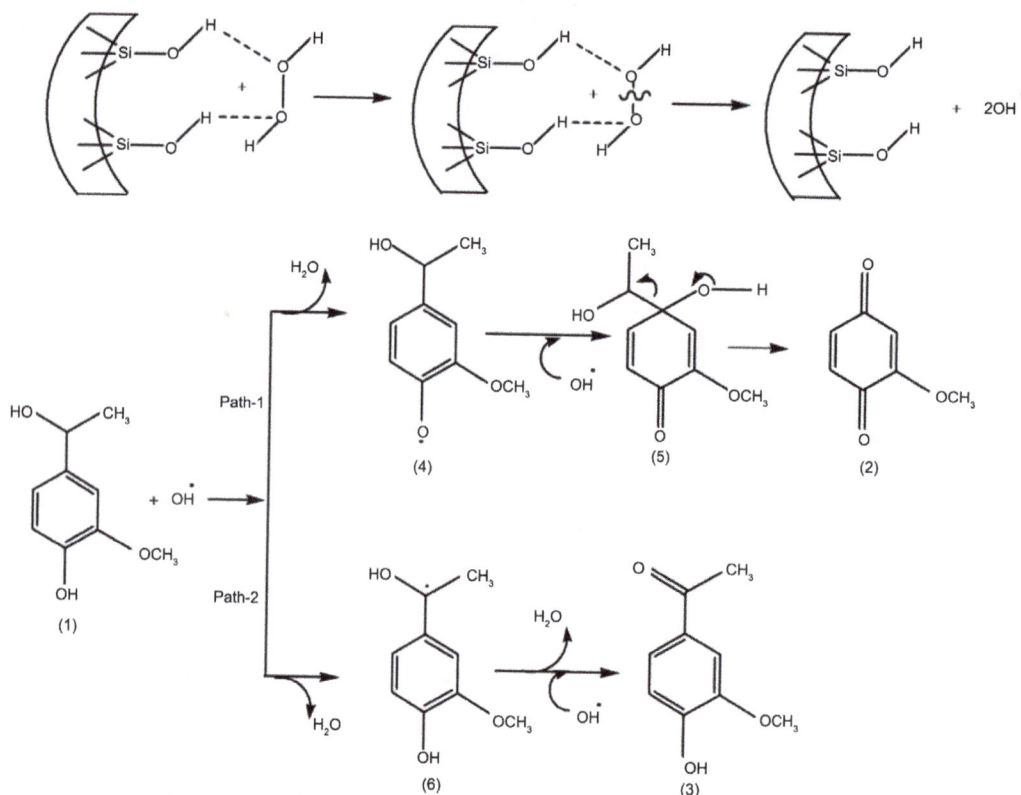

Scheme 2. Probable oxidation pathway of (1) over silica catalysts.

Table 2. Reaction profile of oxidation of apocynol (1) by varying oxidant amount over MCM-48.

Entry	Substrate:oxidant amount	Substrate (1) conv. (%)	Product (2) yield (%)	Product (3) yield (%)
1	In absence of H_2O_2	30	–	–
2	Sub:H_2O_2 = 1:1	58	30	29
3	Sub:H_2O_2 = 1:2	56	70	18
4	Sub:H_2O_2 = 1:3	94	67	–

Reaction conditions: 1.0 mmol apocynol, 5 mL acetonitrile, 100 mg catalyst at 313 K, duration 180 min.

apocynol (**1**) is susceptible to oxidation in presence of H_2O_2 alone; and the reaction proceeds in an uncontrolled way resulting in several high molecular weight products (not identified). On the other hand, mesoporous silicates along with H_2O_2 catalytic system produced 2-methoxybenzoquinone predominantly, and the reaction was much controlled over silicate catalysts and hydrogen peroxide (H_2O_2) oxidant system.

5. Conclusion

In conclusion, we have demonstrated a clean, efficient catalyst system for the selective oxidation of an important lignin model phenolic monomer. Metal free mesoporous silicas along with environmentally benign oxidant, H_2O_2, were found to be highly efficient catalytic system for the chosen oxidation reaction, and quinone was selectively obtained under mild reaction conditions. These results open up for further study to evaluate and understand the oxidative nature of mesoporous amorphous silica materials.

Acknowledgements

We wish to thank Department of Science and Technology, Govt. of India for the financial support under FAST track (SR/FT/CS-93/2007) and PURSE scheme. Authors appreciate many meaningful discussions with Dr. A. Sakthivel, Department of Chemistry, Delhi University, relating the reaction.

References

[1]   Beck, J.S., Vartuli, J.C., Roth, W.J., Leonowicz, M.E., Kresge, C.T., Schmitt, K.D., Chu, C.T.W., Olson, D.H., Sheppard, E.W., McCullen, S.B., Higgins, J.B. and Schlenker, J.L. (1992) A New Family of Mesoporous Molecular Sieves Prepared with Liquid Crystal Templates. *Journal of the American Chemical Society*, **114**, 10834-10843. http://dx.doi.org/10.1021/ja00053a020

[2]   Fryxell, G.E. and Liu, J. (2000) Adsorption on Silica Surfaces. In: Papirer, E., Ed., *Designing Surface Chemistry in Mesoporous Silica*, Marcel Dekker, Inc., New York.

[3]   Cao, G. (2004) Nanostructures and Nanomaterials. Synthesis, Properties, and Applications. Imperial College Press, London. http://dx.doi.org/10.1142/9781860945960

[4]   Zhao, D., Huo, Q., Feng, J., Chmelka, B.F. and Stucky, G.D. (1998) Nonionic Triblock and Star Diblock Copolymer and Oligomeric Surfactant Syntheses of Highly Ordered, Hydrothermally Stable, Mesoporous Silica Structures. *Journal of the American Chemical Society* **120**, 6024-6036. http://dx.doi.org/10.1021/ja974025i

[5]   Selvam, P. and Dapurkar, S.E. (2004) Catalytic Activity of Highly Ordered Mesoporous VMCM-48. *Applied Catalysis A: General*, **276**, 257-265. http://dx.doi.org/10.1016/j.apcata.2004.08.012

[6]   Marasas, L.W. and Harrington, J.S. (1960) Some Oxidative and Hydroxylative Actions of Quartz: Their Possible Relationship to the Development of Silicosis. *Nature*, **188**, 1173-1174. http://dx.doi.org/10.1038/1881173a0

[7]   Schofield, P.J., Ralph, B.J. and Green, J.H. (1964) Mechanisms of Hydroxylation of Aromatics on Silica Surfaces. *Journal of Physical Chemistry*, **68**, 472-476. http://dx.doi.org/10.1021/j100785a006

[8]   Itoh, A., Kodama, T., Masuki, Y. and Inagaki, S. (2006) Oxidative Photo-Decarboxylation in the Presence of Mesoporous Silicas. *Chemical and Pharmaceutical Bulletin*, **54**, 1571-1575. http://dx.doi.org/10.1248/cpb.54.1571

[9]   Badamali, S.K., Luque, R., Clark, J.H. and Breeden, S.W. (2013) Unprecedented Oxidative Properties of Mesoporous Silica Materials: Towards Microwave-Assisted Oxidation of Lignin Model Compounds. *Catalysis Communications*, **31**, 1-4. http://dx.doi.org/10.1016/j.catcom.2012.11.006

[10]   Badamali, S.K., Clark, J.H. and Breeden, S.W. (2008) Microwave Assisted Selective Oxidation of Lignin Model Phenolic Monomer over SBA-15. *Catalysis Communications*, **9**, 2168-2170. http://dx.doi.org/10.1016/j.catcom.2008.04.012

[11]   Holladay, J.E., Bozell, J.J., White, J.F. and Johnson, D. (2007) Top Value-Added Chemicals from Biomass, Results of Screening for Potential Candidates from Biorefinery Lignin, II, USDOE, PNNL-16983.

[12]   Wozniak, J.C., Dimmel, D.R. and Malcom, E.W. (1990) The Generation of Quinones from Lignin and Lignin-Related Compounds, Diels-Adler Reactions of Lignin-Derived Quinones, Lignin-Derived Quinones as Pulping Additives. Institute of Paper Science and Technology Paper Series No. 349, 1-57.

[13]   Cedeno, D. and Bozell, J.J. (2012) Catalytic Oxidation of Para-Substituted Phenols with Cobalt-Schiff Base Complexes/O_2—Selective Conversion of Syringyl and Guaiacyl Lignin Models to Benzoquinones. *Tetrahedron Letters*, **53**, 2380-2383. http://dx.doi.org/10.1016/j.tetlet.2012.02.093

[14]   Canevali, C., Orlandi, M., Pardi, L., Rindone, B., Scotti, R., Sipila, J. and Morazzone, F. (2002) Oxidative Degradation

of Monomeric and Dimeric Phenylpropanoids: Reactivity and Mechanistic Investigation. *Journal of the Chemical Society, Dalton Transactions*, No. 15, 3007-3014. http://dx.doi.org/10.1039/b203386k

[15] Biannic, B. and Bozell, J.J. (2013) Efficient Cobalt-Catalyzed Oxidative Conversion of Lignin Models to Benzoquinones. *Organic Letters*, **15**, 2730-2733. http://dx.doi.org/10.1021/ol401065r

[16] Schmidt, R., Stocker, M., Akporiaye, D., Torstad, E.H. and Olsen, A. (1995) High-Resolution Electron Microscopy and X-Ray Diffraction Studies of MCM-48. *Microporous Materials*, **5**, 1-7. http://dx.doi.org/10.1016/0927-6513(95)00030-D

[17] Sadual, R. (2015) Oxidation of Lignin Model Phenolic Monomer over Cobalt Containing Solid Catalysts. PhD Thesis, North Orissa University, Baripada.

[18] Flanigen, E.M., Khatami, H. and Szymanski, H.A. (1971) Infrared Structural Studies of Zeolite Frameworks. In: Flanigen, E.M. and Sand, L.B., Ed., *Molecular Sieve Zeolites*, Advances in Chemistry 101, American Chemical Society, Washington DC, 201-229. http://dx.doi.org/10.1021/ba-1971-0102

[19] Chen, J.S., Li, Q.H., Xu, R.R. and Xiao, F.S. (1995) Distinguishing the Silanol Groups in the Mesoporous Molecular Sieve MCM-41. *Angewandte Chemie International Edition in English*, **34**, 2694-2696. http://dx.doi.org/10.1002/anie.199526941

[20] Zhao, X.S., Lu, G.Q., Whittaker, A.K., Millar, G.J. and Zhu, H.Y. (1997) Comprehensive Study of Surface Chemistry of MCM-41 Using ^{29}Si CP/MAS NMR, FTIR, Pyridine-TPD, and TGA. *The Journal of Physical Chemistry B*, **101**, 6525-6531. http://dx.doi.org/10.1021/jp971366+

[21] Crestini, C., Pastoni, A. and Taguatesta, P. (2004) Metalloporphyrins Immobilized on Motmorillonite as Biomimetic Catalysts in The Oxidation of Lignin Model Compounds. *Journal of Molecular Catalysis A: Chemical*, **208**, 195-202. http://dx.doi.org/10.1016/j.molcata.2003.07.015

[22] Sadual, R., Badamali, S.K. and Singh, R.K. (2014) Studies on Mesoporous CoSBA-15 Catalysed Selective Oxidation of a Lignin Model Phenolic Monomer. *Advanced Porous Materials*, **2**, 48-53. http://dx.doi.org/10.1166/apm.2014.1044

8

Exact Traveling Wave Solutions for Nano-Solitons of Ionic Waves Propagation along Microtubules in Living Cells and Nano-Ionic Currents of MTs

Emad H. M. Zahran

Department of Mathematical and Physical Engineering, College of Engineering, University of Benha, Shubra, Egypt
Email: e_h_zahran@hotmail.com

Abstract

In this work, the extended Jacobian elliptic function expansion method is used as the first time to evaluate the exact traveling wave solutions of nonlinear evolution equations. The validity and reliability of the method are tested by its applications to nano-solitons of ionic waves propagation along microtubules in living cells and nano-ionic currents of MTs which play an important role in biology.

Keywords

Extended Jacobian Elliptic Function Expansion Method, Nano-Solitons of Ionic Waves Propagation along Microtubules in Living Cells, Nano-Ionic Currents of MTs, Traveling Wave Solutions

1. Introduction

The nonlinear partial differential equations of mathematical physics are major subjects in physical science [1]. Exact solutions for these equations play an important role in many phenomena in physics such as fluid mechanics, hydrodynamics, optics, and plasma physics. Recently many new approaches for finding these solutions have been proposed, for example, tanh-sech method [2]-[4], extended tanh-method [5]-[7], sine-cosine method [8]-[10], homogeneous balance method [11] [12], F-expansion method [13]-[15], exp-function method [16] [17], trigonometric function series method [18], (G'/G) expansion method [19]-[22], Jacobi elliptic function method [23]-[26] and so on.

The objective of this article is to apply the extended Jacobian elliptic function expansion method for finding the exact traveling wave solution of nano-solitons of ionic waves propagate on along microtubules in living cells and nano-ionic currents of MTs which play an important role in biology and mathematical physics.

The rest of this paper is organized as follows: In Section 2, we give the description of the extended Jacobi elliptic function expansion method. In Section 3, we use this method to find the exact solutions of the nonlinear evolution equations pointed out above. In Section 4, conclusions are given.

2. Description of Method

Consider the following nonlinear evolution equation

$$F\left(u, u_t, u_x, u_{tt}, u_{xx}, \cdots\right) = 0,$$
(2.1)

where F is polynomial in $u(x, t)$ and its partial derivatives in which the highest order derivatives and nonlinear terms are involved. In the following, we give the main steps of this method [23]-[26].

Step 1. Using the transformation

$$u = u(\xi), \ \xi = x - ct,$$
(2.2)

where k and c are the wave number and wave speed, to reduce Equation (2.1) to the following ODE:

$$P\left(u, u', u'', u''', \cdots\right) = 0,$$
(2.3)

where P is a polynomial in $u(\xi)$ and its total derivatives, while $u' = du/d\xi$.

Step 2. Making good use of ten Jacobian elliptic functions, we assume that (2.3) have the solutions in these forms:

$$u(\dot{\varepsilon}) = a_0 + \sum_{j=1}^{N} f_i^{j-1}(\dot{\varepsilon})\left[a_j f_i(\dot{\varepsilon}) + b_j g_i(\dot{\varepsilon})\right], \ i = 1, 2, 3, \cdots$$
(2.4)

with

$$f_1(\xi) = sn\xi, \ g_1(\xi) = cn\xi,$$

$$f_2(\xi) = sn\xi, \ g_2(\xi) = dn\xi,$$

$$f_3(\xi) = ns\xi, \ g_3(\xi) = cs\xi,$$

$$f_4(\xi) = ns\xi, \ g_4(\xi) = ds\xi,$$

$$f_5(\xi) = sc\xi, \ g_5(\xi) = nc\xi,$$

$$f_6(\xi) = sd\xi, \ g_6(\xi) = nd\xi,$$
(2.5)

where $sn\xi$, $cn\xi$, $dn\xi$, are the Jacobian elliptic sine function. The jacobian elliptic cosinefunction and the Jacobian elliptic function of the third kind and other Jacobian functions which is denoted by Glaisher's symbols and are generated by these three kinds of functions, namely

$$ns\xi = \frac{1}{sn\xi}, \ nc\xi = \frac{1}{cn\xi}, \ nd\xi = \frac{1}{dn\xi}, \ sc\xi = \frac{cn\xi}{sn\xi},$$

$$cs\xi = \frac{sn\xi}{cn\xi}, \ ds\xi = \frac{dn\xi}{sn\xi}, \ sd\xi = \frac{sn\xi}{dn\xi}$$
(2.6)

that has the relations

$$sn^2\xi + cn^2\xi = 1, \ dn^2\xi + m^2 sn^2\xi = 1, \ ns^2\xi = 1 + cs^2\xi,$$

$$ns^2\xi = m^2 + ds^2\xi, \ sc^2\xi + 1 = nc^2\xi, \ m^2 sd^2 + 1 = nd^2\xi$$
(2.7)

with the modulus m ($0 < m < 1$): In addition we know that

$$\frac{d}{d\xi} sn\xi = cn\xi dn\xi, \ \frac{d}{d\xi} cn\xi = -sn\xi dn\xi, \ \frac{d}{d\xi} dn\xi = -m^2 sn\xi cn\xi.$$
(2.8)

The derivatives of other Jacobian elliptic functions are obtained by using Equation (2.8). To balance the highest order linear term with nonlinear term we define the degree of u as $D[u] = n$ which gives rise to the degrees of other expressions as

$$D\left[\frac{d^q u}{d\xi^q}\right] = n + q, \quad D\left[u^p\left(\frac{d^q u}{d\xi^q}\right)^s\right] = np + s(n+q). \tag{2.9}$$

According the rules, we can balance the highest order linear term and nonlinear term in Equation (2.3) so that n in Equation (2.4) can be determined.

In addition we see that when $m \to 1$, $sn\xi$, $cn\xi$, $dn\xi$ degenerate as $\tanh\xi$, $sech\xi$, $sech\xi$, respectively, while when therefore Equation (2.5) degenerate as the following forms

$$u(\xi) = a_0 + \sum_{j=1}^{N} \tanh_i^{j-1}(\xi)\left[a_j \tanh(\xi) + b_j sech(\xi)\right], \tag{2.10}$$

$$u(\xi) = a_0 + \sum_{j=1}^{N} \coth_i^{j-1}(\xi)\left[a_j \coth(\xi) + b_j \coth(\xi)\right], \tag{2.11}$$

$$u(\xi) = a_0 + \sum_{j=1}^{N} \tan_i^{j-1}(\xi)\left[a_j \tan(\xi) + b_j \sec(\xi)\right], \tag{2.12}$$

$$u(\xi) = a_0 + \sum_{j=1}^{N} \cot_i^{j-1}(\xi)\left[a_j \cot(\xi) + b_j \csc(\xi)\right]. \tag{2.13}$$

Therefore the extended Jacobian elliptic function expansion method is more general than sine-cosine method, the tan-function method and Jacobian elliptic function expansion method.

3. Application

3.1. Example 1: Nano-Solitons of Ionic Waves Propagation along Microtubules in Living Cells [27]

We first consider an inviscid, incompressible and non-rotating flow of fluid of constant depth (h). We take the direction of flow as x-axis and z-axis positively upward the free surface ingravitational field. The free surface elevation above the undisturbed depth h is $\eta(x; t)$, so that the wave surface at height $z = h + \eta(x; t)$, while $z = 0$ is horizontal rigid bottom.

Let $\varphi(x; z; t)$ be the scalar velocity potential of the fluidlying between the bottom ($z = 0$) and free space $\eta(x; t)$, then we could write the Laplace and Euler equation with the boundary conditions at the surface and the bottom, respectively, as follows:

$$\frac{\partial^2 \varphi}{\partial x^2} + \frac{\partial^2 \varphi}{\partial z^2} = 0; \quad 0 < z < h + \eta; \quad -\infty < x < +\infty \tag{3.1}$$

$$\frac{\partial \varphi}{\partial t} + \frac{1}{2}\left(\frac{\partial \varphi}{\partial z}\boldsymbol{i} + \frac{\partial \varphi}{\partial z}\boldsymbol{k}\right)^2 + g\eta = 0, \quad z = h + \eta \tag{3.2}$$

$$\frac{\partial \eta}{\partial t} + \frac{\partial \eta}{\partial x}\frac{\partial \varphi}{\partial x} - \frac{\partial \varphi}{\partial z} = 0, \tag{3.3}$$

$$\frac{\partial \varphi}{\partial z} = 0; \quad z = 0 \tag{3.4}$$

It is useful to introduce two following fundamental dimensionless parameters:

$$\sigma = \frac{\eta_0}{h} < 1; \quad \delta = \left(\frac{h}{l}\right)^2 < 1, \tag{3.5}$$

where η_0 is the wave amplitude, and l is the characteristic length-like wavelength. Accordingly, we also take a complete set of new suitable non-dimensional variables:

$$x = \frac{x}{l}; \quad z = \frac{z}{h}; \quad \tau = \frac{ct}{l}; \quad \psi = \frac{\eta}{\eta_0}; \quad \varnothing = \frac{h}{\eta_0 lc}\varphi, \tag{3.6}$$

where $c = \sqrt{gh}$ is the shallow-water wave speed, with g being gravitational acceleration.

In term of (3.5) and (3.6) the initial system of Equation (3.1)-(3.4) now reads

$$\delta \frac{\partial^2 \varphi}{\partial x^2} + \frac{\partial^2 \varphi}{\partial z^2} = 0;$$

(3.7)

$$\frac{\partial \varnothing}{\partial \tau} + \frac{\sigma}{2} \left(\frac{\partial \varnothing}{\partial x} \right)^2 + \frac{\sigma}{2} \left(\frac{\partial \varnothing}{\partial z} \right)^2 + \psi = 0; \quad Z = 1 + \sigma \psi,$$

(3.8)

$$\frac{\partial \psi}{\partial \tau} + \sigma \left(\frac{\partial \varnothing}{\partial x} \frac{\partial \psi}{\partial x} \right) - \frac{1}{\delta} \frac{\partial \varnothing}{\partial z} = 0; \quad Z = 1 + \sigma \psi,$$

(3.9)

$$\frac{\partial \varphi}{\partial z} = 0; \quad z = 0$$

(3.10)

Expanding $\varnothing(x;t)$ in terms of δ

$$\varnothing = \varnothing_0 + \delta \varnothing_1 + \delta^2 \varnothing_2,$$

(3.11)

and using the dimensionless wave particles velocity in x-direction, by definition $u = \frac{\partial \varnothing}{\partial x}$, then substituting of (3.11) into (3.7)-(3.9), with retaining terms up to linear order of small parameters (σ, δ) in (3.8), and second order in (3.9), we get

$$\frac{\partial \varphi_0}{\partial \tau} - \frac{\delta}{2} \frac{\partial^2 u}{\partial \tau \partial x} + \psi + \frac{1}{2} \sigma u^2 = 0,$$

(3.12)

$$\frac{\partial \psi}{\partial \tau} + \sigma u \frac{\partial \psi}{\partial x} + \frac{1}{\delta} (1 + \sigma \psi) \frac{\partial u}{\partial x} = \frac{\delta}{6} \frac{\partial^3 u}{\partial x^3}.$$

(3.13)

Making the differentiation of (3.12) with respect to x, and rearranging (3.13), we get

$$\frac{\partial u}{\partial \tau} + \sigma u \frac{\partial u}{\partial x} + \frac{\partial \psi}{\partial x} - \frac{1}{2} \delta \frac{\partial^3 u}{\partial x^2 \partial \tau} = 0,$$

(3.14)

$$\frac{\partial \psi}{\partial \tau} + \frac{\partial}{\partial x} \left[u(1 + \sigma \psi) \right] - \frac{1}{6} \delta \frac{\partial^3 u}{\partial x^3} = 0.$$

(3.15)

Returning back to dimensional variables $\eta(x; t)$ and $= d\varphi/dx$, (3.14) now reads

$$\frac{\partial v}{\partial t} + v \frac{\partial v}{\partial x} + g \frac{\partial \eta}{\partial x} = \frac{1}{3} h^2 \frac{\partial^3 u}{\partial x^2 \partial t}.$$

(3.16)

We could define the new function $V(x, t)$ unifying the velocity and displacement of water particles as follows:

$$v = \frac{1}{h} \frac{\partial V}{\partial t}; \quad \eta = -\frac{\partial V}{\partial x}$$

(3.17)

implying that (3.16) becomes

$$\frac{\partial^2 V}{\partial t^2} - gh \frac{\partial^2 V}{\partial x^2} + \frac{1}{2h} \frac{\partial}{\partial x} \left(\frac{\partial V}{\partial t} \right)^2 = \frac{1}{3} h^2 \frac{\partial^4 V}{\partial x^2 \partial t^2}.$$

(3.18)

We seek for traveling wave solutions with moving coordinate of the form $\xi = x - vt$ and with wave speed v, which reduces Equation (3.18) into ordinary nonlinear differential equation as follows:

$$(v^2 - gh) \frac{\partial^2 V}{\partial \zeta^2} + \frac{v^2}{2h} \frac{\partial}{\partial \zeta} \left(\frac{\partial V}{\partial \zeta} \right)^2 = \frac{1}{3} h^2 v^2 \frac{\partial^4 V}{\partial \zeta^4}$$

(3.19)

Integrating Equation (3.19) once, and setting $\frac{\partial V}{\partial \xi} = W$, we get

$$\frac{\partial^2 w}{\partial \xi^2} = \alpha w^2 + \beta w + c_1 \tag{3.20}$$

Balancing w'' and w^2 yields, $N + 2 = 2N \rightarrow N = 2$. Therefore, we can write the solution of Equation (3.20) in the form

$$W(\xi) = a_0 + a_1 sn + b_1 cn + a_2 sn^2 + b_2 sncn \tag{3.21}$$

$$W'(\xi) = \left(a_1 cn - b_1 sn + 2a_2 sncn + b_2 - 2b_2 sn^2 \right) dn \tag{3.22}$$

$$W''(\xi) = 6a_2 m^2 + sn^4 + 2a_1 m^2 sn^3 - 4a_2 \left(m^2 + 1 \right) - a_1 \left(m^2 + 1 \right) sn \tag{3.23}$$
$$+ 6b_2 sm^2 sn^3 cn + 2b_1 m^2 sn^2 cn - b_2 \left(4 + m^2 \right) sncn - b_1 cn + 2a_2$$

Substituting (3.21) into (3.23), setting the coefficients of $(sn^4, sn^3, sn^3 cn, sn^2, sn^2 cn, sncn, sn, cn, sn^0)$ to zero, we obtain the following underdetermined system of algebraic equations for $(a_0, a_1, a_2, b_1, b_2)$:

$$6a_2 m^2 - \alpha \left(a_2^2 - b_2^2 \right) = 0, \tag{3.24}$$

$$2a_1 m^2 - 2\alpha \left(a_1 a_2 - b_1 b_2 \right) = 0, \tag{3.25}$$

$$-4a_2 \left(m^2 + 1 \right) - \alpha \left(a_1^2 - b_1^2 + b_1^2 + 2a_0 a_2 \right) - \beta a_2 = 0, \tag{3.26}$$

$$-a_1 \left(m^2 + 1 \right) - \alpha \left(2a_0 a_1 + 2b_1 b_2 \right) - \beta a_1 = 0, \tag{3.27}$$

$$6b_2 m^2 - 2\alpha a_2 b_2 = 0, \tag{3.28}$$

$$2b_1 m^2 - 2\alpha \left(a_1 b_2 + b_1 a_2 \right) = 0, \tag{3.29}$$

$$-b_2 \left(m^2 + 4 \right) - 2\alpha \left(a_0 b_2 + a_1 b_1 \right) - \beta b_2 = 0, \tag{3.30}$$

$$-b_1 - 2\alpha a_0 b_1 - \beta b_1 = 0, \tag{3.31}$$

$$2a_2 - \alpha a_0^2 - \beta a_0 - c_1 - \alpha b_1^2 = 0. \tag{3.32}$$

Solving the bove system with the aid of Mathematica or Maple, we have the following solution:

$$\alpha = \frac{6m^2}{a_2}, \quad \beta = \frac{4 \left(a_2 m^2 + a_2 + 3a_0 m^2 \right)}{a_2}, \quad a_0 = a_0, \quad a_1 = 0,$$

$$a_2 = a_2, \quad b_1 = 0, \quad b_2 = 0, \quad c_1 = \frac{2 \left(a_2^2 + 3a_0^2 m^2 + 2a_0 a_2 m^2 + 2a_0 a_2 \right)}{a_2} \tag{3.33}$$

Sothat the solution of Equation (3.20) will be in the form:

$$W(\xi) = -\frac{\beta}{2\alpha} - \frac{2}{\alpha} - \frac{2m^2}{\alpha} + 6\frac{m^2}{\alpha} sn^2 (\xi), \tag{3.34}$$

if $m \rightarrow 1$, we have the hyperbolic solution:

$$W(\xi) = -\frac{\beta}{2\alpha} - \frac{2}{\alpha} - \frac{6}{\alpha} \left(\frac{1}{3} - \tanh(\xi) \right). \tag{3.35}$$

3.2. Example 2. Nano-Ionic Currents of MTs

The nano ionic currents are elaborated in [27] take the form

$$\frac{l^2}{3} u_{xxx} + \frac{z^{\frac{3}{2}}}{l} \left(xc_0 - 2ss_0 \right) uu_t + 2u + \frac{zc_0}{l} u_t + \frac{1}{l} \left(Rz^{-1} - G_0 Z \right) u = 0, \tag{3.36}$$

where $R = 0.34 \times 10^9 \ \Omega$ is the resistance of the ER with length, $l = 8 \times 10^{-9}$ m, $c_0 = 1.8 \times 10^{-15}$ F is the maximal capacitance of the ER, $G_0 = 1.1 \times 10^{-13}$ si is conductance of pertaining NPs and $z = 5.56 \times 10^{10} \ \Omega$ is the characteristic impedance of our system parameters δ and x describe nonlinearity of ER capacitor and conductance of NPs in ER, respectively. In order to solve Equation (3.36) we use the travelling wave transformations $u(x,t) = u(\xi)$, $\xi = \frac{1}{l}x - \frac{c}{\tau}t$ with $\tau = Rc_0 = 0.6 \times 10^{-6}$ s, to reduce Equation (3.36) to the following nonlinear ordinary differential equation:

$$\frac{1}{3}u''' - \frac{cz^{\frac{3}{2}}}{\tau}\left(xc_0 - 2sc_0\right)uu' + \left(2 - \frac{zcc_0}{\tau}\right)u' + \left(Rz^{-1} - G_0Z\right)u = 0, \tag{3.37}$$

which can be written in the form

$$\frac{1}{3}u''' + H_1uu' + H_2u' + H_3u = 0, \tag{3.38}$$

$$H_1 = \frac{c}{\tau}B, \ H_2 = \left(2 - \frac{ce}{\tau}\right), \ B = -z^{\frac{3}{2}}\left(xc_0 - 2ss_0\right),$$
$$E = c_{0z}, \ D = H_3. \tag{3.39}$$

Thus Equation (3.38) takes the form

$$\frac{1}{3}u''' + \frac{c}{\tau}Buu' + \left(2 - \frac{ce}{\tau}\right)u' + Du = 0. \tag{3.40}$$

Balancing u''' and uu' yields, $N + 3 = N + N + 1 \rightarrow N = 2$. Consequently, we get

$$u = a_0 + a_1 sn + b_1 cn + a_2 sn^2 + b_2 sncn, \tag{3.41}$$

where a_0, a_1, a_2, b_1, b_2 are arbitrary constants such that $a_2 \neq 0$ or $b_2 \neq 0$. From Equation (3.41), it is easy to see that

$$u' = dna_1cn - dnb_1cn + 2dna_2sn \ cn - 2dnb_2sn^2 + b_2dn, \tag{3.42}$$

$$u'' = -a_1m^2sn + 2a_1sn^3m^2 + 2m^2sn^2b_1cn - 4a_2m^2sn^2 + 6a_2sn^4m^2$$
$$+ 6m^2sn^3cnb_2 - m^2sncnb_2 - a_1sn - b_1cn + 2a_2 - 4a_2sn^2 - 4b_2sncn. \tag{3.43}$$

Substituting Equations (3.41)-(3.43) into Equation (3.40) and equating the coefficients of sn^4dn, sn^3cndn, sn^3dn, sn^2cndn, sn^2dn, $sncndn$, $sndn$, $cndn$ and dn to zero, we obtain

$$-8b_2m^2 - 4\frac{cBa_2b_2}{\tau} = 0, \tag{3.44}$$

$$8a_2m^2 + \frac{cB\left(2a_2^2 - 2b_2^2\right)}{\tau} = 0, \tag{3.45}$$

$$-2b_1m^2 - \frac{cB\left(-3a_1b_2 - 3a_2b_1\right)}{\tau} = 0, \tag{3.46}$$

$$2a_1m^2 + \frac{cB\left(3a_1a_2 - 3b_1b_2\right)}{\tau} = 0, \tag{3.47}$$

$$\frac{4}{3}b_2\left(5m^2 + 2\right) + \frac{cB}{\tau}\left(-2a_0b_2 - 2a_1b_1 + 3a_2b_2\right) + \left(2 - \frac{cE}{\tau}\right)\left(-2b_2\right) + Da_2 = 0, \tag{3.48}$$

$$\frac{-8}{3}a_2\left(m^2+1\right)+\frac{cB}{\tau}\left(2a_0a_2-a_1^2-b_1^2+b_2^2\right)+2\left(2-\frac{cE}{\tau}\right)a_2+Db_2=0, \tag{3.49}$$

$$\frac{1}{3}b_1\left(4m^2+1\right)+\frac{cB}{\tau}\left(-a_0b_1+2a_1b_2+2a_2b_1\right)+\left(2-\frac{cE}{\tau}\right)b_1+Da_1=0, \tag{3.50}$$

$$\frac{-1}{3}a_1\left(m^2+1\right)+\frac{cB}{\tau}\left(2a_0b_1+b_1b_2\right)+\left(2-\frac{cE}{\tau}\right)a_1+Db_1=0, \tag{3.51}$$

$$\frac{-1}{3}b_2\left(m^2+4\right)+\frac{cB}{\tau}\left(2a_0b_2+b_1a_1\right)+\left(2-\frac{cE}{\tau}\right)b_2=0. \tag{3.52}$$

Solving the above system with the aid of Mathematica or Maple, we have the following solution:
Case 1.

$$D=0, m=m, a_0=\frac{1}{3}\frac{4m^2\tau-2\tau+3cE}{cB}, a_1=0, a_2=\frac{-4m^2\tau}{cB}, b_1=0, b_2=0$$

Case 2.

$$D=0, m=\sqrt{\frac{1}{2}-\frac{3cE}{4\tau}}, a_0=0, a_1=0, a_2=\frac{-2\tau+3cE}{cB}, b_1=0, b_2=0$$

So that the solution of Equation (3.40) will be in the form:
Case 1.

$$u\left(\xi\right)=\frac{1}{3}\frac{4m^2\tau-2\tau+3cE}{cB}+\frac{-4m^2\tau}{cB}sn^2 \tag{3.53}$$

Case 2.

$$u\left(\xi\right)=\frac{-2\tau+3cE}{cB}sn^2 \tag{3.54}$$

If $m\to1$, we have the hyperbolic solution:
Case 1.

$$u\left(\xi\right)=\frac{1}{3}\frac{2\tau+3cE}{cB}+\frac{-4\tau}{cB}sn^2 \tag{3.55}$$

Case 2.

$$u\left(\xi\right)=\frac{-2\tau+3cE}{cB}sn^2 \tag{3.56}$$

4. Conclusion

The nano waves propagating along microtubules in living cells play an important role in nano biosciences and cellular signaling where the propagation along microtubules shaped as nanotubes is essential for cell motility, cell division, intracellular trafficking and information processing within neuronal processes. Ionic waves propagating along microtubules in living cells have been also implicated in higher neuronal functions, including memory and the emergence of consciousness and we presented an in viscid, incompressible and non-rotating fluid of constant depth (h). The extended Jacobian elliptic function expansion method has been successfully used to find the exact traveling wave solutions of some nonlinear evolution equations. According to the suggested method we obtained a new and more accurate traveling wave solution of nano ionic-solitons waves' propagation along microtubules in living cells and nano-ionic currents of MTs. Let us compare between our results obtained in the present article with the well-known results obtained by other authors using different methods as follows: Our results of nano-solitons of ionic waves propagation along microtubules in living cells and nano-ionic currents of MTs are new and different from those obtained in [27]. **Figures 1-3** show solitary wave

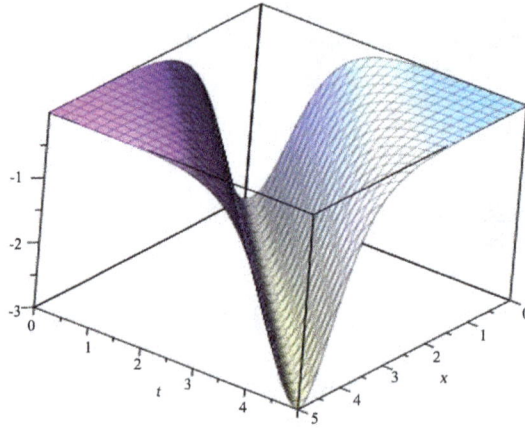

Figure 1. Plot of solution of Equation (3.35).

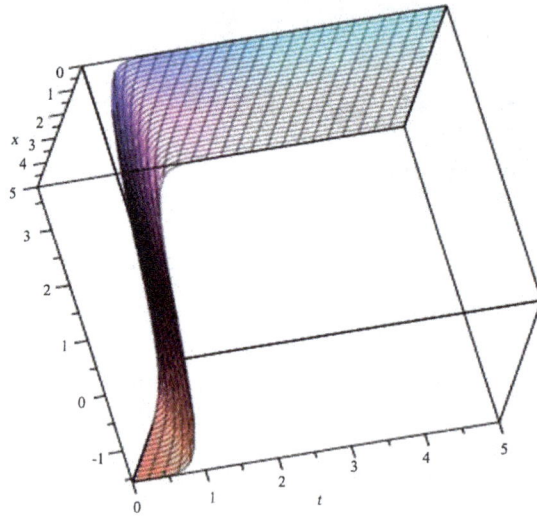

Figure 2. Plot of solution of Equation (3.56).

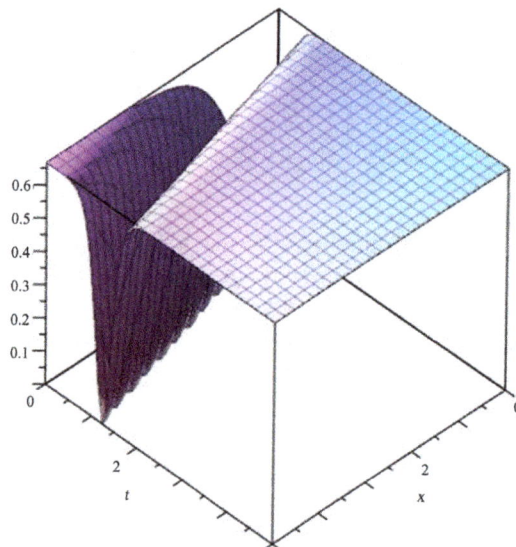

Figure 3. Plot of solution of Equation (3.55).

solution. It can be concluded that this method is reliable and proposes a variety of exact solutions NPDEs. The performance of this method is effective and can be applied to many other nonlinear evolution equations.

References

[1] Ablowitz, M.J. and Segur, H. (1981) Solitions and Inverse Scattering Transform. SIAM, Philadelphia. http://dx.doi.org/10.1137/1.9781611970883

[2] Maliet, W. (1992) Solitary Wave Solutions of Nonlinear Wave Equation. *American Journal of Physics*, **60**, 650-654. http://dx.doi.org/10.1119/1.17120

[3] Maliet, W. and Hereman, W. (1996) The Tanh Method: Exact Solutions of Nonlinear Evolution and Wave Equations. *Physica Scripta*, **54**, 563-568. http://dx.doi.org/10.1088/0031-8949/54/6/003

[4] Wazwaz, A.M. (2004) The Tanh Method for Travelling Wave Solutions of Nonlinear Equations. *Applied Mathematics and Computation*, **154**, 714-723. http://dx.doi.org/10.1016/S0096-3003(03)00745-8

[5] El-Wakil, S.A. and Abdou, M.A. (2007) New Exact Travelling Wave Solutions Using Modified Extended Tanh-Function Method. *Chaos Solitons Fractals*, **31**, 840-852. http://dx.doi.org/10.1016/j.chaos.2005.10.032

[6] Fan, E. (2000) Extended Tanh-Function Method and Its Applications to Nonlinear Equations. *Physics Letters A*, **277**, 212-218. http://dx.doi.org/10.1016/S0375-9601(00)00725-8

[7] Wazwaz, A.M. (2007) The Extended Tanh Method for Abundant Solitary Wave Solutions of Nonlinear Wave Equations. *Applied Mathematics and Computation*, **187**, 1131-1142. http://dx.doi.org/10.1016/j.amc.2006.09.013

[8] Wazwaz, A.M. (2005) Exact Solutions to the Double Sinh-Gordon Equation by the Tanh Method and a Variable Separated ODE Method. *Computers & Mathematics with Applications*, **50**, 1685-1696. http://dx.doi.org/10.1016/j.camwa.2005.05.010

[9] Wazwaz, A.M. (2004) A Sine-Cosine Method for Handling Nonlinear Wave Equations. *Mathematical and Computer Modelling*, **40**, 499-508. http://dx.doi.org/10.1016/j.mcm.2003.12.010

[10] Yan, C. (1996) A Simple Transformation for Nonlinear Waves. *Physics Letters A*, **224**, 77-84. http://dx.doi.org/10.1016/S0375-9601(96)00770-0

[11] Fan, E.G. and Zhang, H.Q. (1998) A Note on the Homogeneous Balance Method. *Physics Letters A*, **246**, 403-406. http://dx.doi.org/10.1016/S0375-9601(98)00547-7

[12] Wang, M.L. (1996) Exact Solutions for a Compound KdV-Burgers Equation. *Physics Letters A*, **213**, 279-287. http://dx.doi.org/10.1016/0375-9601(96)00103-X

[13] Abdou, M.A. (2007) The Extended F-Expansion Method and Its Application for a Class of Nonlinear Evolution Equations. *Chaos, Solitons & Fractals*, **31**, 95-104. http://dx.doi.org/10.1016/j.chaos.2005.09.030

[14] Ren, Y.J. and Zhang, H.Q. (2006) A Generalized F-Expansion Method to Find Abundant Families of Jacobi Elliptic Function Solutions of the (2+1)-Dimensional Nizhnik-Novikov-Veselov Equation. *Chaos, Solitons & Fractals*, **27**, 959-979. http://dx.doi.org/10.1016/j.chaos.2005.04.063

[15] Zhang, J.L., Wang, M.L., Wang, Y.M. and Fang, Z.D. (2006) The Improved F-Expansion Method and Its Applications. *Physics Letters A*, **350**, 103-109. http://dx.doi.org/10.1016/j.physleta.2005.10.099

[16] He, J.H. and Wu, X.H. (2006) Exp-Function Method for Nonlinear Wave Equations. *Chaos, Solitons & Fractals*, **30**, 700-708. http://dx.doi.org/10.1016/j.chaos.2006.03.020

[17] Aminikhad, H., Moosaei, H. and Hajipour, M. (2009) Exact Solutions for Nonlinear Partial Differential Equations via Exp-Function Method. *Numerical Methods for Partial Differential Equations*, **26**, 1427-1433.

[18] Zhang, Z.Y. (2008) New Exact Traveling Wave Solutions for the Nonlinear Klein-Gordon Equation. *Turkish Journal of Physics*, **32**, 235-240.

[19] Wang, M.L., Zhang, J.L. and Li, X.Z. (2008) The (G'/G)-Expansion Method and Travelling Wave Solutions of Nonlinear Evolutions Equations in Mathematical Physics. *Physics Letters A*, **372**, 417-423. http://dx.doi.org/10.1016/j.physleta.2007.07.051

[20] Zhang, S., Tong, J.L. and Wang, W. (2008) A Generalized (G'/G)-Expansion Method for the mKdv Equation with Variable Coefficients. *Physics Letters A*, **372**, 2254-2257. http://dx.doi.org/10.1016/j.physleta.2007.11.026

[21] Zayed, E.M.E. and Gepreel, K.A. (2009) The (G'/G)-Expansion Method for Finding Traveling Wave Solutions of Nonlinear Partial Differential Equations in Mathematical Physics. *Journal of Mathematical Physics*, **50**, Article ID: 013502. http://dx.doi.org/10.1063/1.3033750

[22] Zayed, E.M.E. (2009) The (G'/G)-Expansion Method and Its Applications to Some Nonlinear Evolution Equations in Mathematical Physics. *Journal of Applied Mathematics and Computing*, **30**, 89-103. http://dx.doi.org/10.1007/s12190-008-0159-8

[23] Dai, C.Q. and Zhang, J.F. (2006) Jacobian Elliptic Function Method for Nonlinear Differential-Difference Equations. *Chaos, Solitons & Fractals*, **27**, 1042-1049. http://dx.doi.org/10.1016/j.chaos.2005.04.071

[24] Fan, E. and Zhang, J. (2002) Applications of the Jacobi Elliptic Function Method to Special-Type Nonlinear Equations. *Physics Letters A*, **305**, 383-392. http://dx.doi.org/10.1016/S0375-9601(02)01516-5

[25] Liu, S., Fu, Z., Liu, S. and Zhao, Q. (2001) Jacobi Elliptic Function Expansion Method and Periodic Wave Solutions of Nonlinear Wave Equations. *Physics Letters A*, **289**, 69-74. http://dx.doi.org/10.1016/S0375-9601(01)00580-1

[26] Zhao, X.Q., Zhi, H.Y. and Zhang, H.Q. (2006) Improved Jacobi-Function Method with Symbolic Computation to Construct New Double-Periodic Solutions for the Generalized Ito System. *Chaos, Solitons & Fractals*, **28**, 112-126. http://dx.doi.org/10.1016/j.chaos.2005.05.016

[27] Sataric, M., Dragic, M. and Sejulic, D. (2011) From Giant Ocean Solitons to Cellular Ionic Nano-Solitons. *Romanian Reports in Physics*, **63**, 624-640.

9

Synthesis and Structural Properties of Bismuth Doped Cobalt Nanoferrites Prepared by Sol-Gel Combustion Method

Naraavula Suresh Kumar[1], Katrapally Vijaya Kumar[2*]

[1]Mallareddy Institute of Engineering & Technology, Secunderabad, India
[2]Department of Physics, JNTUH College of Engineering Jagtial, Nachupally (Kondagattu), Karimnagar-Dist, TS, India
Email: *kvkphd@gmail.com

Abstract

A series of Bismuth doped Cobalt nanoferrites of chemical composition $CoBi_xFe_{2-x}O_4$ (where x = 0.00, 0.05, 0.10, 0.15, 0.20 & 0.25) were prepared by sol-gel combustion method and calcinated at 600°C. The structural and morphological studies were carried out by using X-ray diffraction (XRD), Scanning Electron Microscope (SEM), Transmission Electron Microscopy (TEM), Energy Dispersive Spectroscopy (EDS) and Fourier Transform Infrared (FT-IR) spectra showing the single phase spinal structure. The X-ray diffraction (XRD) analysis confirmed a single phase fcc crystal. The crystallite size of all the compositions was calculated using Debye-Scherrer equation and found in the range of 17 to 26 nm. The lattice parameters were found to be decreased as Bi^{3+} ion doping increases. The surface morphology was studied by Scanning Electron Microscope (SEM) and particle size was confirmed by Transmission Electron Microscopy (TEM). The EDS plots revealed existence of no extra peaks other than constituents of the taken up composition. The Fourier Transform Infrared (FT-IR) studies were made in the frequency range 350 - 900 cm^{-1} and observed two strong absorption peaks. The frequency band is found at 596 cm^{-1} where as the lower frequency band at 393 cm^{-1}. It is clearly noticed that the two prominent absorption bands were slightly shifted towards higher frequency side with the increase of Bi^{3+} ion concentration.

Keywords

Bi-Co Nanoferrites, Sol-Gel Combustion Method, XRD, SEM, EDS, TEM and FTIR

---

*Corresponding author.

1. Introduction

In recent years the research on metal nanoferrites has been the subject of much interest due to their unusual structural, magnetic, dielectric and electrical properties. Specifically the materials with nano scale of dimensions in the range 1 - 100 nm show great physical and chemical properties. The nano-size materials attribute increase in relative surface area with decreased particle size leads quantum size effects [1] [2]. The nano particles dramatically change some of the magnetic properties and exhibit super paramagnetic phenomena and quantum tunnelling of magnetisation because each particle can be considered as a single magnetic domain [3]-[5]. In general the nano ferrite materials have high resistivity *i.e.,* low electrical conductivity, moderate saturation magnetization, high coercivity, mechanical hardness and chemical stability [6]-[9]. Any Metal ferrite has the spinal structure belonging to a general formula $M^{2+}Fe_2^{3+}O_4$ (also well known formula AB_2O_4) crystallizing with spinal structure, in which "M" represents tetrahedral site and "Fe" represents the octahedral site respectively and O indicates anion site. Finally, the novel properties of the spinal ferrites result with their ability of distributions among the tetrahedral and octahedral sites [10] [11]. For obtaining the specific properties ferrites can be fabricated by substituting various magnetic and non magnetic ions which greatly affect the magnetic moments, lattice parameters and exchanging interactions. Cobalt Ferrite ($CoFe_2O_4$) is a spinal structured ceramic oxide and well known hard magnetic material with high coercivity and moderate magnetisation. The properties of ferrite materials are known to be strongly influenced by their composition and microstructure. Doping of Cobalt ferrite with one or several metals is the best method to alter its physical and chemical properties. Bismuth Ferrite (BFO) is one of the most important single phase multi ferroic materials due to their high electrical resistivity, low magnetic and dielectric losses [12] [13]. And it is reported that on substituting of bismuth in very small amount does not alter the spinal structure of Cobalt ferrite. From the literature, we found that very less people worked on substitution of Bismuth ferrites [13]-[15]. With the great physicochemical properties and thermal stability, Bismuth doped Cobalt nanoferrites are suitable for magnetic recording applications such as audio, videotapes and high density digital recording discs. Therefore, Bi_2O_3 is a potential substitution for improving the magnetic and electrical properties of ferrites. There are several methods to synthesize nanoferrites such as solid state reaction, co-precipitation, sol-gel method, reverse micelles and hydrothermal method [16]. Among all the sol-gel combustion method is a simple method which speeds up the synthesis of complex materials and brings crystal uniformity [1]. Hence, in this paper we have reported the synthesis and structural characterisations of Bismuth doped Cobalt nonoferrites with the chemical composition $CoBi_xFe_{2-x}O_4$ (where x = 0.00 to 0.25).

2. Experimental

High purity crystalline nanoferrites were prepared by employing sol-gel combustion method. It is probably most effective method for the synthesis of homogenous nano particles which can be prepared at relatively low temperature [17].

2.1. Raw Materials

A series of Bismuth substituted cobalt nanoferrites having the chemical formula $CoBi_xFe_{2-x}O_4$ (where x = 0.00, 0.05, 0.10, 0.15, 0.20, 0.25) were prepared by sol-gel method. The starting materials were cobaltus nitrate $[Co(No_3)_2 \cdot 6H_2O]$ with 99% purity from sd-fine chemicals, bismuth Nitrate $[Bi(No_3)_3 \cdot 5H_2O]$ with 98.9% purity from sd-fine chemicals, ferric nitrate $[Fe(No_3)_3 \cdot 9H_2O]$ with 99.1% purity from sd-fine chemicals, citric acid $[C_6H_8O_7 \cdot H_2O]$ with 99.9% purity from sd-fine chemicals, 25% of aqueous solution of ammonia (NH_3) with A-grade from sd-fine chemicals.

2.2. Synthesis

The stoichiometric amount of all metal nitrates were dissolved in different glass beakers with minimum quantity of double distilled water and mixed together with the help of magnetic stirrer. Later on a solution of citric acid was added to the mixed metal solution with molar ratio of nitrates to citric acid is 1:3. Under constant stirring an aqueous solution of ammonia (NH_3) was added to this nitrate mixture drop wise to adjust the pH value about to 7. And finally a homogeneous solution was obtained. Then the mixed solution was heated about 100°C with uniform stirring and evaporated to obtain a highly viscous gel. The resultant gel was further heated 180°C to 200°C. When all water molecules were removed from the mixture, the gel gave a fast flameless auto combustion reaction with the evolution of gaseous products. And after some time it starts in the hottest zones of the beaker and propagated from the bottom to the top like the eruption of a volcano. After completion of this reaction a dark brown ash powder was yielded at 250°C with a structure similar to branched tree. All these synthesis process was outlined in the following **Figure 1**. At last it is cooled to room temperature and ground in an agate mortar for at least 30 minutes to get a highly dense power. Finally the powder was sintered at 600°C for 5 hours in air and cooled to room temperature. Further the prepared compositions were characterised by XRD, SEM, EDS, TEM and FTIR.

2.3. Characterization

The structural characterization of synthesized samples were carried out XRD by Phillips X-ray Diffractometer (model 3710) using with Cu Kα (λ = 1.54 Å) radiation. This was operated at room temperature by continuous scanning with 2θ values in the range of 10° to 80° to investigate phase, crystallite size and lattice parameters. The surface morphology of the samples was examined by using Scanning Electron Microscope (SEM) and the micro structural analysis of the prepared samples was carried out by Transmission Electron Microscope (TEM). The elemental analysis was carried out by using Energy Dispersive Spectrometer (EDS), which confirm the purity of all prepared samples. The infrared spectra of synthesized powders were recorded by SHIMADZU Fourier Transform Infrared Spectrophotometer (FT-IR), Model P/N-206-73500-38, in the range of 4000 to 250 cm^{-1} with a resolution of 4 cm^{-1}, which reveals the formation of single phase cubic spinal structure.

3. Results and Discussion

3.1. XRD Analysis

X-ray diffraction patterns of composition CoBi$_x$Fe$_{2-x}$O$_4$ (x = 0.0 to 0.25) nano ferrite particles are shown in **Figure 2**. Generally for a pure cobalt ferrite the XRD pattern exhibits eight peaks which were located in between

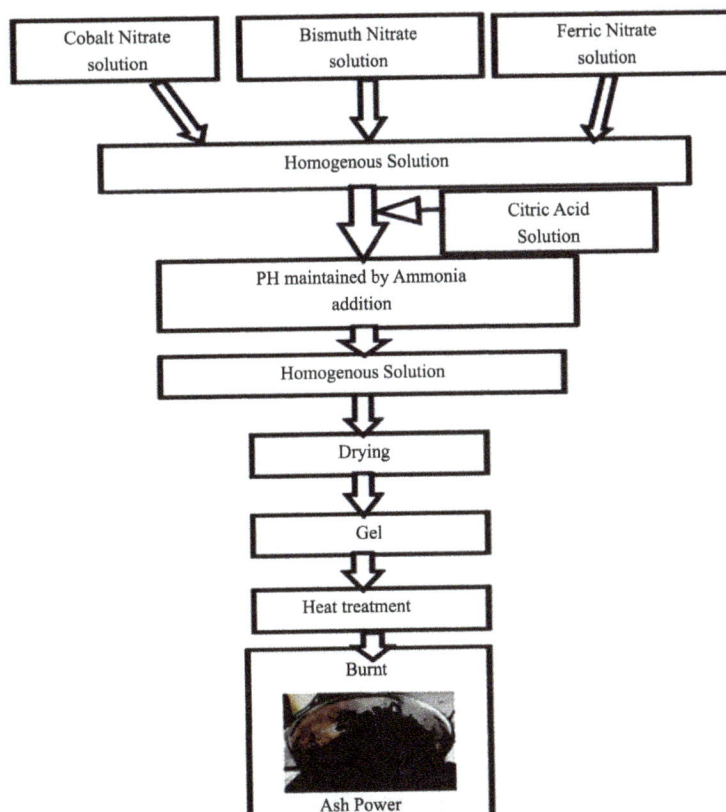

Figure 1. A general flowchart for the synthesis of CoBi$_x$Fe$_{2-x}$O$_4$ nano ferrite particles.

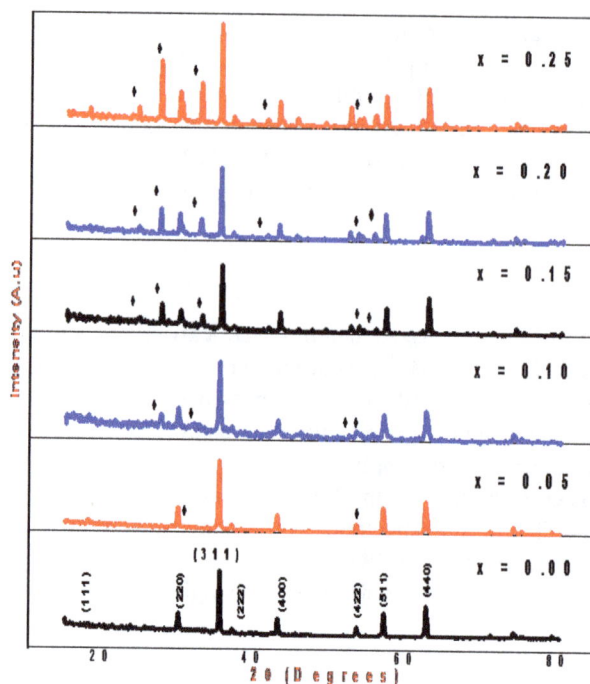

Figure 2. The XRD pattern of $CoBi_xFe_{2-x}O_4$ nano ferrite particles.

$2\theta = 15°$ to $80°$ as follows. In which the corresponding intensities of the peaks displayed as percentage (%) of the most intensive peak located with relative miller indices are shown in **Table 1** [18]. The XRD pattern matches with the (JCPDS-KDD) file number of (22-1086) and the corresponding reflections of peaks (220), (311), (400), (333), (440) were observed in X-ray diffraction pattern. On substitution of Bismuth in very small amount which does not alter the spinal structure of ferrite system [19]. It is also observed that the Bragg's peaks confirm the formation of a single phase fcc spinal structure. From the XRD pattern it is identified for x = 0.00 shows only peaks consistent with cubic spinal phase and rest of all the samples with the composition, x = 0.05 to 0.25, have additional peaks marked with the "♦" sign in **Figure 2**, which corresponds to the element of Bismuth. The crystallite sizes of all compositions were determined from broadening of the peak (311) of XRD pattern using the following Debye-Scherrer formula.

$$D = \frac{0.9\lambda}{\beta \cos\theta} \tag{1}$$

where, D = the average crystallite size of the phase under investigation, λ = wavelength of X-ray beam used, β = full width at half maxima (FWHM) in radians and θ = Bragg's angle.

From **Table 2**, it is observed that the lattice parameter decreases as the Bismuth concentration increases. The decrease in lattice constant (a) is due to the difference in ionic radii of Bi^{3+} (0.74 Å) as compared to Fe^{3+} (0.78 Å). Therefore the smaller Bi^{3+} ions were replaces the larger Fe^{3+} ions completely in octahedral positions [20] and the crystalline size was found to be in the range of 17 to 26 nm which was good agreement with XRD result.

The lattice parameter "a" was calculated using the following equation.

$$a = d\sqrt{h^2 + k^2 + l^2} \tag{2}$$

Figure 3 shows the variation of lattice constant (a) with the composition (X), which indicates that the variation of lattice constant (a) is found to be decrease linearly with increase of Bi^{3+} ion content. The unit cell volume was calculated as $V = a^3$ in atomic units (where a = lattice constant). If we speak in terms of volume, it is clear that the volume of unit cell depends on the lattice constant (a). The increase of bismuth doping in the composition, the lattice constant and the volume of unit cell were decreased linearly [21]. Several investigators were observed the similar behaviour of lattice constants with doping concentration. The decrease in lattice parameter on

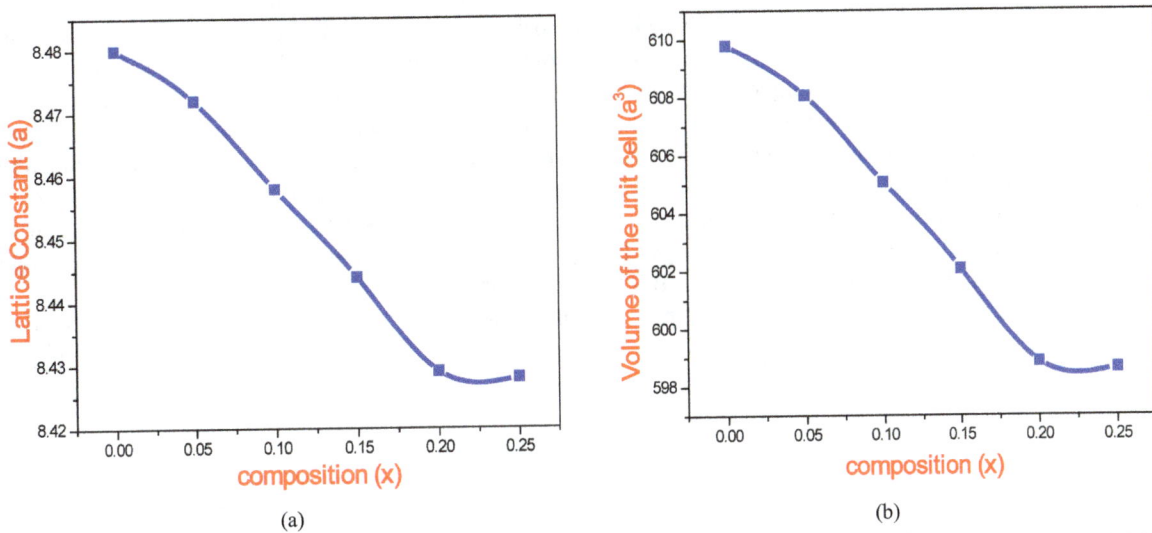

(a)
(b)

Figure 3. (a)-(b) shows the variation of lattice constant (a) and volume of the unit cell with the composition (X) of $CoBi_xFe_{2-x}O_4$ nano ferrite particles.

Table 1. Diffraction angle (2θ), percentage (%) of most intensive peak and Miller indices (hkl) of $CoBi_xFe_{2-x}O_4$ nano ferrite particles.

Diffraction angle (2θ)	% of most intense peak	Miller indices (h k l)
18.289	10%	(111)
30.085	30%	(220)
35.438	100%	(311)
37.057	8%	(222)
43.059	20%	(400)
53.446	10%	(422)
56.975	30%	(511)
62.587	40%	(440)

Table 2. Values of crystallite size (D), lattice constant (a), inter planar distance (d) and unit cell volume (a^3) of $CoBi_xFe_{2-x}O_4$ nano ferrite particles.

X	Composition	Crystallite size (D) nm	Lattice constant (a) Å	Volume of the unit cell (a^3) cm^3	Inter planar spacing (d) Å
0.00	$CoFe_2O_4$	26.458	8.480	609.800	2.550
0.05	$CoBi_{0.05}Fe_{1.95}O_4$	26.450	8.472	608.075	2.541
0.10	$CoBi_{0.10}Fe_{1.90}O_4$	24.020	8.458	605.066	2.546
0.15	$CoBi_{0.15}Fe_{1.85}O_4$	17.631	8.444	602.066	2.541
0.20	$CoBi_{0.20}Fe_{1.80}O_4$	17.630	8.429	598.863	2.554
0.25	$CoBi_{0.25}Fe_{1.75}O_4$	17.627	8.428	598.650	2.557

the basis of relative ionic radii of Cr^{3+} and Fe^{3+} ions [22] [23].

The X-ray densities were calculated using the following Formula (3) and were tabulated in **Table 2**.

$$X-\text{ray density}\left(d_x\right)=\frac{8M}{Na^3}\left(\text{gm/cm}^3\right) \tag{3}$$

where, M = Molecular weight of the sample, N = Avogadro number, a = lattice constant.

The variation of X-ray density (d_x) and porosity (P) with the composition (X) is shown in **Figure 4**. In any ferrite system the variation of X-ray density with composition depends on the lattice constant (a) and total molecular weight (M) of the sample [24]. In this series from the **Table 3**, it was clear that as the Bi^{3+} ion content increases, the total molecular weight of the sample was also increases. This is due to the increase in mass overtakes the decrease in volume of the unit cell. Therefore greater atomic weight of bismuth ions (208.98 gm/mol) were replaces completely with lesser atomic weight of ferric ions (55.85 gm/mol). From **Table 2** and **Table 3** it is clearly shown that, the lattice parameter decreases with the increase of bismuth ion composition. Hence from **Figure 4** shows the X-ray density (d_x) increases with increase of total molecular weight and it also confirms that the porosity decreases linearly with the composition (X). A similar behaviour of X-ray density with Cr-substitution was presented by Md. Javed Iqbal, Mah Rukh, Sidiqyah in Co-Cr ferrite system [25].

The distance between magnetic ions (hopping length) in A-site (tetrahedral) and B-site (octahedral) were calculated using the relations (4) & (5) [23].

$$d_A = 0.25a\sqrt{3} \tag{4}$$

$$d_B = 0.25a\sqrt{2} \tag{5}$$

where, d_A = hopping length for tetrahedral site, d_B = hopping length for octahedral site and a = lattice constant.

The calculated values of d_A and d_B of different compositions were tabulated in **Table 3**. The variation of hopping length for octahedral and tetrahedral sites with Bi^{3+} ion composition is shown in **Figure 5** and it was observed that the hopping length were decreases with the increase of bismuth composition. It may be due to smaller ionic radius of the Bi^{3+} (0.74 Å) than Fe^{3+} (0.78 Å) [21].

3.2. SEM

The Scanning Electron Microscopic (SEM) images of all the synthesized samples were shown in **Figure 6**. The

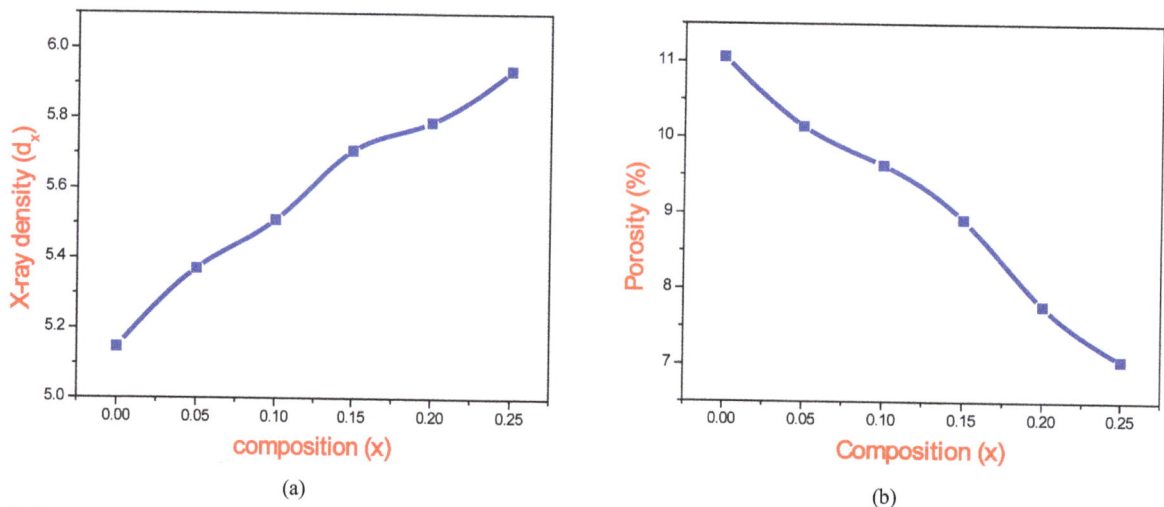

(a)

(b)

Figure 4. (a) (b) shows the variation of X-ray density (d_x) and porosity (P) with the composition (X) of $CoBi_xFe_{2-x}O_4$ nano ferrite particles.

Table 3. Various structural parameters of $CoBi_xFe_{2-x}O_4$ nano ferrite particles.

Sample	Composition	Molecular weight (M) gm/mol	X-ray density (d_x) gm/cc	Bulk density (d_B) gm/cc	Porosity (P) %	A-site (d_A) Å	B-site (d_B) Å
X = 0.00	$CoFe_2O_4$	234.627	5.147	4.57	11.21	3.6718	2.9981
X = 0.05	$CoBi_{0.05}Fe_{1.95}O_4$	242.284	5.371	4.82	10.14	3.6683	2.9952
X = 0.10	$CoBi_{0.10}Fe_{1.90}O_4$	249.942	5.511	4.98	9.63	3.6623	2.9903
X = 0.15	$CoBi_{0.15}Fe_{1.85}O_4$	257.597	5.709	5.20	8.915	3.6562	2.9853
X = 0.20	$CoBi_{0.20}Fe_{1.80}O_4$	265.253	5.790	5.34	7.772	3.6497	2.9800
X = 0.25	$CoBi_{0.25}Fe_{1.75}O_4$	272.910	5.939	5.52	7.055	3.6493	2.9797

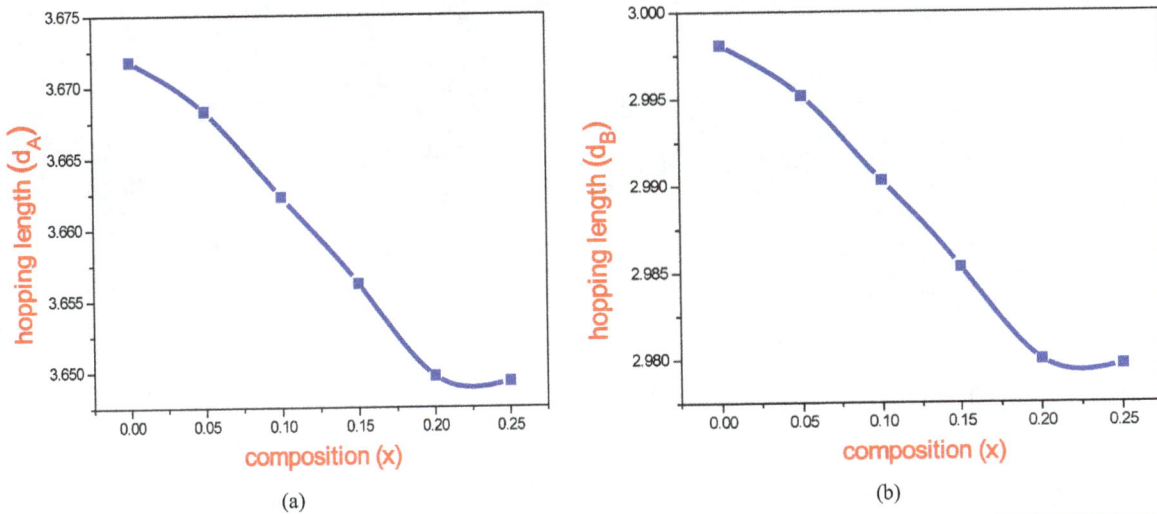

Figure 5. (a)-(b) shows the distance between magnetic ions in both tetrahedral (d_A) and octahedral (d_B) sites as a function of composition (x) of $CoBi_xFe_{2-x}O_4$ nano ferrite particles.

morphology of the prepared samples by sol-gel combustion method was studied by SEM (Scanning Electron Microscope) technique. The prepared samples of composition (x = 0.00 to 0.25) have been presented and secondary electron images were taken at different magnifications. It is clear that the SEM images show that the particles have an almost homogeneous distribution and some of them are agglomerated form. It is evidenced by SEM images that the aggregation of particles lies in nanometre region. The formation of ferrite was during sintering the pores between the particles were removed and formed strong bonds in the agglomeration form [24]. The interaction of grain boundary and porosity is important in determining the limited grain size of the particles. With this study it can be observed that the average grain size of the particles were decreasing trend with Bi^{3+} ion doping, moreover it is slightly lesser than determined by XRD and conforming the synthesized samples were crystalline structure.

3.3. TEM

Figures 7 (a)-(c) Show images of Transmission Electron Microscope (TEM) and Selected Area Electron Diffraction (SAED) patterns of cobalt ferrite of the sample x = 0.10 sintered at 600˚C. The TEM micrographs of these samples show the complete view about crystallite size, morphology and micro structure. From **Figure 7(a)**, it can be observed that the particles were rounded in cubic shape and formed loose aggregates and the particle size was found 50 nm which was a good agreement to XRD result. Some separated particles are also seen in those samples. However some moderately agglomerated particles as well as separated particles also present in the images. Which is due to increases with sintering temperature and hence some degree of agglomeration at this 600˚C appears which unavoidable [25] [26].

The corresponding selective area electron diffraction (SAED) analysis of the sample x = 0.10 were shown in **Figure 7(b)** and **Figure 7(c)**. This indicates that Bismuth doped cobalt ferrite nano particles were found well-crystalline in nature. According to this diffraction pattern the measured lattice constant and inter planar spacing $d_{(hkl)}$ agrees well with the XRD result.

3.4. EDS

The elemental analysis of the Co-Bi nano ferrite samples with different compositions was done by Energy dispersive spectrometer (EDS). The product of $CoBi_xFe_{2-x}O_4$ of all the compositions has been determined by the EDS and the patterns obtained are shown in **Figure 8**. All the EDS images show the presence of Co, Bi, Fe and Oxygen in the sample which did not contain the elements of Na and other elements. So the results indicated that the cation Na^+ did not take part in the reaction. Most of the undesired precursor materials like chloride ions have been completely removed from quantification of the peaks were found to be values of 1:1:1 and the elemental % and atomic % of different elements were shown in **Table 4**.

Figure 6. Shows the SEM image of all Co-Bi ferrite (CBF) nano particles which are sintered at 600°C of $CoBi_xFe_{2-x}O_4$ nano ferrite particles.

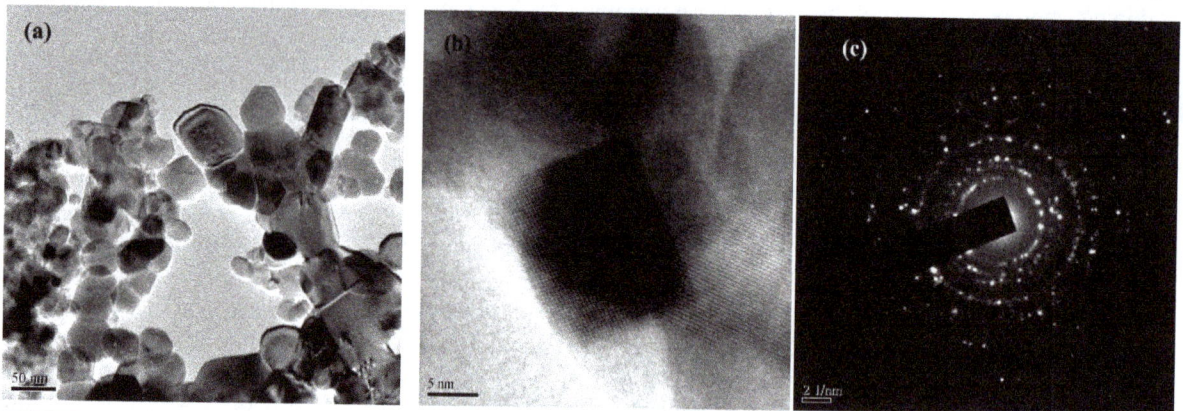

Figure 7. (a)-(c) shows TEM and SAED images of $CoBi_xFe_{2-x}O_4$ nano ferrite particles of x = 0.10.

3.5. FTIR

FTIR spectroscopic analysis is an additional tool for the structural characterisation of the spinal structure of Co-Bi ferrite system. The **Figure 9** shows FT-IR spectra of $CoBi_x Fe_{2-x}O_4$ (x = 0.00 to 0.25) nano ferrite particles at room temperature in the range of 350 to 900 cm^{-1}. The spectra of all the ferrites have been used to locate the band positions which are summarized in the **Table 5**. In the present study two prominent absorption bands v_1 and v_2 are found at around 596.15 and 393.50 cm^{-1} respectively for all the compositions. The absorption bands observed within the specific limits reveal the formation of single phase spinal structure having two sub lattices, *i.e.*, tetrahedral (A-site) and octahedral (B-site) [27]. The high frequency band v_1 called strongest absorption lies in the range 591.33 to 596.15 cm^{-1} while the low frequency band v_2 called weakest absorption lies in the range of 384.09 to 393.50 cm^{-1} and the both the frequency values for all the composition are furnished in **Table 5**.

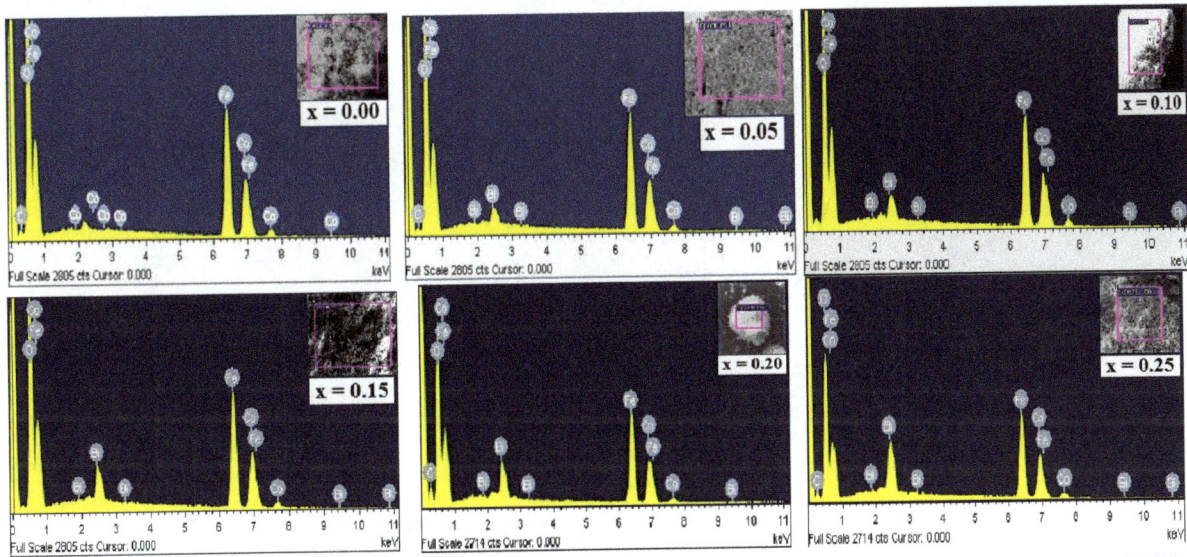

Figure 8. The EDS pattern of $CoBi_xFe_{2-x}O_4$ nano ferrite particles sintered at $600°C$ with the composition (x = 0.00 to 0.25).

Table 4. The relative elemental (%) and atomic (%) of composition of $CoBi_xFe_{2-x}O_4$ nano ferrite particles.

Element	O		Co		Bi		Fe	
Ferrite composition	Element %	Atomic %	Element %	Atomic %	Element %	Atomic %	Element %	Atomic %
$CoFe_2O_4$	26.34	54.76	26.27	16.23	--	--	47.39	29.01
$CoBi_{0.05}Fe_{1.95}O_4$	25.57	55.85	25.26	15.33	4.25	0.71	44.92	28.11
$CoBi_{0.10}Fe_{1.90}O_4$	27.01	57.31	24.90	15.23	6.14	1.14	41.95	26.32
$CoBi_{0.15}Fe_{1.85}O_4$	24.85	58.12	24.32	15.12	8.94	1.42	41.89	25.34
$CoBi_{0.20}Fe_{1.80}O_4$	26.63	59.14	23.13	14.22	10.49	1.92	39.75	24.72
$CoBi_{0.25}Fe_{1.75}O_4$	23.65	58.91	22.87	13.98	15.46	2.94	38.02	24.17

Table 5. The relative frequency bands of lower and higher of the $CoBi_xFe_{2-x}O_4$ nano ferrite particles.

Ferrite composition	Lower frequency bands v_1 (cm^{-1})	Higher frequency bands v_2 (cm^{-1})
$CoFe_2O_4$	592.55	393.50
$CoBi_{0.05}Fe_{1.95}O_4$	591.33	369.10
$CoBi_{0.10}Fe_{1.90}O_4$	593.65	384.09
$CoBi_{0.15}Fe_{1.85}O_4$	596.15	385.68
$CoBi_{0.20}Fe_{1.80}O_4$	591.63	368.50
$CoBi_{0.25}Fe_{1.75}O_4$	592.55	385.60

It is observed that the differences in the band positions is expected because of the difference in bond length of $Fe^{3+} - O^{2-}$ ions at the octahedral and tetrahedral sites [28]-[30]. Because of tetrahedral dimensions are less with compared to octahedral site dimensions, the absorption bands have inverse relationship with the bond length. It is also observed that, the bands v_1 and v_2 were slightly shifted towards higher frequency side with the increase of bismuth content. Due to decrease in site radius, that increase the fundamental frequency and hence the central frequency should shift towards the higher frequency side [31] [32].

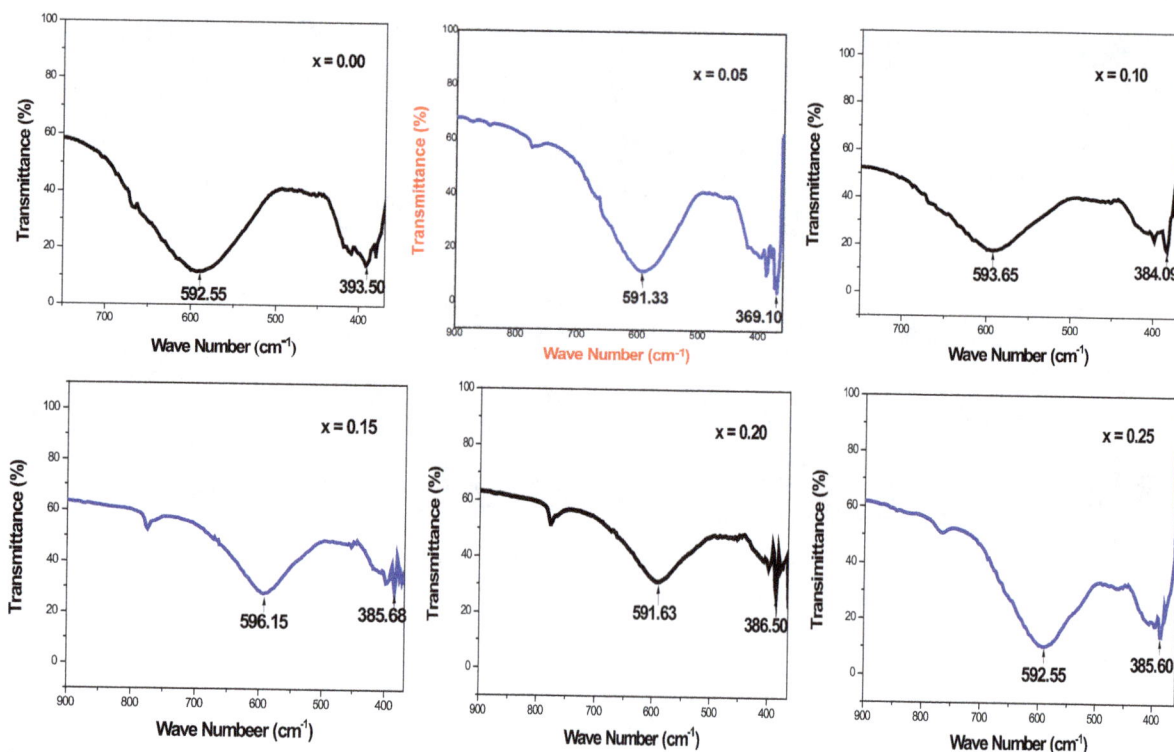

Figure 9. FTIR spectra of all Co-Bi ferrite (CBF) nano particles which are sintered at 600°C of the composition Co-$Bi_xFe_{2-x}O_4$ (x = 0.00 to 0.25).

4. Conclusions

On the basis of above discussion, the following conclusions are drawn

➤ A series of Bismuth doped Cobalt ferrite with the compositional formula $CoBi_x Fe_{2-x}O_4$ (where x = 0.00 to 0.25) were prepared by sol-gel method successfully and sintered at 600°C for 5 hrs.

➤ It is observed that the sol-gel combustion method is a convenient way for obtaining a homogeneous nano sized mixed nanoferrites.

➤ The X-ray diffraction patterns confirm the formation of cubic spinal structure in single phase without any impurity peaks.

➤ Due to very small amount of Bi^{3+} ion doping, the molecular weight increases and the unit cell dimensions were shrinkage. Therefore the lattice parameters also decrease with the increase of doping concentration.

➤ In this system the crystallite size was found to be in the range of 17 nm to 26 nm.

➤ The relation between hopping length for tetrahedral site (A-site) and octahedral site (B-site) was investigated as a function of "X" and it was found to be decrease with Bi^{3+} ion content. This is due to the difference in ionic radii of Fe^{3+} and Bi^{3+} ions.

➤ The SEM and TEM images show the morphology of the prepared samples at different magnifications. And it is clear that, the SEM images have an almost homogeneous distribution and some of them are agglomerated.

➤ The particle size is evidenced by TEM images that the aggregation of particles lies in nano scale and conform all the samples were crystalline structure.

➤ The EDS data gives the elemental % and atomic % of Bi doped Cobalt Ferrite system. And which shows the presence of Co, Fe, O and Bi (except x = 0) participating cations.

➤ The FT-IR spectra of the compositions under investigation reveal the formation of a single phase cubic spinal structure. This shows two significant absorption bands which confirm the characteristics of ferrite sample.

Acknowledgements

The authors are thankful to Dr. K.E. Balachandrudu, Principal, MRIET, Secunderabad for providing the necessary facilities to bring out this research work. Dr. K.V.K. is grateful to Prof. N. V. Ramana, Principal, JNTUH

College of Engineering Jagtial, Nachupally (Kondagattu), Karimnagar-Dist for his encouragement.

References

[1]   Pallai, V. and Shah, D.O. (1996) Synthesis of High-Coercivity Cobalt Ferrite Particles Using Water-in-Oil Microemulsions. *Journal of Magnetism and Magnetic Materials*, **163**, 243-248. http://dx.doi.org/10.1016/S0304-8853(96)00280-6

[2]   Skomski, R. (2003) Nanomagnetics. *Journal of Physics*: *Condensed Matter*, **15**, R1. http://dx.doi.org/10.1088/0953-8984/15/20/202

[3]   Kumar, S., Shinde, T.J. and Vasambekar, P.N. (2013) Microwave Synthesis and Characterization of Nano Crystalline Mn-Zn Ferrites. *Advanced Materials Letters*, **4**, 373-377. http://dx.doi.org/10.5185/amlett.2012.10429

[4]   Dixit, G., Singh, J.P., Srivastava, R.C., Agrawal, H.M. and Chaudhary, R.J. (2012) Structural, Magnetic and Optical Studies of Nickel Ferrite Thin Films. *Advanced Materials Letters*, **3**, 21-28. http://dx.doi.org/10.5185/amlett.2011.6280

[5]   Giannakopoulou, T., Kompotiatis, L., Kontogeorgakos, A. and Kordas, G. (2002) Microwave Behavior of Ferrites Prepared via Sol-Gel Method. *Journal of Magnetism and Magnetic Materials*, **246**, 360-365. http://dx.doi.org/10.1016/S0304-8853(02)00106-3

[6]   Pardavi-Horvath, M. (2000) Microwave Applications of Soft Ferrites. *Journal of Magnetism and Magnetic Materials*, **215-216**, 171-183. http://dx.doi.org/10.1016/S0304-8853(00)00106-2

[7]   Gupta, N., Dimri, M.C. Kashyap, S.C. and Dube, D.C. (2005) Processing and Properties of Cobalt-Substituted Lithium Ferrite in the GHz Frequency Range. *Ceramics International*, **31**, 171-176. http://dx.doi.org/10.1016/j.ceramint.2004.04.004

[8]   Bate, G. (1975) In: Craik, D.J., Ed., *L: Magnetic Oxides, Part* 2, Wiley Interscience, New York, 703.

[9]   Sharrock, M.P. (1989) Particulate Magnetic Recording Media: A Review. *IEEE Transactions on Magnetics*, **25**, 4374-4389. http://dx.doi.org/10.1109/20.45317

[10]   Handley, R.C.O. (2000) Modern Magnetic Materials: Principles and Applications. John Wiley & Sons, New York.

[11]   Chen, D., Zhang, Y. and Tu, C. (2012) Preparation of High Saturation Magnetic $MgFe_2O_4$ Nanoparticles by Microwave-Assisted Ball Milling. *Materials Letters*, **82**, 10-12. http://dx.doi.org/10.1016/j.matlet.2012.05.034

[12]   Amighian, J., Mozaffari, M. and Nasr, B. (2006) Preparation of Nano-Sized Manganese Ferrite ($Mn-Fe_2O_4$) via Coprecipitation Method. *Journal of Solid State Physics*, **3**, 3188-3192.

[13]   Meenakshisundaram, A., Gunasekaran, N. and Srinivasan, V. (1982) Distribution of Metal Ions in Transition Metal Manganites AMn_2O_4 (A: Co, Ni, Cu, or Zn). *Physica Status Solidi(a)*, **69**, K15-K19. http://dx.doi.org/10.1002/pssa.2210690149

[14]   Pallai, V. and Shah, D.O. (1996) Synthesis of High-Coercivity Cobalt Ferrite Particles Using Water-in-Oil Microemulsions. *Journal of Magnetism and Magnetic Materials*, **163**, 243-248. http://dx.doi.org/10.1016/S0304-8853(96)00280-6

[15]   Skomski, R. (2003) Nanomagnetics. *Journal of Physics*: *Condensed Matter*, **15**, R841-R896. http://dx.doi.org/10.1088/0953-8984/15/20/202

[16]   Yang, H., Zhang, X.C., Tang, A.D. and Qiu, G.Z. (2004) Cobalt Ferrite Nanoparticles Prepared by Coprecipitation/Mechanochemical Treatment. *Chemistry Letters*, **33**, 826-827. http://dx.doi.org/10.1246/cl.2004.826

[17]   Brinker, C.J. and Scherer, G.W. (1990) Sol-Gel Science: The Physics and Chemistry of Sol-Gel Processing. Academic Press, San Diego.

[18]   Ai, L.H. and Jiang, J. (2010) Influence of Annealing Temperature on the Formation, Microstructure and Magnetic Properties of Spinel Nanocrystalline Cobalt Ferrites. *Current Applied Physics*, **10**, 284-288. http://dx.doi.org/10.1016/j.cap.2009.06.007

[19]   Bensebaa, F., Zavaliche, F., L'Ecuyer, P., Cochrane, R.W. and Veres, T. (2004) Microwave Synthesis and Characterization of Co-Ferrite Nanoparticles. *Journal of Colloid and Interface Science*, **277**, 104-110. http://dx.doi.org/10.1016/j.jcis.2004.04.016

[20]   El Haiti, M.A. (1994) DC Conductivity for $Zn_xMg_{0.8-x}Ni_{0.2}Fe_2O_4$ Ferrites. *Journal of Magnetism and Magnetic Materials*, **136**, 138-142. http://dx.doi.org/10.1016/0304-8853(94)90457-X

[21]   Khan, H.M., Misbah-ul-Islam, Ali, I. and Rana, M. (2011) Electrical Transport Properties of Bi_2O_3-Doped $CoFe_2O_4$ and $CoHo_{0.02}Fe_{1.98}O_4$ Ferrites. *Materials Sciences and Application*, **2**, 1083-1089.

[22]   Kawade, V.B., Bichile, G.K. and Jadhav, K.M. (2000) X-Ray and Infrared Studies of Chromium Substituted Magnesium Ferrite. *Material Letters*, **42**, 33-37. http://dx.doi.org/10.1016/S0167-577X(99)00155-X

[23]   Arulmurugan, R., Jeyadevan, B., Vaidyanathan, G. and Sendhilnathan, S. (2005) Effect of Zinc Substitution on Co-Zn and Mn-Zn Ferrite Nanoparticles Prepared by Coprecipitation. *Journal of Magnetism and Magnetic Materials*, **288**,

470-477. http://dx.doi.org/10.1016/j.jmmm.2004.09.138

[24] Goldman, A. (1990) Modern Ferrite Technology. Van No Strand Reinhold, New York.

[25] Iqbal, M.J. and Siddiquah, M.R. (2008) Electrical and Magnetic Properties of Chromium Substituted Cobalt Ferrite Nano Materials. *Journal of Alloys and Compounds*, **453**, 513-518. http://dx.doi.org/10.1016/j.jallcom.2007.06.105

[26] Maaz, K., Mumtaz, A., Hasanain, S.K. and Ceylan, A. (2007) Synthesis and Magnetic Properties of Cobalt Ferrite ($CoFe_2O_4$) Nanoparticles Prepared by Wet Chemical Route. *Journal of Magnetism and Magnetic Materials*, **308**, 289-295. http://dx.doi.org/10.1016/j.jmmm.2006.06.003

[27] Zhou, B., Zhang, Y.W., Liao, C.S., Yan, C.H., Chen, L.Y. and Wang, S.Y. (2004) Rare-Earth-Mediated Magnetism and Magneto-Optical Kerr Effects in Nano-Crystalline $CoFeMn_{0.9}RE_{0.1}O_4$ Thin Films. *Journal of Magnetism and Magnetic Materials*, **280**, 327-333. http://dx.doi.org/10.1016/j.jmmm.2004.03.031

[28] Zaki, H.M. and Dawoud, H.A. (2010) Far-Infrared Spectra for Copper-Zinc Mixed Ferrites. *Physica B*, **405**, 4476-4479. http://dx.doi.org/10.1016/j.physb.2010.08.018

[29] Kambale, R.C., Song, K.M., Koo, Y.S. and Hur, N. (2011) Low Temperature Synthesis of Nano Crystalline Dy3p Doped Cobalt Ferrite: Structural and Magnetic Properties. *Journal of Applied Physics*, **110**, Article ID: 053910. http://dx.doi.org/10.1063/1.3632987

[30] Srivastava, M., Chaubey, S. and Ojha, A.K. (2009) Investigation on Size Dependent Structural and Magnetic Behavior of Nickel Ferrite Nanoparticles Prepared by Sol-Gel and Hydrothermal Methods. *Materials Chemistry and Physics*, **118**, 174-180. http://dx.doi.org/10.1016/j.matchemphys.2009.07.023

[31] Maensiri, S., Masingboon, C., Boonchom, B. and Seraphin, S. (2007) A Simple Route to Synthesize Nickel Ferrite ($NiFe_2O_4$) Nanoparticles Using Egg White. *Scripta Materialia*, **56**, 797-800. http://dx.doi.org/10.1016/j.scriptamat.2006.09.033

[32] Srivastava, M., Ojha, A.K., Chaubey, S., Sharma, P.K. and Pandey, A.C. (2010) Influence of pH on Structural Morphology and Magnetic Properties of Ordered Phase Cobalt Doped Lithium Ferrites Nanoparticles Synthesized by Sol-Gel Method. *Materials Science and Engineering B*, **175**, 14-21. http://dx.doi.org/10.1016/j.mseb.2010.06.005

10

In the Heart of Femtosecond Laser Induced Nanogratings: From Porous Nanoplanes to Form Birefringence

R. Desmarchelier, B. Poumellec, F. Brisset, S. Mazerat, M. Lancry

ICMMO, UMR CNRS-UPSud 8182, Université Paris Sud (in Université Paris Saclay), Orsay, France
Email: matthieu.lancry@u-psud.fr

Abstract

It is demonstrated that the form birefringence related to the so-called nanogratings is quantitatively correlated to the porosity-filling factor of these nanostructures. We reveal that matters surrounding the nanopores exhibit significant refractive index decrease which is likely due to the fictive temperature increase and/or the presence of a significant amount of interstitial O_2. The control of the porosity was achieved by adjusting the laser pulse energy and the number of pulses/micron *i.e.* the overlapping rate. Applications can be numerous in fast material processing by the production of nanoporous matter, and photonics by changing the optical properties.

Keywords

Silica, Glasses, Nanoporous, Femtosecond Laser Processing, Birefringence, Nanogratings

1. Introduction

One of the main advantages of using femtosecond pulses to induce 3D refractive index changes is that energy can be rapidly deposited before thermal build-up which can occur. This restricts thermal spreading so simple, the relaxation kinetics associated with quenched glass do not easily explain changes beyond the damage region when low repetition rate lasers are used. These attributes allow compact 2-D and 3-D multi-component photonic devices [1]-[4] to be fabricated in a single step within a wide variety of transparent materials [5]. For many applications, silica glasses are the preferred material, providing excellent physical and chemical properties such as optical transparency from IR to UV, a low thermal expansion coefficient, long term stability and a high resistance to laser induced damage.

Depending on the pulse parameters, we may define a variety of different laser-material regimes. Typically, the key parameters that can be tuned are the laser repetition rate, the pulse energy and the polarization, with the laser pulse duration and wavelength usually being fixed [6]. In terms of repetition rate, two regimes are categorized: (1) the *low repetition rate regime* where the material changes are caused by individual pulses [7]; and (2) the *high repetition rate regime* where changes arise due to cumulative thermal effects [8] since the time between each pulse is less than the thermal diffusion time of silica. In silica, the border between these regimes occurs around a repetition rate of 1 MHz although this is not easily defined since the thermal dissipation depends on volume and surface area. The investigations of this paper will focus in the low repetition rate regime. Depending on the exposure parameters, three qualitatively different types of structural changes can be induced in fused silica as recently review in [6]: (1) an isotropic positive refractive index change (type-I); (2) a form birefringence with negative index change [9] (type-II); and (3) voids (type-III). *Type-I modifications* occur above an energy of ~ 0.1 μJ ($\lambda = 800$ nm, $\Delta t_p = 160$ fs, 200 kHz and NA = 0.5) where the index change is permanent and isotropic with a slight stress birefringence [10] [11]. The maximum index change is $\Delta n \sim (3 - 6) \times 10^{-3}$ in fused silica [7] which is relatively large when compared with the index change achievable with nanosecond lasers [12] [13].

In contrast, *Type-II modifications* are significantly different and happen above a higher energy threshold of 0.31 μJ ($\lambda = 800$ nm, $\Delta t_p = 160$ fs, 200 kHz and NA = 0.5) [11]. The index change can be as large as $\Delta n \sim -2 \times 10^{-2}$ [14] and exhibits impressive thermal durability, exhibiting no signs of decay after two hours at 1000°C [15]. The most striking features of type-II modifications are that the index change is highly anisotropic. The origins of this anisotropy lie in the formation of sub-wavelength features of the so-called "nanogratings" or "nanoplanes" [16]. The nanoplanes are oriented perpendicular to the direction of /2n but it is the laser polarization. Their spacing is found to be dependent significantly on the pulse energy, and the number of laser pulses [17]. Recently, we have shown that these nanoplanes consisted of porous matter most likely produce as a result of decomposition of SiO_2 into $SiO_{2(1-x)} + x \cdot O_2$ under the intense plasma generation [18] [19]. The formation of nanoporous silica likely explains the refractive index contrast of these nanostructures and thus the observed form birefringence. In this paper, we investigate in details the form birefringence model and the related porous nanoplanes produced by the interaction of the femtosecond laser with silica glass. We reveal that the so-called nanogratings and porous nanoplanes are correlated and we study their changes with both the pulse energy and the pulse-to-pulse overlap.

2. Experimental Details

The direct writing procedure using infrared femtosecond laser pulses has been already described extensively in other work [10] [20] and is reminded in **Figure 1**. In this work, Ge-doped (12 w% in GeO_2) silica glass plates of 1 mm thickness and a refractive index equal to 1.4722 at 550 nm are used. Processing is undertaken with a femtosecond fibre laser ($\lambda = 1030$ nm, $\Delta t_p = 300$ fs). The single mode output is focused below the surface of the silica plate using a 0.6 NA aspheric lens with the k vector of the beam being perpendicular to the surface of the plate. The sample can then be moved in three dimensions using computer-controlled stages. The linear polarization of the laser output was

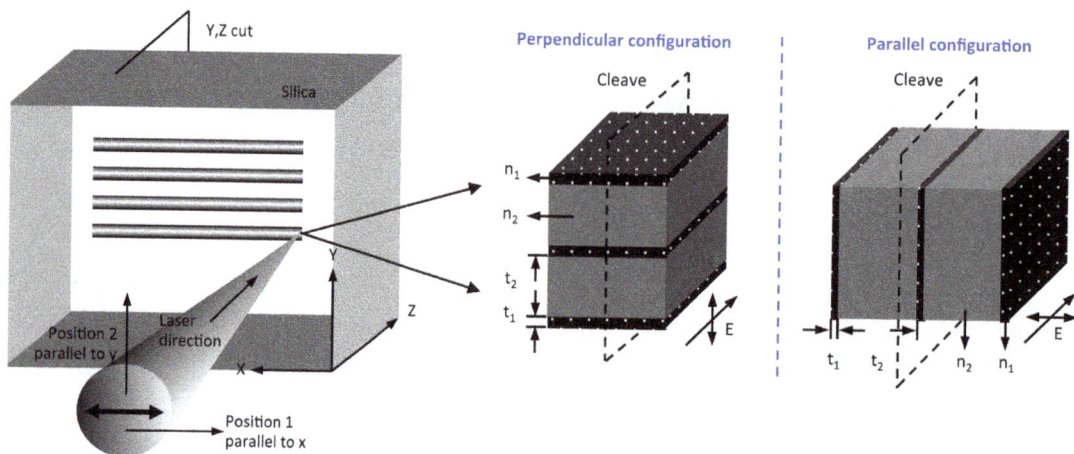

Figure 1. Experimental setup scheme for configuration for writing (from the left side) and schematic of sub-wavelength periodic structure formed in cross-section of the irradiated region. n_1 et n_2: local refractive indices of the nanoplates; t_1 and t_2, are their respective thicknesses.

usually kept parallel to the sample translation direction (parallel configuration) although the polarization transverse configuration state was also explored. The laser pulse energy was varied over (0.05 – 1) μJ; *i.e.* above the second damage threshold where nanostructures are formed. The scanning speed was varied from 10 up to 1000 μm/s and the repetition rate from 10 kHz up to 500 kHz. This allows varying the pulse-to-pulse overlapping rate from 1 up to 2×10^5 pulses/μm. After irradiation, QPm (Quantitative Phase Microscopy) and quantitative birefringence measurements were performed. The birefringence measurement is based on a Sénarmont compensator that couples a highly precise quarter waveplate with a 180-degree rotating analyzer to provide retardation measurements having an accuracy that approaches $\lambda/100$. The device is utilized for retardation measurements over one wavelength range (up to 551.5 nm) for the quantitative analysis of birefringence. Then in order to observe the intimate structure of the nanoplanes, it has been decided to use an original approach. The samples were cleaved using a diamond pen as described in the right side of **Figure 1**. The laser tracks have been observed using a Field-Emission Gun Scanning Electron Microscope (FEG-SEM, ZEISS SUPRA 55 VP). This allows to examine uncoated dielectric specimens using low accelerating voltage (typ. in the range of 1 kV) and very low current (a few pA) because they can keep an image resolution good enough even in these extreme conditions) and thus the original characteristics of the samples may be preserved for further testing or manipulation (no conductive coating in particular).

3. Results

In order to correlate the birefringence to the existence of porous nanogratings, we have analyzed the laser tracks cross sections using SEM images in two different writing configurations noted Xx and Xy, namely parallel and perpendicular polarization to writing direction. In addition for better investigation, we have studied two parameters dependence: the pulse energy and the overlapping rate that is defined by the ratio of the laser repetition rate to the writing speed.

The FEG-SEM secondary electrons images shown in **Figure 2** highlight the modifications morphology occur

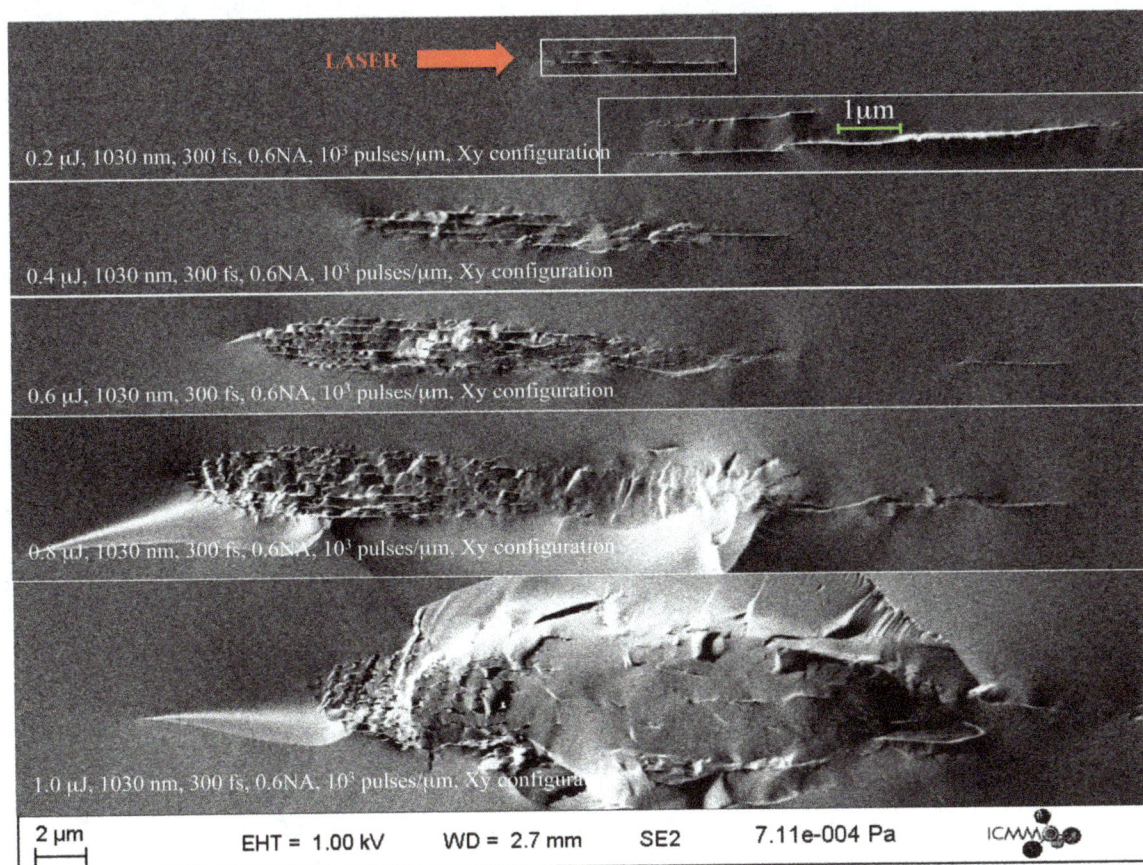

Figure 2. FEG-SEM, secondary electrons images of laser tracks cross-section for perpendicular writing configuration. The laser parameters were: 1030 nm, 300 fs, 0.6 NA and 10^3 pulses/μm and Xy configuration.

ring above T2 threshold for different pulse energy. The laser polarization was perpendicular to writing direction in order to reveal the nanoplanes or so-called nanogratings in the literature. Firstly, there is a trend towards increasing widths and lengths of the modified regions as the pulse energy increases from 0.2 to 1 µJ. In particular, we can observe that the number of nanoplanes in increasing along the width of the laser tracks. At low energy, we observe only 2 nanoplanes in the laser propagation direction for the studied overlapping rate that has been fixed at 10^3 pulses/micron. Then, above 0.6 µJ, disruptive modulation of the nanoplanes builds up in the head of the modified regions and becomes massive as the pulse energy is increased up to 1 µJ and above.

For a deeper investigation of the nanoplane structure, we turned the laser polarization 90° into a parallel configuration Xx (*i.e.* laser polarization is parallel to the scanning direction). As shown in **Figure 3**, when the laser polarization is parallel to the scanning direction, one can directly image the modified region inside the nanoplanes of the head of the interaction volume. The image is not uniform in term of nanoporosity and we observed what appears to be a "white" layer arising from surface topography variations that corresponds to material between the nanolayers. In all our samples, we observed that these pores always occur along with nanoplane formation for pulse energies higher than 0.2 µJ. The observed morphology is similar to the perpendicular configuration in terms of laser track lengths and widths.

The retardance of those laser tracks was then measured by using the Sénarmont method as shown in **Figure 4(a)**. The pulse energy used in these experiments varied from 0.025 to 1 µJ. First, measurements highlight no retardance produced below 0.2 µJ/pulse, which confirms the previous measurement in the same irradiation conditions [6]. Then, the retardance increases up to 0.7 µJ/pulse and decreases for higher pulse energies. We observe the opposite phenomena with the QPm measurements, which consist in the difference of phase in non-irradiated substrate compared to the irradiated one. The phase difference is zero below 0.2 µJ/pulse, then decreases down to

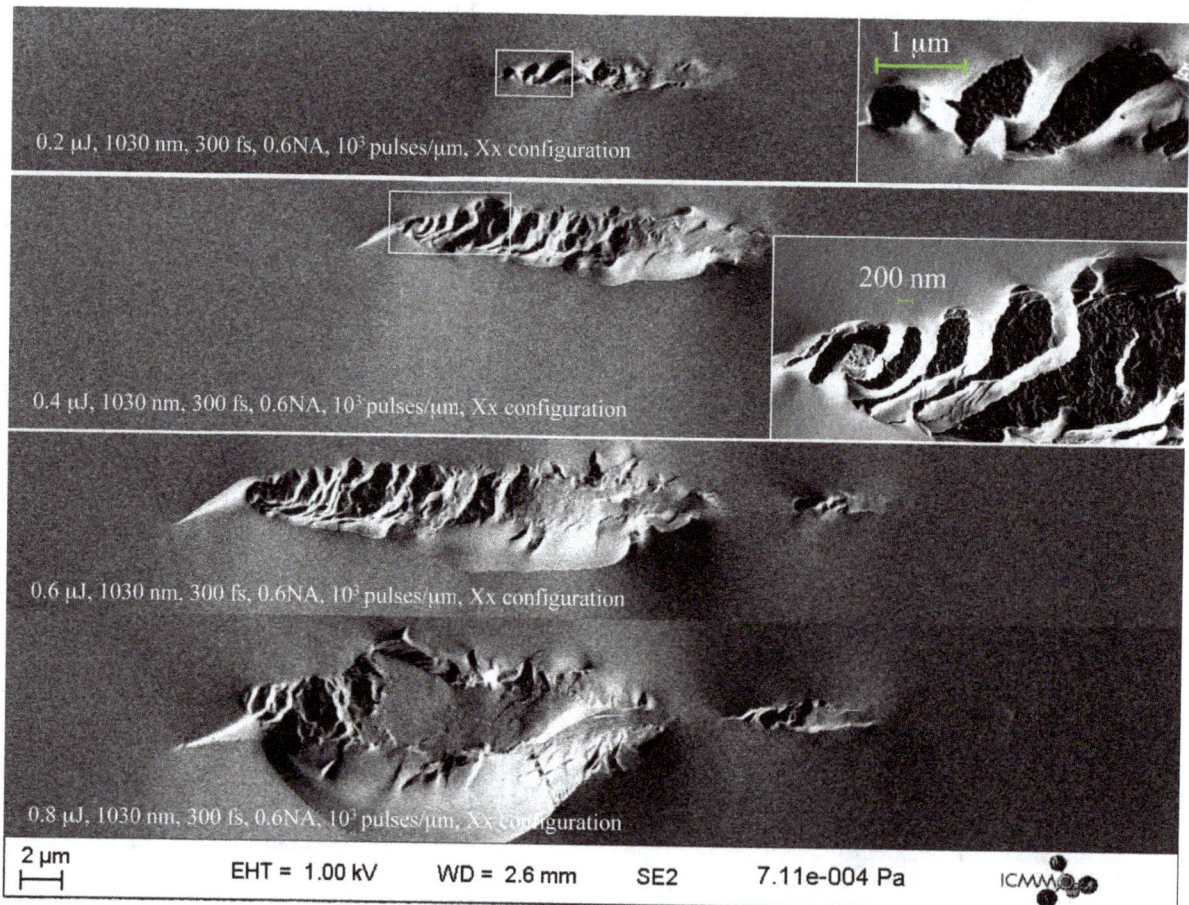

Figure 3. FEG-SEM, Secondary electrons images of laser tracks cross-section for parallel writing configuration. The laser parameters were: 1030 nm, 300 fs, 0.6 NA and 10^3 pulses/µm and Xx configuration.

-2p at 0.7 µJ/pulse and increases at higher pulse energy.

In order to deduce the quantitative birefringence from the retardance measurements, we need to determine the length (or the thickness) of the birefringent zone from the SEM images. The total length L_t of the laser tracks and the length L_{nano} of the nanostructured area (*i.e.* the porous nanoplane zone) are reported in **Figure 4(b)**. The whole laser track length (including head and tail) in increasing monotonously with the pulse energy whereas the length of the nanostructured area increases up to 0.6 µJ and then decreases for higher energy. We can notice that L_{nano} curve exhibits the same qualitative shape as for the measured retardance shown in **Figure 4(a)**.

Next, in order to correlate the birefringence to the porous nanogratings characteristics, SEM images where analysed in order to deduce the nanoplanes and porosity changes according to the laser parameters. In perpendicular configuration Xy, we can see that the number of nanoplanes within the laser track width increases linearly from 0.2 µJ up to 1 µJ as reported in **Figure 5(a)**. At the same time, measurements extracted from SEM images reveal that the average spacing between nanoplanes is decreasing from 450 nm down to 330 nm when increasing the pulse energy up to 1 µJ. This last observation is in agreement with the literature [20]. From the data gathered in parallel configuration Xx we can deduce that the average pore size increases from 16 to 50 nm to 0.4 µJ/pulse and then decrease to 35 nm at higher energy as shown in **Figure 5(b)**. In contrast, the 2D porosity-filling factor of the nanoplans $f_\parallel$, defined by ratio between area of pores and the remaining area within a

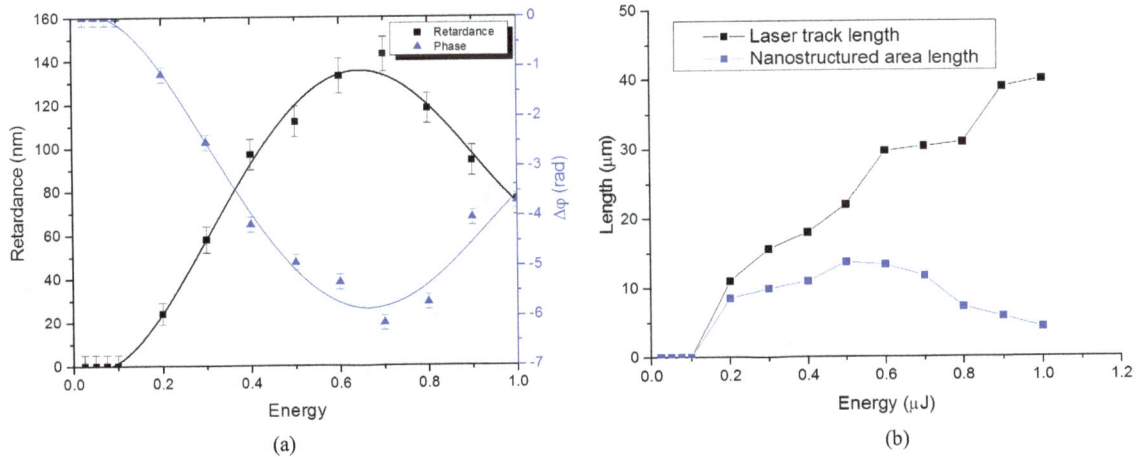

(a)   (b)

Figure 4. Detailed analyses of laser track (a) Plot of the retardance (black curve) and phase (blue curve) according to pulse energy for parallel writing configuration. (b) Plot of laser tracks length and nanostructured area length (the error bar ±1.0 µm). The laser parameters were: 1030 nm, 300 fs, 0.6 NA, 10^3 pulses/µm and Xx configuration.

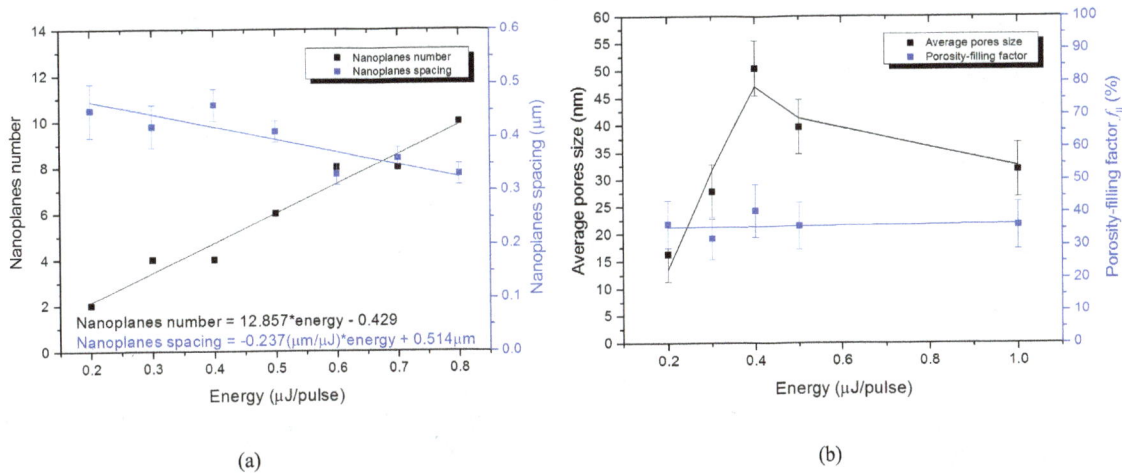

(a)   (b)

Figure 5. Structural characterization by means of SEM images: (a) Plot of the nanoplanes number and spacing according pulse energy for perpendicular writing configuration. (b) Plot of the porosity-filling factor $f_\parallel$ and average pores size according pulse energy for parallel writing configuration. The laser parameters were: 1030 nm, 300 fs, 0.6 NA, 10^3 pulses/µm.

nanoplane, is independent of the laser pulse energy within the studied range. The porosity-filling factor is found to be around 36% ± 7% whatever the pulse energy may be.

The FEG-SEM images in **Figure 6** demonstrate the modification in the transformation morphology occurring for a parallel configuration Xx and for different overlapping rates (or pulse densities). The pulse energy was fixed to 0.5 μJ. For low or no overlapping rate *i.e.* 1 pulse/μm, only isotropic index changes are detected by optical microscopy and no specific nanostructure can be observed using SEM apart a slight contrast related to volume change. Then for overlapping rate in the range of 2 - 20 pulses/μm we detect a topographic contrast e.g. see the white area. FEG-SEM also detects some kind of bubbles that evolves into elongated ones (1 micron in length and 100 nm wide) when increasing the overlapping from 2 to 5 pulses/μm. Finally the nanoplanes are porous for overlapping rate higher than 100 pulses/micron and up to 2×10^5 pulses/μm. Notice that for low repetition rate *i.e.* 1 - 10 kHz, porous nanoplanes are also observed if the writing speed is small enough *i.e.* below 10 μm/s. **Figure 7** highlights a specific laser track in SEM and its corresponding surface topography recorded in AFM as shown in **Figure 7(b)**. The area appearing white is dense matter, although having been irradiated, whereas the parts that present "roughness" are the cleaved nanoplanes.

After SEM images analysis we have plotted in **Figure 8** the average pore size together with the porosity-filling factor $f_{\parallel}$ according to the overlapping rate. The average pore size is strongly decreasing from 1 - 1000 pulses/μm and it decreases from 62 nm down to 45 nm for higher overlapping rate. In contrast, the porosity-filling factor of 24% ± 5% is rather independent of overlapping rate for more than 100 pulses/μm.

4. Discussion

For an irradiated area larger than the probe beam, it is well known that the strength of form birefringence de-

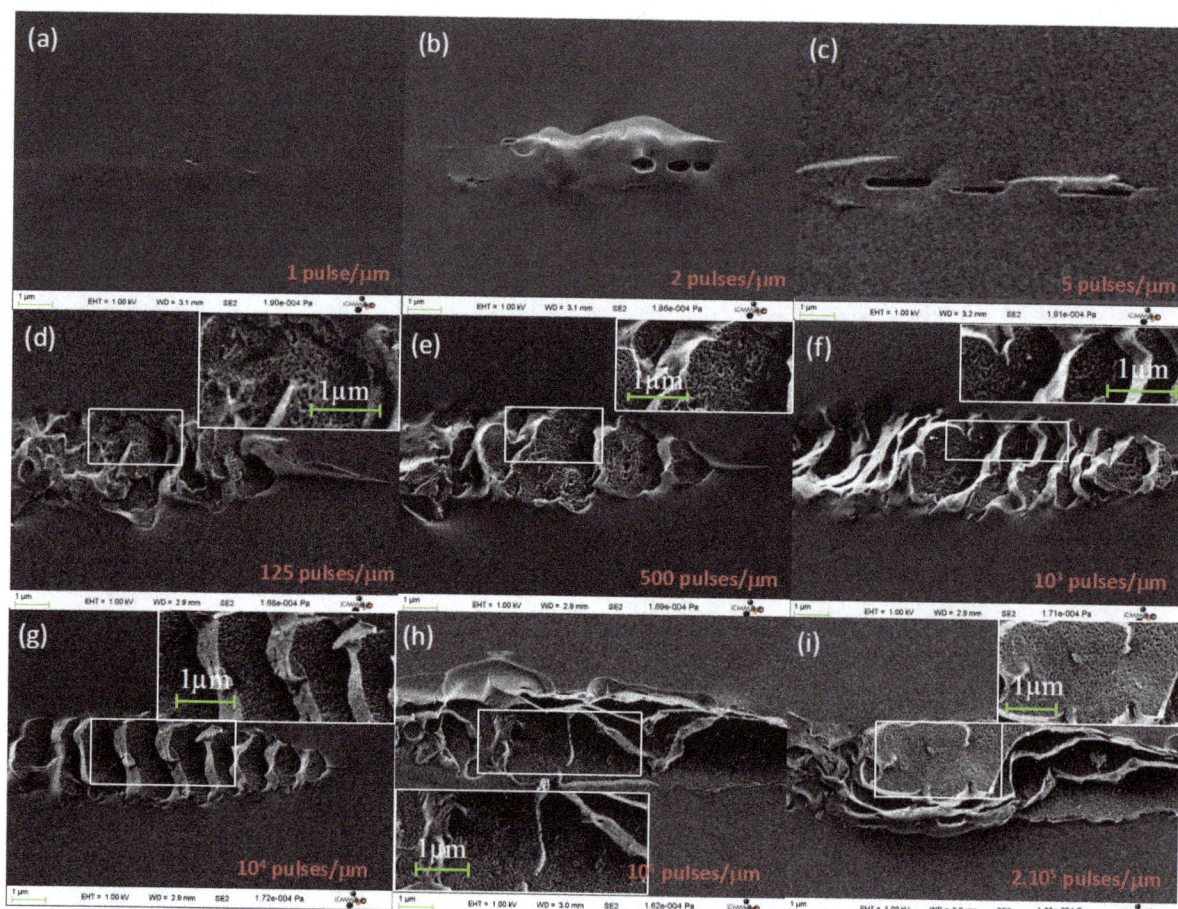

Figure 6. FEG-SEM, Secondary electrons images of laser tracks cross-section for parallel writing configuration, according to overlapping rate in pulses/μm. The laser parameters were: 0.5 μJ/pulse, 1030 nm, 300 fs, 0.6 NA and Xx configuration.

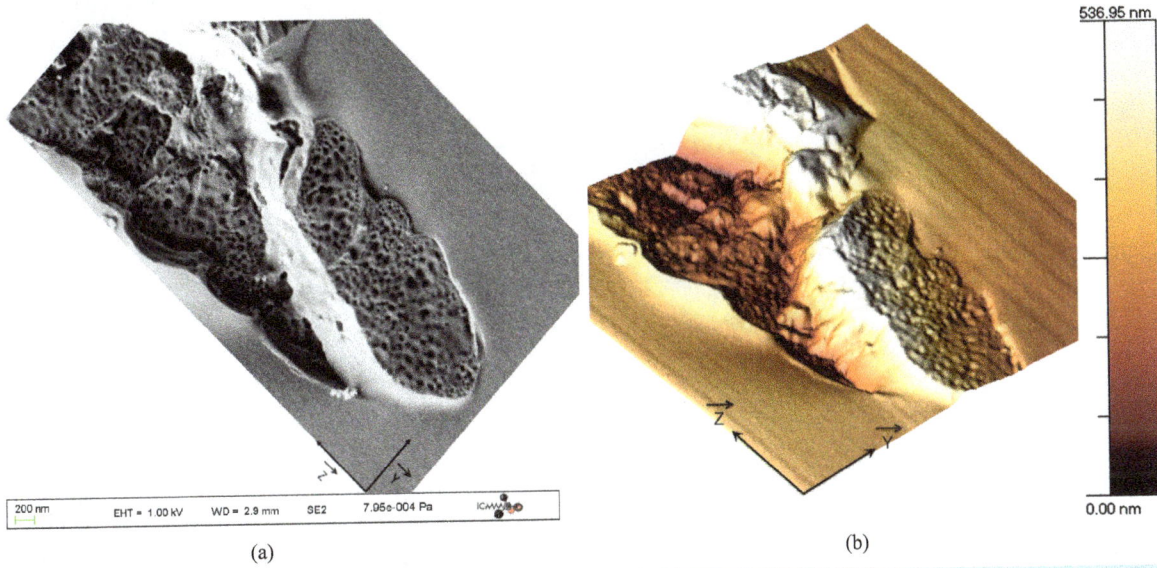

Figure 7. FEG-SEM, Secondary electrons image (a) and 3D AFM image 3.6 μm × 3.6 μm (b) of a laser track cross-section for parallel writing configuration. The laser parameters were: 0.4 μJ/pulse, 1030 nm, 300 fs, 0.6 NA, 10^3 pulses/μm and Xx configuration.

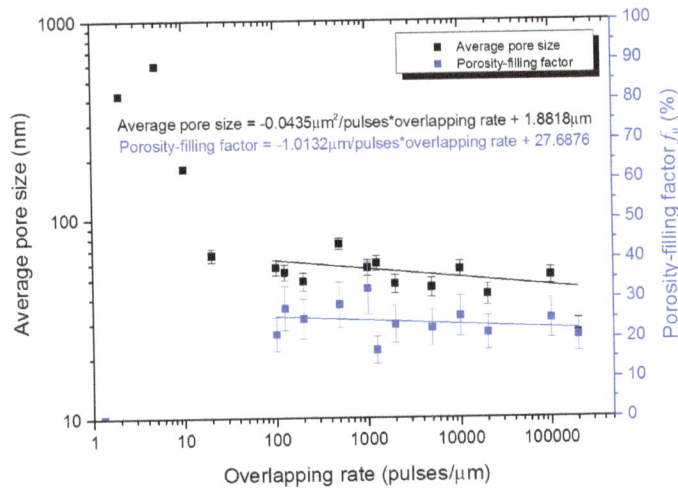

Figure 8. Plot of porosity-filling factor $f_{\parallel}$ (black points) and average pores size (blue points) according to overlapping rate for parallel Xx writing configuration. The laser parameters were: 1030 nm, 0.5 μJ/pulse, 0.6 NA.

pends on the nanoplanes spacing, the nanolayers thicknesses (t_1 and t_2) and the refractive indices (n_1 and n_2) of these two nanolayers. The uniaxial birefringence Dn due to the nanogratings can be written as a refractive index difference between ordinary (n_0) and extraordinary (n_e) wave [21]:

$$\Delta n = n_e - n_0 = \underbrace{\frac{n_1 n_2}{\sqrt{f_\perp n_2^2 + (1 - f_\perp) n_1^2}}}_{n_e} - \underbrace{\sqrt{f_\perp n_1^2 + (1 - f_\perp) n_2^2}}_{n_0} \tag{1}$$

where $f_\perp = t_1/(t_1 + t_2) = t_1/\Lambda$ is the 1D nanogratings filling factor $f_\perp$ and Λ is the nanoplanes spacing, n_1, n_2, t_1 and t_2 are defined in **Figure 1** and **Figure 10**. This relation, valid for nanogratings, can be generalised for any anisotropic structures like series of nanoplanes, in the case where the probe beam diameter w is larger than nanoplanes area:

$$\Delta n = \frac{\Lambda - \sqrt{\left(t_1 n_1^2 + t_2 n_2^2\right)\left(\dfrac{t_1}{n_1^2} + \dfrac{t_2}{n_2^2}\right)}}{\Lambda \left(\dfrac{t_1}{n_1^2} + \dfrac{t_2}{n_2^2}\right)} \tag{2}$$

and, in the case where w is smaller than nanoplane area:

$$\frac{w - \sqrt{\left[l_t n_2^2 + N_p\left(t_1 n_1^2 + t_2 n_2^2\right)\right]\left[\dfrac{l_t}{n_2^2} + N_p\left(\dfrac{t_1}{n_1^2} + \dfrac{t_2}{n_2^2}\right)\right]}}{\sqrt{w\left[\dfrac{l_t}{n_2^2} + N_p\left(\dfrac{t_1}{n_1^2} + \dfrac{t_2}{n_2^2}\right)\right]}} \tag{3}$$

where N_p is the number of period Λ and w is probe beam diameter (equal to 1.2 µm in our experimental conditions) and l_t is defined by $l_t = w - N_p \Lambda$.

On the other hand, QPm gives a measurement that is proportional to the mean refractive index defined by $\overline{n} = (n_e + n_0)/2$. The measured value $\Delta\varphi$ is the phase difference between non-irradiated and irradiated material. So, after measuring the retardance and the thickness of L_{nano} of the nanogratings structures in the direction of the light propagation, it is possible to calculate birefringence by the relation $\Delta n = R/L_{\text{nano}}$ and $\overline{n} = \dfrac{\lambda}{2\pi L_{\text{nano}}} \Delta\varphi + n_{\text{Ge-doped silica}}$. We deduce a system (4) of 2 equations with 2 unknown variables n_e and n_0 that we can solve:

$$\begin{cases} \Delta n = n_e - n_0 \\ \overline{n} = (n_e + n_0)/2 \end{cases} \tag{4}$$

The **Figure 9(a)** shows the result after solving the above equations set. We observe that both ordinary and extraordinary refractive indices decrease down to 0.7 µJ/pulse and seem increase to higher energies. The maximum decrease is around -4×10^{-2} and maximum birefringence Dn is around -7×10^{-3}. This order of magnitude is in agreement with previous publication from E. Bricchi et al. [21].

From these experimental results, we can extract information about the matter within the porous nanoplanes

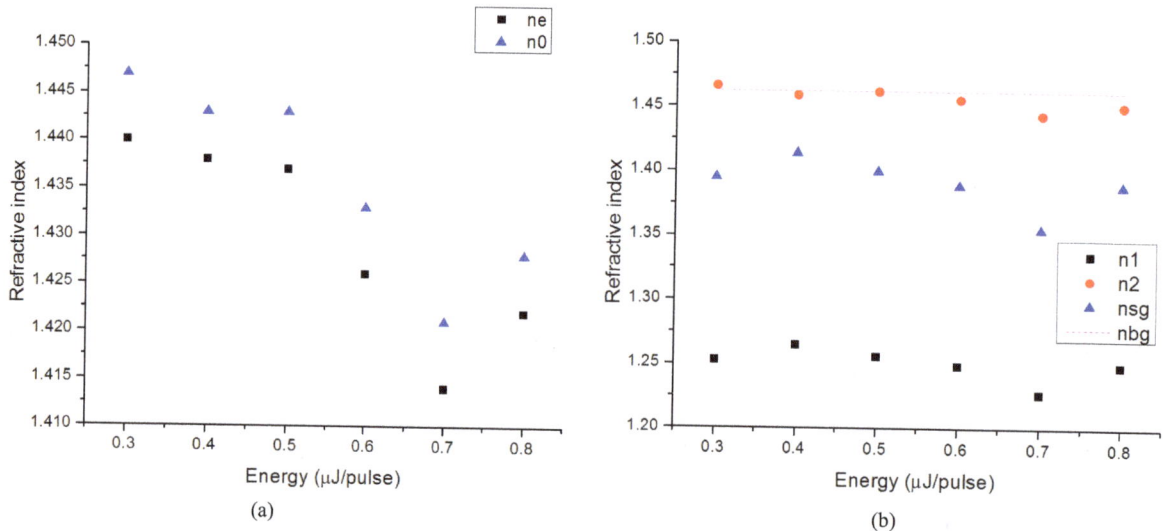

Figure 9. Refractive indices deduced from the measurements: (a) Ordinary n_o and extraordinary n_e indices, (b) n_1 and n_2, local indices of the plates of thickness t_1 and t_2, respectively and n_{sg}, the local refractive index of region surrounding nanopores and n_{bg} the refractive index of region surrounding laser track before irradiation.

and the matter around them. As shown in **Figure 10**, these periodic structures are ruled in the direction perpendicular to the polarization of the writing laser and consist of thin regions of refractive index n_1, characterized by a strong oxygen deficiency and oxide decomposition [6], surrounded by larger regions of index n_2. Using Equation (1) we can thus deduce n_0 and n_e from n_1, n_2 and based on the SEM measurements of the nanogratings filling factor $f_\perp$. Then we can extract information related to the nanoplanes themselves. Indeed, based on the Maxwell-Garnett theory, the effective refractive index (n_1) of the mesoporous nanoplanes can be decomposed as follow:

$$n_1 = f_{\parallel} n_{pore} + \left(1 - f_{\parallel}\right) n_{sg} \tag{5}$$

where n_{pore} and n_{sg} is the local refractive index for nanopores and for surrounding regions, respectively. Thus assuming that the nanopores refractive index is equal to 1, we can deduce n_{sg}.

Figure 9(b) shows that the refractive index n_2 (red dots), corresponding to the matter between nanoplanes, seems to remain unchanged after laser irradiation, whereas the mesoporous nanoplanes exhibit a significant decrease of their refractive index n_1 (black squares) down to 1.25 that is mainly due to its porous nature. It's more difficult to explain the fact that $n_{sg} < n_{bg}$ i.e. the matter in-between nanopores (blue triangles) exhibits a refractive index decrease! Whereas, we would expect to have under-stoichiometric silica and thus a refractive index larger than the background material n_{bg}. We can interpret this refractive index decrease by an irreversible volume expansion $\Delta V > 0$ of the matter as a normal glass (12w% GeO_2-doped silica) due to increase of a local fictive temperature. A second possibility may be that this matter in-between nanopores is composed of silica with a significant amount of Frenkel oxygen defects n_{bg} (an interstitial oxygen and a vacancy) resulting in a refractive index decrease. Raman microspectroscopy has indeed revealed the presence of molecular oxygen dissolved in the glass matrix [18] [19]. In addition, since fluctuations of n_1, n_2 and n_{sg}, remain quite small, we can consider that the refractive indices are constant according to the laser pulse energy. So we can deduce that the birefringence changes are mainly due to structural modifications of the laser tracks namely the number of nanoplans, their length and their spacing.

5. Conclusions

Here, we analyzed the laser tracks in the condition for obtaining porous nanogratings and related anisotropic optical properties at the nanoscopic scale. We revealed that the magnitude of the birefringence could be accurately modeled based on SEM observations with the following input parameters: number of nanoplanes, nanoplanes thickness, nanostructuration length, and the porosity-filling factor. This study reveals definitely that modification of birefringence is mainly due to the structural modification (number of nanoplanes, their thickness and length) and not related to refractive index changes after laser irradiation according to energy or overlapping rate. It is

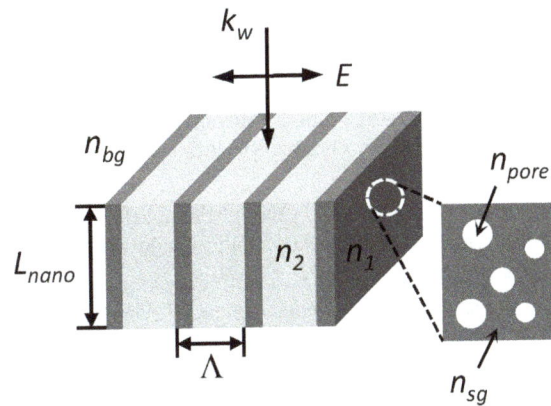

Figure 10. Schematic of sub-wavelength periodic structure formed in cross-section of the irradiated region. n_1 and n_2: local refractive indices of the plates of the thickness t_1 and t_2, respectively, n_{pore} (=1) and n_{sg}: local refractive index for nanopores and for surrounding oxygen defect regions, respectively, $\Lambda = t_1 + t_2$: period of nanogratings and n_{bg}: refractive index of the surrounding material, L_{nano}: thickness of nanogratings.

worth noticing that the matter in-between nanopores exhibits significant refractive index decrease which is likely due to fictive temperature increase and/or the presence of a significant amount of interstitial O_2 in the silica surrounding nanopores.

The average pore size increases with the pulse energy and slightly decreases at high repetition rate or high overlapping rate. This should provide a way to reduce the light scattering related to the porous nanoplanes formation. From a practical perspective, such control allows the fabrication of mesoporous nanolayers that can be arranged in a nearly regular array, leading to many novel applications not only for birefringent devices (such as waveplates [22], micro-patterned waveplates, polarization converters, 5D optical data storage) but we can also expect the catalysts, molecular sieves, encapsulants, and composites. Such mesoporous glass could also be used for filtration and separation of compounds. For example, by controlling the pore diameters and homogeneity, the mesoporous network allows permeability on a size-selective basis and can be integrated with high precision into waveguides and patterned components, a potentially more robust and efficient top-down alternative to methods based on bottom-up self-assembly.

Acknowledgements

This work has been performed in the framework of FLAG (Femtosecond Laser Application in Glasses) consortium project with the support of several organisations: the Agence Nationale pour la Recherche (ANR-09-BLAN-0172-01), the RTRA Triangle de la Physique (Réseau Thématique de Recherche Avancée, 2008-056T) and FP7-PEOPLE-IRSES (e-FLAG 247635).

References

[1]   Ams, M., Marshall, G., Dekker, P., Dubov, M., Mezentsev, V., Bennion, I. and Withford, M. (2008) Investigation of Ultrafast Laser—Photonic Material Interactions: Challenges for Directly Written Glass Photonics. *IEEE Journal of Selected Topics in Quantum Electronics*, **14**, 1370-1381. http://dx.doi.org/10.1109/JSTQE.2008.925809

[2]   Gattass, R.R. and Mazur, E. (2008) Femtosecond Laser Micromachining in Transparent Materials. *Nature Photonics*, **2**, 219-225. http://dx.doi.org/10.1038/nphoton.2008.47

[3]   Itoh, K., Watanabe, W., Nolte, S. and Schaffer, C. (2006) Ultrafast Processes for Bulk Modification of Transparent Materials. *MRS BULLETIN*, **31**, 620-625. http://dx.doi.org/10.1557/mrs2006.159

[4]   Qiu, J., Miura, K. and Hirao, K. (2008) Femtosecond laser-Induced Microfeatures in Glasses and Their Applications. *Journal of Non-Crystalline Solids*, **354**, 1100-1111. http://dx.doi.org/10.1016/j.jnoncrysol.2007.02.092

[5]   Lancry, M., Poumellec, B., Chahid-Erraji, A., Beresna, M. and Kazansky, P. (2011) Dependence of the Femtosecond Laser Refractive Index Change Thresholds on the Chemical Composition of Doped-Silica Glasses. *Optical Materials Express*, **1**, 711-723. http://dx.doi.org/10.1364/OME.1.000711

[6]   Poumellec, B., Lancry, M., Chahid-Erraji, A. and Kazansky, P. (2011) Modification Thresholds in Femtosecond Laser Processing of Pure Silica: Review of Dependencies on Laser Parameters. *Optical Materials Express*, **1**, 766-782. http://dx.doi.org/10.1364/OME.1.000766

[7]   Eaton, S., Zhang, H., Herman, P., Yoshino, F., Shah, L., Bovatsek, J. and Arai, A. (2005) Heat Accumulation Effects in Femtosecond Laser-Written Waveguides with Variable Repetition Rate. *Optics Express*, **13**, 4708-4716. http://dx.doi.org/10.1364/OPEX.13.004708

[8]   Schaffer, C., Brodeur, A., Garcia, J. and Mazur, E. (2001) Micromachining Bulk Glass by Use of Femtosecond Laser Pulses with Nanojoule Energy. *Optics Letters*, **26**, 93-95. http://dx.doi.org/10.1364/OL.26.000093

[9]   Bricchi, E., Klappauf, B. and Kazansky, P. (2004) Form Birefringence and Negative Index Change Created by Femtosecond Direct Writing in Transparent Materials. *Optics Letters*, **29**, 119-121. http://dx.doi.org/10.1364/OL.29.000119

[10]   Poumellec, B., Lancry, M., Poulin, J. and Ani-Joseph, S. (2008) Non Reciprocal Writing and Chirality in Femtosecond Laser Irradiated Silica. *Optics Express*, **16**, 18354-18361. http://dx.doi.org/10.1364/OE.16.018354

[11]   Sudrie, L., Franco, M., Prade, B. and Mysyrowicz, A. (1999) Writing of Permanent Birefringent Microlayers in Bulk Fused Silica with Femtosecond Laser Pulses. *Optics Communications*, **171**, 279-284. http://dx.doi.org/10.1016/S0030-4018(99)00562-3

[12]   Lancry, M., Niay, P. and Douay, M. (2005) Comparing the Properties of Various Sensitization Methods in H2-Loaded, UV Hypersensitized or OH-Flooded Standard Germanosilicate Fibers. *Optics Express*, **13**, 4037-4043. http://dx.doi.org/10.1364/OPEX.13.004037

[13]   Lancry, M. and Poumellec, B. (2013) UV Laser Processing and Multiphoton Absorption Processes in optical Tele-

communication Fibers Materials. *Physics Reports*, **523**, 207-229. http://dx.doi.org/10.1016/j.physrep.2012.09.008

[14] Eaton, S.M., Ng, M.L., Osellame, R. and Herman, P.R. (2010) High Refractive Index Contrast in Fused Silica Waveguides by Tightly Focused, High-Repetition Rate Femtosecond Laser. *Journal of Non-Crystalline Solids*.

[15] Bricchi, E. and Kazansky, P. (2006) Extraordinary Stability of Anisotropic Femtosecond Direct-Written Structures Embedded in Silica Glass. *Applied Physics Letters*, **88**, 111119-111119. http://dx.doi.org/10.1063/1.2185587

[16] Shimotsuma, Y., Kazansky, P., Qiu, J. and Hirao, K. (2003) Self-Organized Nanogratings in Glass Irradiated by Ultrashort Light Pulses. *Physical Review Letters*, **91**, Article ID: 247405. http://dx.doi.org/10.1103/PhysRevLett.91.247405

[17] Bhardwaj, V., Simova, E., Rajeev, P., Hnatovsky, C., Taylor, R., Rayner, D. and Corkum, P. (2006) Optically Produced Arrays of Planar Nanostructures Inside Fused Silica. *Physical Review Letters*, **96**, Article ID: 057404. http://dx.doi.org/10.1103/PhysRevLett.96.057404

[18] Canning, J., Lancry, M., Cook, K., Weickman, A., Brisset, F. and Poumellec, B. (2011) Anatomy of Femtosecond Laser processed Silica Waveguide. *Optical Materials Express*, **1**, 998-1008. http://dx.doi.org/10.1364/OME.1.000998

[19] Lancry, M., Poumellec, B., Canning, J., Cook, K., Poulin, J.C. and Brisset, F. (2013) Ultrafast Nanoporous Silica Formation Driven by Femtosecond Laser Irradiation. *Laser & Photonics Reviews*, **7**, 953-962. http://dx.doi.org/10.1002/lpor.201300043

[20] Yang, W.J., Bricchi, E., Kazansky, P.G., Bovatsek, J. and Arai, A.Y. (2006) Self-Assembled Periodic Sub-Wavelength Structures by Femtosecond Laser Direct Writing. *Optics Express*, **14**, 10117-10124. http://dx.doi.org/10.1364/OE.14.010117

[21] Bricchi, E., Klappauf, B.G. and Kazansky, P.G. (2004) Form Birefringence and Negative Index Change Created by Femtosecond Direct Writing in Transparent Materials. *Optics letters*, **29**, 119-121. http://dx.doi.org/10.1364/OL.29.000119

[22] Lancry, M., Desmarchelier, R., Cook, K., Canning, J. and Poumellec, B. (2014) Compact Birefringent Waveplates Photo-Induced in Silica by Femtosecond Laser. *Micromachines*, **5**, 825-838. http://dx.doi.org/10.3390/mi5040825

11

Hubble Scale Dark Energy Meets Nano Scale Casimir Energy and the Rational of Their T-Duality and Mirror Symmetry Equivalence[*]

M. S. El Naschie

Department of Physics, University of Alexandria, Alexandria, Egypt
Email: Chaossf@aol.com

Abstract

We establish that ordinary energy, Casimir energy and dark energy are not only interlinked but are basically the same thing separated merely by scale and topology. Casimir energy is essentially a nano scale spacetime phenomenon produced by the boundary condition of the two Casimir plates constituting the Casimir experimental set up for measuring the Casimir force. By contrast dark energy is the result of the cosmic boundary condition, *i.e.* the boundary of the universe. This one sided Möbius-like boundary located at vast cosmic distance and was comparable only to the Hubble radius scales of the universe. All the Casimir energy spreads out until the majority of it reaches the vicinity of the edge of the cosmos. According to a famous theorem due to the Ukrainian-Israeli scientist I. Dvoretzky, almost 96% of the total energy will be concentrated at the boundary of the universe, too far away to be measured directly. The rest of the accumulated Casimir energy density is consequently the nearly 4% to 4.5%, the existence of which is confirmed by various sophisticated cosmic measurements and observations. When all is said and done, the work is essentially yet another confirmation of Witten's T-duality and mirror symmetry bringing nano scale and Hubble scale together in an unexpected magical yet mathematically rigorous way.

Keywords

Mirror Symmetry, Casimir Energy, Dark Energy, Zero Point Vacuum Energy, T-Duality, Nano Scale-Hubble Scale, Möbius Holographic Boundary, Dvoretzky's Theorem, Banach-Tarski Theorem

1. Introduction

The present paper is concerned with a wide range of problems connecting high energy quantum physics with

[*]Dedicated to the memory of Sir Herman Bondi and Prof. Thomas Barta—two very memorable teachers and fatherly friends.

relativity, quantum gravity and cosmology [1]-[96].

One of the many memorable words of wisdom that Prof. Sir Herman Bondi was known for is to the effect that once a great truth about nature is uncovered, it becomes so obvious and self evident that we are bound to wonder how we did not notice it for such a long time and how it could have eluded us all despite its compelling simple logic [1]. Sir Herman used to cite Einstein' relativity as his favourite parade example [1]-[10].

Referring to the present work the author would like to think that the thesis that we are about to reveal belongs in the same category which Prof. Bondi would have liked [1]-[10]. We are pointing here at our discovery that we perceived a few months ago as tantalizing because it implies the realization that the Casimir energy involved in the famous Casimir effect as well as dark and ordinary energy density of the cosmos are nothing but different forms of the same physical reality by virtue of some mathematical theorem which goes by the name of mirror symmetry and T-duality [93]-[95]. With that we do not only mean that they are different manifestations of the zero point energy of vacuum fluctuation only [11]-[31]. It is far more than that. How much more is the subject of the present paper and we will, without much ado, turn our attention now to the present main task of explaining it in as simple a manor as possible with the only provision which Einstein would have said, "put not more".

2. An Exact Analytical-Topological Picture of Spacetime

Before reading a single word or equation of the present paper, our advise to the prospective reader is to first have a long contemplative look at **Figure 1** and **Figure 2** of the present work. The first figure is actually an artist impression of quantum or fractal-Cantorian spacetime (**Figure 1**). The second is an exact mathematical-topological picture in the sense of the scientific philosophy of people like Wittgenstein (**Figure 2**). Such a scientific picture is not the product of wild imagination but of stringent application of the dimensional function of von Neumann-Connes' continuous and noncommutative geometry, namely [71]

$$D = a + b\phi; \quad a,b \in Z \quad \text{and} \quad \phi = \left(\sqrt{5}-1\right)\big/2 .$$

In particular it can be shown that $D(O)$ is the dimensionality of the zero set which models the pre-quantum particle while $D(-1)$ is the dimensionality of the empty set which models the quantum wave. More specifically the pre-quantum particle is accurately represented by the bidimension [64] [71]

$$D(O) = \left(D_T; D_H\right) = \left(O, \phi\right)$$

where $D = O$ is the Menger-Urysohn deductive topological dimension and $D_H = \phi$ is the corresponding Hausdorff dimension which happens to be the same dimension as that of a triadic Cantor set that is constructed randomly using a uniform distribution. In other words, we could use the follow shorthand notation [64] [71]

$$D(QP) \equiv \left(O, \phi\right) \equiv D(O) \equiv \text{zero set}$$

for the pre-quantum particle and

$$D(QW) \equiv \left(-1, \phi^2\right) \equiv D(-1) \equiv \text{empty set}$$

for the pre-quantum wave. Now using Fibonacci's growth law and starting with $D(O)$ using $(a, b) = (0, 1)$ and $D(1)$ $(a, b) = (1, 0)$ as said, one obtains for positive dimension the following series

$$D(O) = 0 + \phi = \phi$$
$$D(1) = 1 + 0 = 1$$
$$D(2) = 1 + \phi = 1 + \phi = \left(1/\phi\right)$$
$$D(3) = 2 + \phi = 2 + \phi = \left(1/\phi\right)^2$$
$$D(4) = 3 + 2\phi = 4 + \phi^3 = \left(1/\phi\right)^3$$
$$D(5) = 5 + 3\phi = 6.854101 = \left(1/\phi\right)^4$$
$$\vdots$$
$$D(n) = \cdots = \cdots = \left(1/\phi\right)^{n-1} .$$

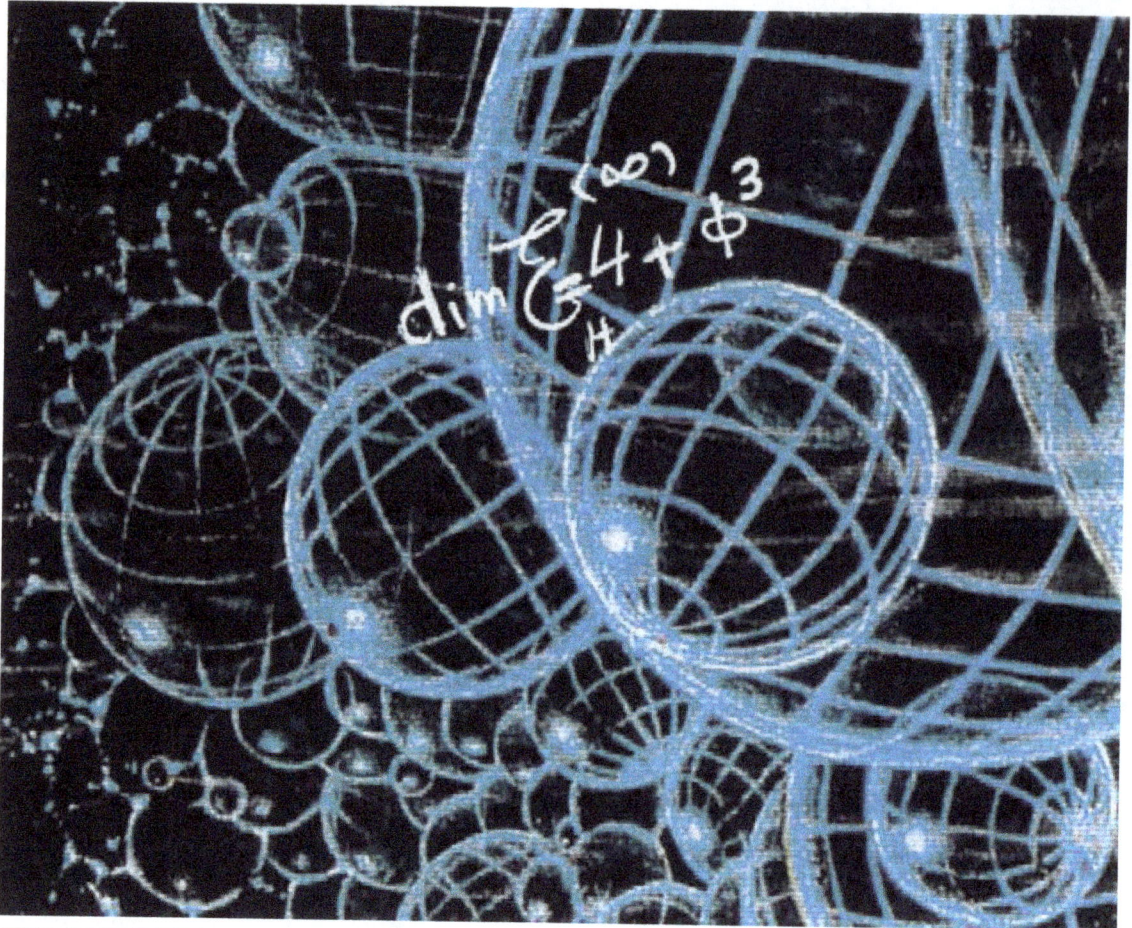

Figure 1. T-duality and Banach-Tarski sphere decomposition Cantorian spacetime of E-infinity theory that is considered here to model our actual spacetime may be envisaged advantageously as in this artist impression. This is basically a two dimensional projection in which each of the larger balls (circles) are a zero set $(0;\phi)$ representing the quantum particle while the surface (circumference) represents the empty set $(-1,\phi^2)$ which in turn represents the quantum wave [1] [17]. This wave is then surrounded by an infinite hierarchy of smaller (fractal) spheres (surfaces), which may be seen as the emptier set $(-2;\phi^3)$, *i.e.* the surface of the empty set quantum wave. Remarkably the average set of all zero and empty sets is an expectation value equal $\langle -2;\phi^3 \rangle$. In other words $\langle -2;\phi^3 \rangle$ is our quantum spacetime, which is the cobordism of the quantum wave, which in turn is the cobordism of the quantum particle, floating and propagating with the help of its wave in our Cantorian E-infinity spacetime [1] [2] [10] [11]. It is likewise remarkable that ϕ^3 is simultaneously equal to the topological Casimir force as well as the topological mass of the ordinary energy of spacetime [96]. Thus all matter and energy manifestations in our cosmos are essentially a manifestation of the zero point energy of the vacuum of spacetime. To obtain Einstein maximal energy density we just need to find first the topological energy density by adding Kaluza-Klein D = 5 to ϕ^3 of the spacetime vacuum and find the fractal Kaluza-Klein dimension $5+\phi^3$ then multiply this with the average Cantorian interval speed of light c = ϕ squared. The result is $(5+\phi^3)\phi^2 = 2$. Inserting in Newton's kinetic energy one finds

$$E(Einstein) = \frac{1}{2}m(v \to c)^2(2) = mc^2$$

exactly as should be. The preceding explanation amounts to a paradigm shift in physics where the totally empty vacuum of spacetime is taken as fundamental and everything else is derivable from it. To prove this point was a dream of Serbian American inventor *N.* Tesla who died in 1943 as well as Soviet physicist A. Zakharov. In fact in his later years Nobel Laureate J. Schwinger was a champion of cold fusion [12] which comes very near to our present concept of a Casimir-nano energy reactor [10] [11]. We also stress that we are making tacit use of the Banach-Tarski decomposition theorem as a Schwinger-like source [18] [21] [34]. The main conclusion that jumps out of this picture is that scale in physics is not a trivial idea and can only be deeply comprehended via mathematical tools such as P-Adic quantum mechanics and Witten's T-duality [93]-[95].

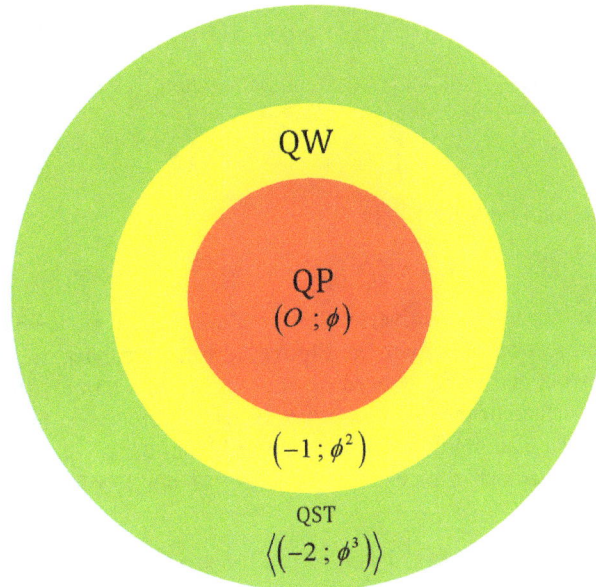

Figure 2. The quantum spacetime E-infinity hierarchy [12] [13].

Remarkably our Fibonacci-like dimension series could be extended into the negative side using the same logic as follows [64] [71]

$$D(1) = 1 + 0$$

$$D(O) = 0 + \phi$$

$$D(-1) = (1-0) + (0-\phi) = 1 - \phi = \phi^2$$

$$D(-2) = (0-1) + \phi - (-\phi) = -1 + 2\phi = \phi^3$$

$$D(-3) = 1 - (-1) + (\phi) - 2\phi = (1+1) - 3\phi = \phi^4$$

$$D(-4) = \cdots = \phi^5$$

$$\vdots$$

$$D(-n) = \cdots = \phi^{n+1}.$$

Although the preceding computation is mechanistic and essentially elementary, it is quite tedious because of the heavy and slightly confusing notation for negative dimensions. It is at this point that one discovers the advantage of the equivalent notation used in E-infinity theory [69]. The equivalent formula to that of the dimensional function is simply the bijection formula [33] [65] [69]

$$d_c^{(n)} = (1/\phi)^{n-1}.$$

Here n is the Menger-Urysohn topological dimension while $d_c^{(n)}$ denotes the Hausdorff dimension corresponding to n. Consequently for the zero set and the empty set we have [69] [70]

$$d_c^{(0)} = (1/\phi)^{0-1} = \phi$$

and

$$d_c^{(-1)} = (1/\phi)^{-1-1} = \phi^2$$

correspondingly. For the core Hausdorff dimension of E-infinity we just need to set $n = 4$ and find [69]-[71]

$$d_c^{(4)} = (1/\phi)^{4-1} = (1/\phi)^3 = 4 + \phi^3$$

We note for later use that for $n = 2$, which means the surface of the pre-quantum wave we find

$$d_c^{(-2)} = (1/\phi)^{-2-1} = (1/\phi)^{-3} = \phi^3.$$

Now we ask an important question, namely what happens when $n \to \pm\infty$? In the positive domain we find of course that

$$d_c^{(n)} = (1/\phi)^{\infty-1} = (1/\phi)^{\infty} = \infty.$$

By contrast for $n \to -\infty$ one finds [69]-[71]

$$d_c^{(n)} = (1/\phi)^{-\infty-1} = (1/\phi)^{-\infty} = \phi^{\infty} = \text{zero}.$$

This is clearly the end of our and any universe. Said in a mundane way, we basically have a Cantorian onion in the two dimensional projection as shown in **Figure 2**. First we have the zero set quantum pre-particle $D(O) \equiv (O, \phi)$. This is surrounded by its cobordism, *i.e.* its surface, namely the empty set quantum pre-wave $D(1) \equiv (-1, \phi^2)$. Next the surface of the quantum wave is also an empty set but not only simply empty. It is an emptier set than the empty set. The highly interesting point is that the inversion of this ϕ^3 empty set given by $D(-2) = \phi^3$ is simply the Hausdorff dimension of the core of our E-infinity Cantorian spacetime. This is so because [69]-[71]

$$1/(D(-2)) = 1/\phi = 4 + \phi^3$$

while the average dimension of the entire spacetime is given by

$$\sum_{n=1}^{\infty} \phi^n + \phi^0 + \phi^{-1} = 1/\phi + 1 + 1/\phi = 1 + \phi + 1 + 1 + \phi$$

$$= 3 + 2\phi = 3 + 1 + \phi^3 = 4 + \phi^3$$

which is precisely the same value as that of the inversion of the cobordism of the empty set quantum wave. Thus we conclude that the picture of **Figure 2** is accurate in the most stringent mathematical and transfinite topological sense and we just have to be put $D(-2, \phi^3)$ between averaging brackets to read $\overline{D(-2, \phi^3)}$. That way the picture is complete and will play a pivotal role in understanding the mirror symmetry and T-duality of Casimir and dark energy [93]-[95].

3. The Geometry and Topology of the Casimir Effect [89] [91]

Consider the classical Casimir set up and the hydromechanical analogy that goes with it. Bringing the two Casimir plates as near as a few nano units, we can consider the space between the plates practically totally empty of any "waves" or "particles" pushing them apart. Outside the plates spacetime resembles a stormy sea, which pushes the plates towards each other. Now the inside of the plates has the ϕ^2 topological "pressure" of the empty set inherited from its Hausdorff dimension. The outside will have at least the topological pressure of quantum particles which is the zero set represented by ϕ. The pressure difference between the outside and the inside of the two plates is given by the difference namely $\phi - \phi^2 = \phi^3$. In other words in the limit the exact Casimir topological pressure is simply ϕ^3. Recalling that Einstein spacetime $D = 4$ is a smooth spacetime while the core is $D = 4 + \phi^3$ indicating its fractal character we see that the extra ϕ^3 fractal part is equal to the limit of the topological Casimir pressure. To find the density of the Casimir energy in a way comparable to that of the dark energy density of the cosmos we have to transfer the result to a five dimensional K-K spacetime or a five dimensional anti de Sitter spacetime. This is easily done because it is nothing but the union of all the four dimensional fusion algebra dimension function given by $d(1) = d(\epsilon) = 1$ and $d(\alpha) = d(\beta) = \phi$ which means: [92]

$$D = 1 + 1 + 1/\phi + 1/\phi = 2(1 + 1/\phi) = 4 + 2\phi = 5 + \phi^3$$

The Casimir energy density is therefore given by the following ratio [35]-[51]

$$\gamma = \frac{\phi^3}{5 + \phi^3} = \phi^5/2 \approx 1/22.$$

This is nothing but the familiar ordinary energy density of the universe that agrees with all previous analysis of the problem as well as accurate observations and measurements. Now this is only about 4.5 percent of the expected energy, so where is the rest is a natural question. However this question was answered some time ago. The bulk of the energy density is due to the quantum wave and is given by [33]-[51]

$$\gamma(\text{dark}) = (D = 5)(\phi^2)/2 \simeq (21)/(22).$$

This is about 95.5 percent of the energy and is stored in the integer part of $D = 5 + \phi^3$, namely $D = 5$. Consequently

$$\gamma(\text{dark}) = 5/5 + \phi^3 \simeq 21/22.$$

The result is in complete agreement with that obtained using various methods. The fact that this 95.5 percent is located at the edge of the universe is explained within our five dimensional spacetime theory via Dvoretzky's theorem [59] [63]. On reflection the result is obvious. To see how obvious let us imagine that the two Casimir plates of the famous experiment are taken apart in two opposite directions until each one reaches the cosmological boundary. Thus instead of having ϕ^2 of the empty set between them they have now the entire topological pressure of the universe between them, that is to say $5 + \phi^3$. Outside there is nothing to balance this pressure because on the other side there is zero pressure, which is natural for a one sided Möbius-like boundary. The topological density is thus $(5 + \phi^3)/(5 + \phi^3) = 1$ which is Einstein's theoretical maximal topological energy density. To find the dark energy density we just have to subtract ϕ^3 of the Casimir density which is pushing intrinsically in the opposite direction and find that $\left[(5 + \phi^3) - \phi^3\right]/(5 + \phi^3) = 0.954915$ percent is the relevant energy density. The analysis could be extended to find the Immirzi parameter connecting loop quantum gravity with string theory by looking at $(4)/4 + \phi^3 = \phi^6$ which is the value found using different methods.

4. The Mirror Symmetry and T-Duality Connecting Casimir Energy with Dark Energy

There is a fascinating history behind the subject of this section testifying to the unity of science and showing that at the end, physicists and mathematicians of all moulds must think and work together on the most fundamental level to be able to push the borders of science forward [95]. The most word economical way to explain mirror symmetry for someone—like the author—living between physics and mathematics, is to say mirror symmetry means that two Calabi-Yau manifolds with different topologies give the same conformal quantum field theory [95]. A more general and probably stronger statement of the same fact can be formulate as a duality which in our particular mirror symmetry case is Witten's T-duality [93]-[95].

Let me be more specific even at the risk of appearing elementary or even trivial. Let us start with Hardy's quantum entanglement. The exact solution for two quantum particles entanglement is given by the well known accurate result ϕ^5. Now quantum entanglement experiments may be performed on a billiard table rather than using the entire universe [96] [97]. However the inversion of ϕ^5 gives us the amazing fractal M-theory dimensionality of the entire universe, namely [94]

$$1/\phi^5 = 11 + \cfrac{1}{11 + \cfrac{1}{11 + \cdots}}.$$

This is an implicit application of T-duality which is clear cut and in view of the experimental verification of Hardy's entanglement and the theoretical success of M-theory points clearly to being a physical reality [9] [12] [15] [30] [94]. It is really hard to believe that all these results, pertinent to the dark energy density of the cosmos $E(D) = (\phi^5/2)mc^2$ and the complimentary result of ordinary energy density $E(O) = 1 - E(D) = (5\phi^2/2)mc^2$ was not immediately noticed by the author that it means $E(O) = \left[\phi^3/(5 + \phi^3)\right]mc^2$ and consequently implies mirror symmetry [95] and T-duality transformation of dark energy to ordinary measurable energy which is nothing but the intrinsic spacetime energy of our universe [94]. Even more astonishing was the initial failure of the present author to recognize that the $(22 + k)/2 = 11 + \phi^5$ factor is a super symmetric isomorphic length for a super symmetric Klein-Penrose fractal universe given by $(4 + \phi^3)(5 + \phi^3)$ and implies a topological Planck length equal this isomorphic radius and by mirror symmetry and T-duality ϕ^5 is a topological Planck energy [6] [7] [10] [15] [22] [23] [69] [94] [97]. The author is sure Sir Herman Bondi would have loved this conclusion. We urge the reader to study carefully Figures 2-4 of Ref. [94] to gain an intuitive grasp for the power and simplicity of Witten's T-duality connecting the quantum with the cosmic scale in the way presented here. We see here the interaction not only of physics and mathematics but also mathematics and engineering leading to a design of a Casimir-dark energy nano reactor [96].

5. Conclusions

To put things into perspective we should not forget that we always referred to the Casimir effect and dark energy as different forms of quantum vacuum zero point fluctuation. Nevertheless the fact that ϕ^3 is the intrinsic energy of spacetime and is equal to the global part of Hardy's quantum entanglement was not clear at all let alone that the Casimir energy density is equal to the ordinary energy density of the universe. Consequently we see that the Casimir energy density plus the energy of the quantum pre-wave in five dimensional spacetime leads to Einstein energy density. We see clearly that Casimir energy density and the quantum wave dark energy density lead to Einstein energy. This introduces to our modern physical and cosmological theory a hitherto unknown unit and elegance and all wrapped up in some of the most beautiful mathematical theorems like that of Dvoretzky and mirror symmetry of Witten's T-duality [93]-[95]. At the end we cannot find a rational reason for overlooking these results for a long time except Sir Herman Bondi's wise words about the nature of certain fundamental laws of nature.

It consists of three main layers [41]-[45]. We have first an infinite number of zero and empty sets with an average bi-dimension $\left\langle\left(-2;\phi^3\right)\right\rangle$. This is the outer circle representing quantum spacetime. Inside this we have the quantum wave given by the bi-dimension $\left(-1;\phi^2\right)$ which is the empty set. Finally inside the quantum wave as its inner eye, we have the zero set quantum particle with the bi-dimension $\left(O;\phi\right)$. The above picture also gives us an almost trivial resolution for quantum wave collapse. This is so because to "locate" QP we must somehow penetrate QW. Since QW is the empty set, the slightest touch would convert it to a non-empty set. Consequently QW disappears and metamorphose into QP. This is the observed mysterious state vector reduction which as the reader sees, is not mysterious at all within this topological set theoretical picture. The above picture gives us also an almost trivial resolution for quantum wave collapse. This is so because to "locate" QP we must penetrate somehow QW. Since QW is the empty set the slightest touch would convert it to a non-empty set. Consequently QW disappears and metamorphs into QP. This is the observed mysterious state vector reduction which as the reader sees, is not mysterious at all within this topological set theoretical picture.

Acknowledgements

For a young man who said more than half a century ago "never trust anyone above thirty", to become himself over seventy it is a time to pause and thank all those who really fundamentally influenced him and helped him in so many ways that he feels today grateful to them. On the top of the list his father, Major General Salah El Naschie (Alnashaee), then his teacher and friends, Prof. Dr. Alf Pflüger, Prof. T. Lehmann, Prof. E. Stein, Prof. R. Deboer, Prof. J.M.T. Thompson, Prof. T. Barta, Prof. Sir Herman Bondi, Prof. W. Martienssen, Prof. W. Greiner and Prof. IlyaPigogine. Last but not least, working with His Excellency Prof. Salah Al Athel of Saudi Arabia was one of the most important scientific experiences in the author's life. There are of course many other names which the fading memory of an old man is not helping to remember but to all of them, I am truly indebted.

References

[1]   Bondi, S.H. (1964) Relativity and the Commonsense. A New Approach to Einstein. Heinemann, London.

[2]   Rindler, W. (2004) Relativity (Special, General and Cosmological). Oxford University Press, Oxford.

[3]   Okun, L.B. (2009) Energy and Mass in Relativity Theory. World Scientific, Singapore.

[4]   Rindler, W. (1991) Introduction to Special Relativity. Oxford Science Publications, Oxford.

[5]   El Naschie, M.S. (2014) On a New Elementary Particle from the Disintegration of the Symplectic 't Hooft-Veltman-Wilson Fractal Spacetime. *World Journal of Nuclear Science and Technology*, **4**, 216-221.
http://dx.doi.org/10.4236/wjnst.2014.44027

[6]   Helal, M.A., Marek-Crnjac, L. and He, J.-H. (2013) The Three Page Guide to the Most Important Results of M. S. El Naschie's Research in E-Infinity Quantum Physics and Cosmology. *Open Journal of Microphysics*, **3**, 141-145.
http://dx.doi.org/10.4236/ojm.2013.34020

[7]   Marek-Crnjac, L. and He, J.-H. (2013) An Invitation to El Naschie's Theory of Cantorian Space-Time and Dark Ener-

Hubble Scale Dark Energy Meets Nano Scale Casimir Energy and the Rational of Their T-Duality...

107

gy. *International Journal of Astronomy and Astrophysics*, **3**, 464-471. http://dx.doi.org/10.4236/ijaa.2013.34053

[8] Auffray, J.-P. (2014) E-Infinity Dualities, Discontinuous Spacetimes, Xonic Quantum Physics and the Decisive Experiment. *Journal of Modern Physics*, **5**, 1427-1436. http://dx.doi.org/10.4236/jmp.2014.515144

[9] El Naschie, M.S. (2011) Quantum Entanglement as a Consequence of a Cantorian Micro Spacetime Geometry. *Journal of Quantum Information Science*, **1**, 50-53. http://dx.doi.org/10.4236/jqis.2011.12007

[10] El Naschie, M.S. (2013) A Resolution of Cosmic Dark Energy via a Quantum Entanglement Relativity Theory. *Journal of Quantum Information Science*, **3**, 23-26. http://dx.doi.org/10.4236/jqis.2013.31006

[11] El Naschie, M.S. (2013) A Unified Newtonian-Relativistic Quantum Resolution of the Supposedly Missing Dark Energy of the Cosmos and the Constancy of the Speed of Light. *International Journal of Modern Nonlinear Theory and Application*, **2**, 43-54. http://dx.doi.org/10.4236/ijmnta.2013.21005

[12] El Naschie, M.S. (2013) Quantum Entanglement: Where Dark Energy and Negative Gravity plus Accelerated Expansion of the Universe Comes From. *Journal of Quantum Information Science*, **3**, 57-77. http://dx.doi.org/10.4236/jqis.2013.32011

[13] El Naschie, M.S. (2013) A Fractal Menger Sponge Space-Time Proposal to Reconcile Measurements and Theoretical Predictions of Cosmic Dark Energy. *International Journal of Modern Nonlinear Theory and Application*, **2**, 107-121. http://dx.doi.org/10.4236/ijmnta.2013.22014

[14] El Naschie, M.S. (2013) The Hydrogen Atom Fractal Spectra, the Missing Dark Energy of the Cosmos and Their Hardy Quantum Entanglement. *International Journal of Modern Nonlinear Theory and Application*, **2**, 167-169. http://dx.doi.org/10.4236/ijmnta.2013.23023

[15] El Naschie, M.S. (2013) A Rindler-KAM Spacetime Geometry and Scaling the Planck Scale Solves Quantum Relativity and Explains Dark Energy. *International Journal of Astronomy and Astrophysics*, **3**, 483-493. http://dx.doi.org/10.4236/ijaa.2013.34056

[16] El Naschie, M.S. (2013) What Is the Missing Dark Energy in a Nutshell and the Hawking-Hartle Quantum Wave Collapse. *International Journal of Astronomy and Astrophysics*, **3**, 205-211. http://dx.doi.org/10.4236/ijaa.2013.33024

[17] El Naschie, M.S. (2013) The Missing Dark Energy of the Cosmos from Light Cone Topological Velocity and Scaling of the Planck Scale. *Open Journal of Microphysics*, **3**, 64-70. http://dx.doi.org/10.4236/ojm.2013.33012

[18] El Naschie, M.S. (2013) From Yang-Mills Photon in Curved Spacetime to Dark Energy Density. *Journal of Quantum Information Science*, **3**, 121-126. http://dx.doi.org/10.4236/jqis.2013.34016

[19] El Naschie, M.S. (2014) Calculating the Exact Experimental Density of the Dark Energy in the Cosmos Assuming a Fractal Speed of Light. *International Journal of Modern Nonlinear Theory and Application*, **3**, 1-5. http://dx.doi.org/10.4236/ijmnta.2014.31001

[20] El Naschie, M.S. (2014) Cosmic Dark Energy Density from Classical Mechanics and Seemingly Redundant Riemannian Finitely Many Tensor Components of Einstein's General Relativity. *World Journal of Mechanics*, **4**, 153-156. http://dx.doi.org/10.4236/wjm.2014.46017

[21] El Naschie, M.S. (2014) Capillary Surface Energy Elucidation of the Cosmic Dark Energy—Ordinary Energy Duality. *Open Journal of Fluid Dynamics*, **4**, 15-17. http://dx.doi.org/10.4236/ojfd.2014.41002

[22] El Naschie, M.S. (2014) Einstein's General Relativity and Pure Gravity in a Cosserat and De Sitter-Witten Spacetime Setting as the Explanation of Dark Energy and Cosmic Accelerated Expansion. *International Journal of Astronomy and Astrophysics*, **4**, 332-339. http://dx.doi.org/10.4236/ijaa.2014.42027

[23] El Naschie, M.S. (2014) Electromagnetic—Pure Gravity Connection via Hardy's Quantum Entanglement. *Journal of Electromagnetic Analysis and Applications*, **6**, 233-237. http://dx.doi.org/10.4236/jemaa.2014.69023

[24] El Naschie, M.S. (2014) Cosmic Dark Energy from 't Hooft's Dimensional Regularization and Witten's Topological Quantum Field Pure Gravity. *Journal of Quantum Information Science*, **4**, 83-91. http://dx.doi.org/10.4236/jqis.2014.42008

[25] El Naschie, M.S. (2014) Entanglement of E8E8 Exceptional Lie Symmetry Group Dark Energy, Einstein's Maximal Total Energy and the Hartle-Hawking No Boundary Proposal as the Explanation for Dark Energy. *World Journal of Condensed Matter Physics*, **4**, 74-77. http://dx.doi.org/10.4236/wjcmp.2014.42011

[26] El Naschie, M.S. (2014) The Meta Energy of Dark Energy. *Open Journal of Philosophy*, **4**, 157-159. http://dx.doi.org/10.4236/ojpp.2014.42022

[27] El Naschie, M.S. (2014) Pinched Material Einstein Space-Time Produces Accelerated Cosmic Expansion. *International Journal of Astronomy and Astrophysics*, **4**, 80-90. http://dx.doi.org/10.4236/ijaa.2014.41009

[28] El Naschie, M.S. (2014) From Chern-Simon, Holography and Scale Relativity to Dark Energy. *Journal of Applied Mathematics and Physics*, **2**, 634-638. http://dx.doi.org/10.4236/jamp.2014.27069

[29] El Naschie, M.S. (2014) Why E Is Not Equal to mc^2. *Journal of Modern Physics*, **5**, 743-750.

http://dx.doi.org/10.4236/jmp.2014.59084

[30] El Naschie, M.S. (2013) Nash Embedding of Witten's M-Theory and the Hawking-Hartle Quantum Wave of Dark Energy. *Journal of Modern Physics*, **4**, 1417-1428. http://dx.doi.org/10.4236/jmp.2013.410170

[31] El Naschie, M.S. (2013) Dark Energy from Kaluza-Klein Spacetime and Noether's Theorem via Lagrangian Multiplier Method. *Journal of Modern Physics*, **4**, 757-760. http://dx.doi.org/10.4236/jmp.2013.46103

[32] El Naschie, M.S. (2013) The hyperbolic Extension of Sigalotti-Hendi-Sharifzadeh's Golden Triangle of Special Theory of Relativity and the Nature of Dark Energy. *Journal of Modern Physics*, **4**, 354-356. http://dx.doi.org/10.4236/jmp.2013.43049

[33] El Naschie, M.S. (2013) Topological-Geometrical and Physical Interpretation of the Dark Energy of the Cosmos as a "Halo" Energy of the Schrödinger Quantum Wave. *Journal of Modern Physics*, **4**, 591-596. http://dx.doi.org/10.4236/jmp.2013.45084

[34] El Naschie, M.S. (2014) From Modified Newtonian Gravity to Dark Energy via Quantum Entanglement. *Journal of Applied Mathematics and Physics*, **2**, 803-806. http://dx.doi.org/10.4236/jamp.2014.28088

[35] He, J.H. and Marek-Crnjac, L. (2013) Mohamed El Naschie's Revision of Albert Einstein's $E = m_0c^2$: A Definite Resolution of the Mystery of the Missing Dark Energy of the Cosmos. *International Journal of Modern Nonlinear Theory and Application*, **2**, 55-59. http://dx.doi.org/10.4236/ijmnta.2013.21006

[36] El Naschie, M.S. and Marek-Crnjac, L. (2012) Deriving the Exact Percentage of Dark Energy Using a Transfinite Version of Nottale's Scale Relativity. *International Journal of Modern Nonlinear Theory and Application*, **1**, 118-124. http://dx.doi.org/10.4236/ijmnta.2012.14018

[37] El Naschie, M.S. and Helal, A. (2013) Dark Energy Explained via the Hawking-Hartle Quantum Wave and the Topology of Cosmic Crystallography. *International Journal of Astronomy and Astrophysics*, **3**, 318-343. http://dx.doi.org/10.4236/ijaa.2013.33037

[38] Marek-Crnjac, L. and El Naschie, M.S. (2013) Chaotic Fractal Tiling for the Missing Dark Energy and Veneziano Model. *Applied Mathematics*, **4**, 22-29. http://dx.doi.org/10.4236/am.2013.411A2005

[39] Marek-Crnjac, L. and El Naschie, M.S. (2013) Quantum Gravity and Dark Energy Using Fractal Planck Scaling. *Journal of Modern Physics*, **4**, 31-38. http://dx.doi.org/10.4236/jmp.2013.411A1005

[40] Auffray, J.P. (2015) E Infinity, the Zero Set, Absolute Space and Photon Spin. *Journal of Modern Physics*, **6**, 536-545. http://dx.doi.org/10.4236/jmp.2015.65058

[41] El Naschie, M.S., Olsen, S., He, J.H., Nada, S., Marek-Crnjac, L. and Helal, A. (2012) On the Need for Fractal Logic in High Energy Quantum Physics. *International Journal of Modern Nonlinear Theory and Application*, **1**, 84-92. http://dx.doi.org/10.4236/ijmnta.2012.13012

[42] El Naschie, M.S., Marek-Crnjac, L., Helal, A.M. and He, J.-H. (2014) A Topological Magueijo-Smolin Varying Speed of Light Theory, the Accelerated Cosmic Expansion and the Dark Energy of Pure Gravity. *Applied Mathematics*, **5**, 1780-1790. http://dx.doi.org/10.4236/am.2014.512171

[43] El Naschie, M.S. (2014) Compactified Dimensions as Produced by Quantum Entanglement, the Four Dimensionality of Einstein's Smooth Spacetime and "tHooft"s 4-ε Fractal Spacetime. *American Journal of Astronomy & Astrophysics*, **2**, 34-37. http://dx.doi.org/10.11648/j.ajaa.20140203.12

[44] El Naschie, M.S. (2014) Hardy's Entanglement as the Ultimate Explanation for the Observed Cosmic Dark Energy and Accelerated Expansion. *International Journal of High Energy Physics*, **1**, 13-17. http://dx.doi.org/10.11648/j.ijhep.20140102.11

[45] El Naschie, M.S. (2014) Deriving E = mc²/22 of Einstein's Ordinary Quantum Relativity Energy Density from the Lie Symmetry Group SO (10) of Grand Unification of All Fundamental Forces and without Quantum Mechanics. *American Journal of Mechanics & Applications*, **2**, 6-9. http://dx.doi.org/10.11648/j.ajma.20140202.11

[46] El Naschie, M.S. (2014) Cosserat-Cartan Modification of Einstein-Riemann Relativity and Cosmic Dark Energy Density. *American Journal of Modern Physics*, **3**, 82-87. http://dx.doi.org/10.11648/j.ajmp.20140302.17

[47] El Naschie, M.S. (2014) Asymptotically Safe Pure Gravity as the Source of Dark Energy of the Vacuum. *International Journal of Astrophysics & Space Science*, **2**, 12-15. http://dx.doi.org/10.11648/j.ijass.20140201.13

[48] El Naschie, M.S. (2014) Logarithmic Running of 't Hooft-Polyakov Monopole to Dark Energy. *International Journal of High Energy Physics*, **1**, 1-5. http://dx.doi.org/10.11648/j.ijhep.20140101.11

[49] El Naschie, M.S. (2013) Experimentally Based Theoretical Arguments That Unruh's Temperature, Hawking's Vacuum Fluctuation and Rindler's Wedge Are Physically Real. *American Journal of Modern Physics*, **2**, 357-361. http://dx.doi.org/10.11648/j.ajmp.20130206.23

[50] Marek-Crnjac, L. (2013) Modification of Einstein's E = mc² to E = 1/22 mc². *American Journal of Modern Physics*, **2**, 255-263. http://dx.doi.org/10.11648/j.ajmp.20130205.14

[51] El Naschie, M.S. (2013) The Quantum Gravity Immirzi Parameter—A General Physical and Topological Interpretation. *Gravitation and Cosmology*, **19**, 151-155. http://dx.doi.org/10.1134/S0202289313030031

[52] El Naschie, M.S. (2013) Determining the Missing Dark Energy Density of the Cosmos from a Light Cone Exact Relativistic Analysis. *Journal of Physics*, **2**, 19-25.

[53] El Naschie, M.S. (2013) The Quantum Entanglement behind the Missing Dark Energy. *Journal of Modern Physics and Applications*, **2**, 88-96.

[54] El Naschie, M.S. (2014) Dark Energy via Quantum Field Theory in Curved Spacetime. *Journal of Modern Physics and Applications*, **2**, 1-7.

[55] El Naschie, M.S. (2014) Rindler Space Derivation of Dark Energy. *Journal of Modern Physics and Applications*, **6**, 1-10.

[56] Tang, W., *et al.* (2014) From Nonlocal Elasticity to Nonlocal Spacetime and Nano Science. *Bubbfil Nanotechnology*, **1**, 3-12.

[57] El Naschie, M.S. (2014) To Dark Energy Theory from a Cosserat-Like Model of Spacetime. *Problems of Nonlinear Analysis in Engineering Systems*, **20**, 79-98.

[58] El Naschie, M.S. (2012) Revising Einstein's $E = mc^2$. A Theoretical Resolution of the Mystery of Dark Energy. *Proceedings of the 4th Arab International Conference in Physics and Material Science*, Egypt, 1-30 October 2012, 1.

[59] Ball, K.M. (1991) Volume Ratios and a Reverse Isoperimetric Inequality. *Journal of London Mathematical Society*, **44**, 351-359. http://dx.doi.org/10.1112/jlms/s2-44.2.351

[60] Pisier, G. (1989) The Volume of Convex Bodies and Banach Space Geometry. Tracts in Math 94, Cambridge University Press, Cambridge. http://dx.doi.org/10.1017/CBO9780511662454

[61] Kasin, B.S. (1977) The Width of Certain Finite-Dimensional Sets and Classes of Smooth Functions. *Izvestiya Akademii Nauk SSSR. Seriya Matematicheskaya*, **41**, 334-351. (In Russian)

[62] Guedon, O. (2013) Concentration Phenomena in High Dimensional Geometry. arXiv:1310.1204V1 [math.FA].

[63] He, J.-H. (Guest Editor) (2013) Special Issue on Recent Developments on Dark Energy and Dark Matter. *Fractal Spacetime and Noncommutative Geometry in Quantum and High Energy Physics*, **3**, 1-62.

[64] He, J.-H. and El Naschie, M.S. (2012) On the Monadic Nature of Quantum Gravity as a Highly Structured Golden Ring, Spaces and Spectra. *Fractal Spacetime and Noncommutative Geometry in Quantum and High Energy Physics*, **2**, 94-98.

[65] El Naschie, M.S. (2012) Towards a General Transfinite Set Theory for Quantum Mechanics. *Fractal Spacetime and Noncommutative Geometry in Quantum and High Energy Physics*, **2**, 135-142.

[66] El Naschie, M.S., He, J.-H., Nada, S., Marek-Crnjac, L. and Helal, M. (2012) Golden Mean Computer for High Energy Physics. *Fractal Spacetime and Noncommutative Geometry in Quantum and High Energy Physics*, **2**, 80-92.

[67] El Naschie, M.S. (2012) The Minus One Connection of Relativity, Quantum Mechanics and Set Theory. *Fractal Spacetime and Noncommutative Geometry in Quantum and High Energy Physics*, **2**, 131-134.

[68] El Naschie, M.S. Dark Energy and Its Cosmic Density from Einstein's Relativity and Gauge Fields Renormalization Leading to the Possibility of a New 'tHooft Quasi Particle. *The Open Journal of Astronomy*, in Press.

[69] El Naschie, M.S. (2004) A Review of *E* Infinity and the Mass Spectrum of High Energy Particle Physics. *Chaos, Solitons & Fractals*, **19**, 209-236. http://dx.doi.org/10.1016/S0960-0779(03)00278-9

[70] El Naschie, M.S. (2009) The Theory of Cantorian Spacetime and High Energy Particle Physics (An Informal Review). *Chaos, Solitons & Fractals*, **41**, 2635-2646. http://dx.doi.org/10.1016/j.chaos.2008.09.059

[71] Connes, A. (1994) Noncommutative Geometry. Academic Press, San Diego.

[72] Krantz, S.G. and Parks, H.R. (2008) Geometric Integration Theory. Birkhauser, Boston. http://dx.doi.org/10.1007/978-0-8176-4679-0

[73] El Naschie, M.S. (1997) Remarks on Super Strings, Fractal Gravity, Nagasawa's Diffusion and Cantorian Spacetime. *Chaos, Solitons & Fractals*, **8**, 1873-1886. http://dx.doi.org/10.1016/S0960-0779(97)00124-0

[74] El Naschie, M.S. (1997) Introduction to Nonlinear Dynamics, General Relativity and the Quantum—The Uneven Flow of Fractal Time. *Chaos, Solitons & Fractals*, **8**, vii-x. http://dx.doi.org/10.1016/S0960-0779(97)88695-X

[75] El Naschie, M.S. (1999) Hyper-Dimensional Geometry and the Nature of Physical Spacetime. *Chaos, Solitons & Fractals*, **10**, 155-158. http://dx.doi.org/10.1016/S0960-0779(98)00235-5

[76] El Naschie, M.S. *E* Infinity—High Energy Communications Nos. 1 to 90. April 2010 to December 2012.

[77] Chen, N.X. (2010) Möbius Inversion in Physics. World Scientific, Singapore.

[78] He, J.-H. (2014) A Tutorial Review on Fractal Spacetime and Fractional Calculus. *International Journal of Theoretical*

Physics, **53**, 3698-3718. http://dx.doi.org/10.1007/s10773-014-2123-8

[79]   El Naschie, M.S. (2001) On Twistors in Cantorian *E* Infinity Space. *Chaos, Solitons & Fractals,* **12**, 741-746.
http://dx.doi.org/10.1016/S0960-0779(00)00193-4

[80]   Rössler, O.E. (1998) Endophysics. World Scientific, Singapore. http://dx.doi.org/10.1142/3183

[81]   El Naschie, M.S. (2001) On a General Theory for Quantum Gravity. In: Diebner, H., Druckrey, T. and Weibel, P., Eds.,
Science of the Interface, Genista Verlag, Tübingen.

[82]   Li, M. (2004) A Model of Holographic Dark Energy. *Physics Letters B,* **603**, 1-5.
http://dx.doi.org/10.1016/j.physletb.2004.10.014

[83]   El Naschie, M.S. (2006) Holographic Dimensional Reduction. Center Manifold Theorem and *E* Infinity. *Chaos, Solitons & Fractals,* **29**, 816-822. http://dx.doi.org/10.1016/j.chaos.2006.01.013

[84]   Balachandran, A.P., Kürkcüoglu, S. and Vaidya, S. (2007) Lectures on Fuzzy and Fuzzy Susy Physics. World Scientific, Singapore.

[85]   Bahcall, J., Piran, T. and Weinberg, S. (2004) Dark Matter in the Universe. World Scientific, Singapore.

[86]   Amendola, L. and Tsujikawa, S. (2010) Dark Energy: Theory and Observations. Cambridge University Press, Cambridge.

[87]   Ruiz-Lapuente, P. (2010) Dark Energy, Observational and Theoretical Approaches. Cambridge University Press, Cambridge.

[88]   El Naschie, M.S. (2014) From $E = mc^2$ to $E = mc^2/22$—A Short Account of the Most Famous Equation in Physics and Its Hidden Quantum Entangled Origin. *Journal of Quantum Information Science,* **4**, 284-291.
http://dx.doi.org/10.4236/jqis.2014.44023

[89]   El Naschie, M.S. (2014) Casimir-Like Energy as a Double Eigenvalue of Quantumly Entangled System Leading to the Missing Dark Energy Density of the Cosmos. *International Journal of High Energy Physics,* **1**, 55-63.
http://dx.doi.org/10.11648/j.ijhep.20140105.11

[90]   Perlmutter, S., *et al.* (1999) Supernova Cosmology Project Collaboration: Measurements of Ω and Λ from 42 High Redshift Supernova. *Astrophysics Journal,* **517**, 565-585. http://dx.doi.org/10.1086/307221

[91]   El Naschie, M.S. (2015) On a Non-Perturbative Quantum Relativity Theory Leading to a Casimir-Dark Energy Nanotech Reactor Proposal. *Open Journal of Applied Science,* in Press.

[92]   Kodiyalam, V. and Sunder, V.S. (2001) Topological Quantum Field Theories from Subfactors. Chapman & Hall/CRC, London.

[93]   El Naschie, M.S. (2005) A Few Hints and Some Theorem about Witten's M Theory and T-Duality. *Chaos, Solitons & Fractals,* **25**, 545-548. http://dx.doi.org/10.1016/j.chaos.2005.01.009

[94]   Marek-Crnjac, L., El Naschie, M.S. and He, J.-H. (2013) Chaotic Fractals at the Root of Relativistic Quantum Physics and Cosmology. *International Journal of Modern Nonlinear Theory and Application,* **2**, 78-88.
http://dx.doi.org/10.4236/ijmnta.2013.21A010

[95]   Yau, S.T. and Nadis, S. (2010) The Shape of Inner Space. Basic Book, Persens Group, New York.

[96]   Bell, J.S. (1991) Speakable and Unspeakable in Quantum Mechanics. Cambridge University Press, Cambridge.

[97]   Penrose, R. (2004) The Road to Reality. A Complete Guide to the Laws of the Universe. Jonathan Cape, London.

12

Nanoscale Stiffness Distribution in Bone Metastasis

Ludovic Richert[1*], Laetitia Keller[2,3*], Quentin Wagner[2,3], Fabien Bornert[2,3,4],
Catherine Gros[2,3,4], Sophie Bahi[2,3,4], François Clauss[2,3,4], William Bacon[2,3,4],
Philippe Clézardin[5], Nadia Benkirane-Jessel[2,3,4], Florence Fioretti[2,3,4]

[1]Centre National de la Recherche Scientifique (CNRS), UMR, Faculté de Pharmacie de l'Université de Strasbourg (UdS), Illkirch, France
[2]Institut National de la Santé et de la Recherche Médicale (INSERM), Osteoarticular and Dental Regenerative Nanomedicine, UMR, Faculté de Médecine de l'Université de Strasbourg and FMTS, Strasbourg, France
[3]Faculté de Chirurgie Dentaire de l'Université de Strasbourg (UdS), Strasbourg, France
[4]Hôpitaux Universitaires de Strasbourg, Strasbourg, France
[5]Institut National de la Santé et de la Recherche Médicale (INSERM), UMR, Faculté de Médecine Laënnec de l'Université Claude Bernard Lyon 1, Lyon, France

Email: f.fioretti@unistra.fr

Abstract

Nanomechanical heterogeneity is expected to have an effect on elasticity, injury and bone remodelling. In normal bone, we have two types of cells (osteoclasts and osteoblasts) working together to maintain existing bone. Bone cancers can produce factors that make the osteoclasts work harder. This means that more bone is destroyed than rebuilt, and leads to weakening of the affected bone. We report here the first demonstration of the nanoscale stiffness distribution in bone metastases before and after treatment of animals with the bisphosphonate Risedronate, a drug which is currently used for the treatment of bone metastases in patients with advanced cancers. The strategy used here is applicable to a wide class of biological tissues and may serve as a new reflection for biologically inspired scaffolds technologies.

Keywords

Bone Metastasis, Stiffness, Risedronate

1. Introduction

As cancer becomes more advanced, it tends to spread throughout the body, with the bones being a common site

---

*These authors contributed equally to this work.

of spread for many cancers [1]. Spread of cancer to the bone from its original site is referred to as bone metastases. Treatment may consist of radiation therapy, bisphosphonates, hormone therapy and/or chemotherapy, depending upon the type of cancer from which the metastasis originated [2] [3].

Researchers are evaluating ways to avoid or decrease the ache or break caused by bone metastasis, not just provide treatment once they occur. Myeloma and some secondary bone cancers can produce factors that make the osteoclasts work harder [1]-[4]. This means that more bone is destroyed than rebuilt, and leads to weakening of the affected bone. This can cause pain and means that the bone can fracture or break more easily; in this case, ablation of metastasis bone and implantation of biomaterials are needed.

As many natural materials, bone is heterogeneous. Mechanical heterogeneity is expected to exist at different length scales. It has become evident that the nanoscale properties of bone participate in its macroscopic biomechanical funcion [4]-[7]. At the macroscopic level, considerable variations in mechanical properties have been detected for different tissues locations [8], as well as for regions within a specific location [9]. Microscopically, indentation has further identified differences in modulus and hardness for specific features such as lamellae (thin and thick) in osteons of bone which have been recognized as collagen fibril orientation, as well as variations in mineral composition [10] [11]. The cellular remodelling process resulting in a combination of new and old bone is a result of this heterogeneity at this length scale [12]. Atomic force microscopy (AFM)-based nanoindentation has been used to distinguish mechanically heterogeneous microscale regions in bone tissue from genetically modified mice [13]-[16]. These studies also raise important issues as to whether heterogeneity is advantageous or disadvantageous to the mechanical function of bone [17]. Recently, the detailed study of the consequences of heterogeneity, in particular at the nanoscale, was reported [18].

An inherent feature of bones is their heterogeneity in the collagen fibril orientation and the mineral content. This particular spatial structure has a direct incidence on their elasticity even on the nanometer scale [18]. The elasticity, or equivalently the stiffness, of a material can be quantified using nano-indentation. It consists in fixing a probe (most often a sphere, a cone or a pyramid) at the end of the cantilever of an atomic force microscope (AFM) and to measure the deflection of the cantilever when the material to be characterized is pushed into contact with the probe. The applied force is derived from the deflection, and the indentation is derived from the deflection and the material displacement. In this way, we obtain a so-called force curve. If the material behaves as an elastic solid, a force curve can be interpreted using Hertzian mechanics [19]. This characterization method was used to compare healthy bone, bone with metastases, and bone after treatment with bisphosphonates.

Bisphosphonates drugs are used for the treatment of cancer-related hypercalcemia and treatment of bone metastases in patients with advanced cancers [2]. Bisphosphonates decrease the rate of bone destruction in patients with bone metastases and clinical studies have demonstrated that bisphosphonates can significantly decrease the pain and number of fractures induced by bone metastases.

Bisphosphonates are established in the treatment of skeletal metastases [12]. They have a common P-C-P structural feature consisting in a central carbon atom correlated to two phosphonate moieties, and two substituents (R1, R2) (**Figure 1**). Phosphonate and R1 (preferably hydroxyl) groups allow bisphosphonates to bind

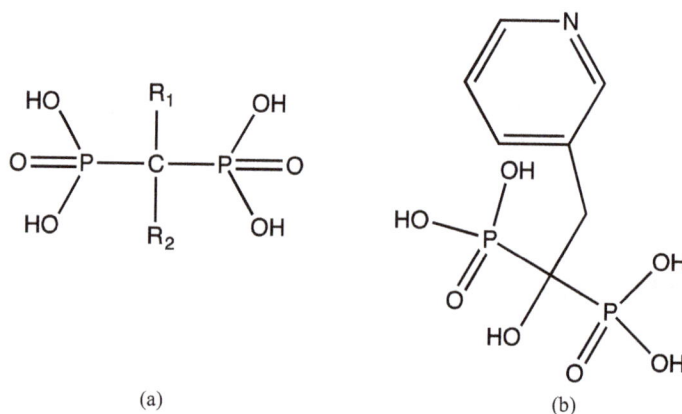

(a)   (b)

Figure 1. (a) Chemical structure of bisphosphonate drugs. P-C-P common structural feature of bisphosphonates and location of the specific R1 and R2; (b) Chemical structure of the Risedronate (RIS) antitumor drug (2-(3-pyridinyl)-1-hydroxyethylidene-bisphosphonic acid).

avidly to bone hydroxyapatite, while R2 determines their effectiveness to inhibit osteoclast-mediated bone resorption [3]. Clinically, bisphosphonates significantly diminish bone destruction rate, pain and fractures associated with bone metastases [2]. Moreover, there is extensive *in vivo* preclinical evidence that bisphosphonates reduce skeletal tumor burden and inhibit bone metastasis formation in animals [12]. Among proposed mechanisms of actions, bisphosphonates may render bone a less favorable microenvironment for metastasis development by reducing osteoclast-mediated bone resorption, which, in turn, deprives tumor cells of bone-derived growth factors required for their proliferation. Additionally, bisphosphonates were shown to exert direct antitumor action, as they inhibited tumor cell adhesion, invasion, and proliferation, and induced apoptosis of various human tumor cell lines *in vitro* [3]. However, to date this clear direct antitumor potential was not verified *in vivo* [10], due to the high affinity of bisphosphonates for bone mineral which must obstruct their availability for tumor cells. We undeniably observed a significantly higher potency of soluble compared to mineral-bound bisphosphonates at inhibiting tumor cell adhesion to bone *in vitro* [1]. Therefore strategies are needed to optimize bisphosphonates bioavailability and direct antitumor activity *in vivo*.

As well, cells are constantly changing their mechanical environment [20] and nanomechanical heterogeneity detection could facilitate damage detection in the extracellular matrix and improved remodelling responses. The heterogeneous nanomechanical prototypes measured experimentally could in turn induce local heterogeneous strains when loaded macroscopically. Such strains are expected to be increased by the softer surrounding cellular matrix of osteocytes [21] and also expected to influence interstitial fluid flow.

The objective of this work was to analyze the nanomechanical distribution of breast cancer bone metastasis before and after treatment of metastatic animals with Risedronate, the usually administrated drug in clinic. In this study, we have first analyzed by radiography, histology and histomorphometry the legs from metastatic mice and compared them with those from animals that had not been injected with breast cancer cells (naïve mice).

By radiography, we have analyzed and compared the healthy bone without metastasis (**Figure 2(a)**), bone with metastasis induced after inoculation (**Figure 2(b)**) and the treated bone after treatment by Risedronate (**Figure 2(c)**).

After radiography on day 32 after tumor cell injection, we have shown that metastatic animals had large osteolytic lesions in hind limbs when compared to the bones of naïve mice (**Figure 2**, panels b vs a). Metastatic animals were then treated either with the bisphosphonate Risedronate (administered by subcutaneous injection in

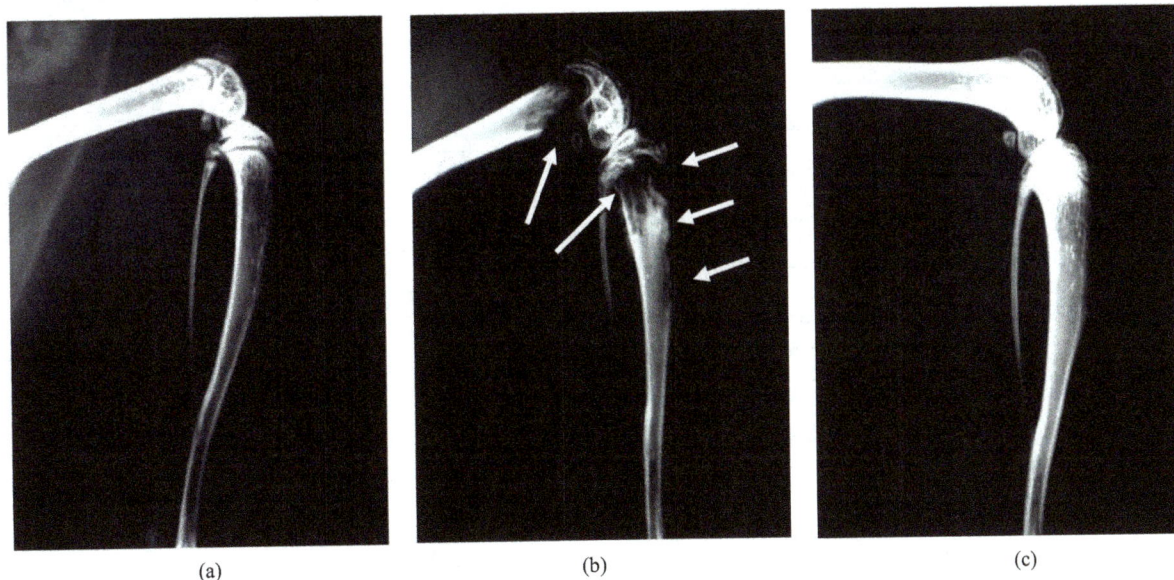

(a)   (b)   (c)

Figure 2. Identification of bone metastasis on radiographs of hind limbs from mice not injected with tumor cells with healthy bone, Naïve (a); b and c radiographs were obtained from different mice at day 32 after CHO-β3 tumor cells inoculation. PBS was used as vehicle and corresponds to bone with metastasis (b); Risedronate was used to treat mice and the radiographs represent the treated bone (c). The radiographs displayed are examples that best illustrate the effects of treatments. Arrows indicate osteolytic lesions.

PBS, used here as the vehicle) or the vehicle only.

After radiographic analysis, we examined the effect of Risedronate on the extent of bone destruction and determined the size of bone metastasis (**Table 1**).

Our results indicate clearly the antitumor efficacy of the treatment by Risedronate (0.6 mm^2 for treated bone, compared to 5 mm^2 for bone without treatment). We also performed histology and compared tumor burden and soft tissue volume ratio before and after treatment by Risedronate (**Figure 3** and **Table 2**).

The images shown in **Figure 3** are examples that best illustrate the effects of the treatments with Risedronate (**Figure 3(c)**) compared to no treated bone with tumor cells (asterisk in **Figure 3(b)**). To get more information

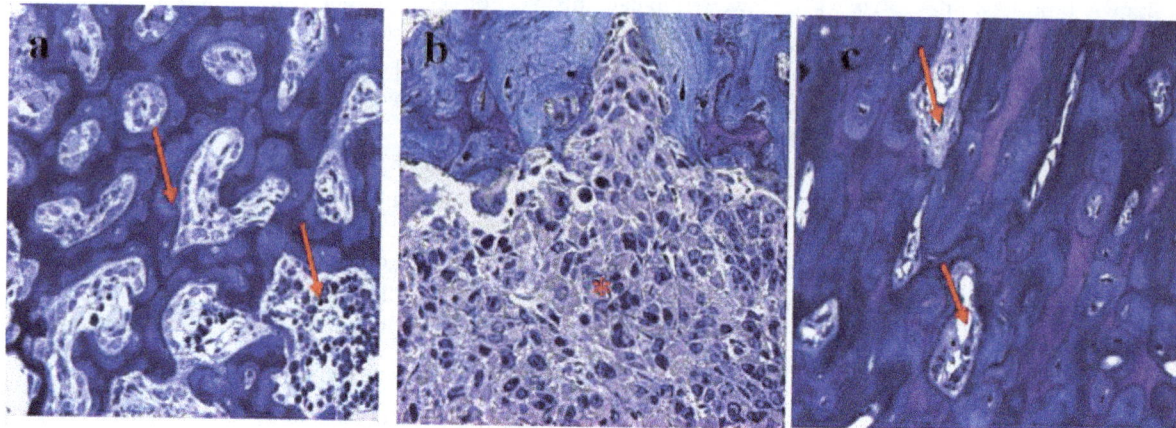

Figure 3. Optical microscopy visualization after semi-thin section (7 μm) Histologic analysis of hind limbs from mice naïve with healthy bone (a); b and c were obtained from different mice at day 32 after CHO-β3 tumor cells inoculation. PBS was used as vehicle and corresponds to bone with metastasis (b); Risedronate was used to treat mice and represent the treated bone (c). After toluidine blue stain, bone is stained blue and tumor cells zone were represented by asterisk. The arrows indicate the bone marrow.

Table 1. Tumor size determination after radiography.

Treatment[*]	Radiography (Tumor size (mm^2/mouse)
Naive "healthy bone"	0
Vehicle "bone metastasis"	5 ± 0.6
Risedronate "treated bone"	$0.6 \pm 0.08^{\S}$

[*]Drug administration was initiated from the time of tumor cell inoculation (day 0) to the end of the protocol (day 32). All measurements were made 32 days after tumor cell injection. Naive animals had not been injected with tumor cells and correspond to the healthy bone. PBS was used as vehicle for animals injected with tumor cells and correspond to bone metastasis. Risedronate animals treated by daily dose of 150 μg/kg, s.c. (clinical dose for patient with bone metastasis). Results are expressed as the mean ± SD (2 Naive mice, 3 Vehicle mice and 3 Risedronate mice). $^{\S}P < 0.01$, compared to the vehicle-treated group. Statistical pairwise comparisons were made using Mann-Whitney U test.

Table 2. Histomorphometry from histology analysis.

Treatment[*]	Histomorphometry[ξ]	
	BV/TV (%)	TB/STV (%)
Naive "healthy bone"	$22.7 \pm 1.8^{\S}$	0
Vehicle "bone metastasis"	9.5 ± 2.6	42.4 ± 3.6
Risedronate "treated bone"	$33.2 \pm 2.4^{\S}$	$6.4 \pm 3.4^{\S}$

Naive animals are mice that had not been injected with tumor cells. Risedronate (150 μg/kg, daily) was administered to metastatic animals by subcutaneous injection in 0.1 mL PBS (vehicle). Control mice bearing metastatic lesions received a daily treatment with the vehicle only. [*]Drug administration was initiated from the time of tumor cell inoculation (day 0) to the end of the protocol (day 32). [ξ]Histomorphometry was performed on legs from naïve and metastatic animals. BV/TV: bone volume/tissue volume ratio (a measurement of the bone volume). TB/STV: tumor burden/soft tissue volume ratio (a measurement of the tumor volume). Results are expressed as the mean ± SD (2 Naive mice, 3 Vehicle mice and 3 Risedronate mice). $^{\S}P < 0.01$, compared to the vehicle-treated group. Statistical pairwise comparisons were made using Mann-Whitney U test.

concerning the quantification of bone, we analyzed finely by histomorphometry the histologic data and determined the ratio bone volume/tissue volume and tumor burden/soft tissues volume (**Table 2**).

The beneficial effect of the treatment is demonstrated by the increase of the ratio BV/TV from 9.5% for bone without treatment to 33.2% for the treated bone. We have also shown that after treatment the ratio TB/STV was 6.4% in comparison to 42.4% without treatment.

Metastatic animals treated with Risedronate had osteolytic lesions that were 95% smaller than those of tumor-bearing animals treated with the vehicle only (**Table 1** and **Figure 2(c)**). Histologic analysis of hind limbs with metastases from vehicle-treated mice showed that tumor cells completely filled the bone marrow cavity and bone trabeculae were almost completely destroyed when compared to the bone histology of legs from naive mice (**Figure 2**, panels b vs a). By contrast, the extent of bone trabeculae in legs from Risedronate-treated mice was markedly increased, indicating a complete prevention of bone loss by the bisphosphonate (**Figure 2(c)**). In this respect, histomorphometric analysis of metastatic hind limbs from mice treated with Risedronate showed that the bone volume (BV) to tissue volume (TV) ratio was statistically significantly higher than that corresponding to vehicle-treated animals and naive mice (**Table 2** and **Figure 2**). Risedronate also decreased the tumor burden (TB) to soft tissue volume (STV) ratio by 85% compared with vehicle (**Table 2** and **Figure 2**).

In order to get more information about the structure of bone after Risedronate treatment, we analyzed and compared by electron microscopy the bones from naive animals with those obtained from metastatic animals treated with the bisphosphonate. By optical microscopy, we visualized different cells (**Figure 3**) after semi-thin sections (7 μm) of the different bone and toluidine blue stain. **Figure 3(b)** shows the invasion of bone by tumor cells working harder to destroy bone. In **Figure 3(a)** and **Figure 3(c)**, we have a normal bone with bone marrow section and mineralized bone with osteoblasts and osteoclasts. In **Figure 3(c)**, we can also see more mineralization than in **Figure 3(a)**. This result could be explained by the capacity of Risedronate to induce mineralization.

Bones from naive animals and metastatic animals treated or not with Risedronate were next analyzed by Atomic Force Microscopy (AFM) in order to examine the ultrastructure and nanomechanical spatial heterogeneity of the bone tissue (**Figure 4**).

The raw indentation curves were obtained by atomic force microscopy with diamond-like tip (**Figure 4(a)**). We report here the first demonstration of the ultrastructure and nanomechanical spatial heterogeneity of bone

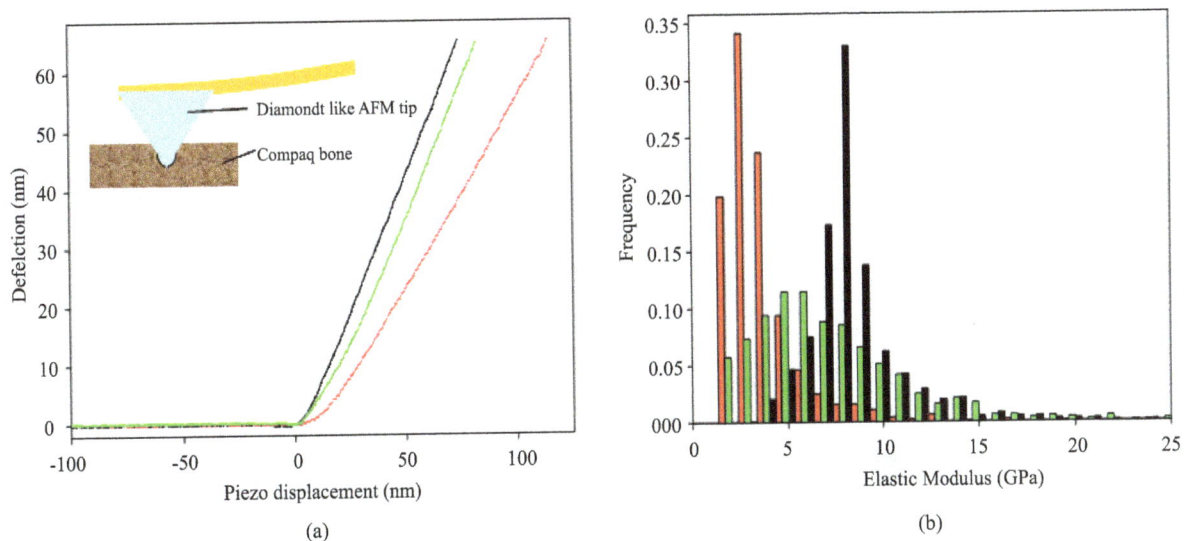

(a)

(b)

Figure 4. Representative raw indentation curves (a) and comparison of bone stiffness distribution (b). (a) Raw indentation curves obtained by atomic force microscopy with diamond like tip (100 nm of radius) for naïve healthy bone (black bar), vehicle bone with metastasis (red bar) and Risedronate treated bone (green bar) selected in the highest value of distribution. (b) Bone stiffness distributions for naive healthy bone (black bars) with a median of 8.14 GPa (first quartile = 6.39 GPa, third quartile = 10.33 GPa), for bone with metastasis and Risedronate treated bone (green bars) with a median of 5.45 GPa (first quartile = 3.49 GPa, third quartile = 8.08 GPa), and vehicle bone with metastasis (red bars) with a median of 2.38 GPa (first quartile = 1.60 GPa, third quartile = 3.39 GPa), obtained after computation of the elastic modulus from nanoindentation data.

stiffness with metastasis and the effect of the treatment on this nanomechanical distribution. The distribution of elastic modulus highlights the major decrease of elastic modulus in bone with metastasis (see histograms above). The median value is divided by a factor larger than 3 (8.14 GPa for healthy bone, 2.38 GPa for bone with metastasis). The Risedronate treatment reduces this decrease with an intermediate median value of 5.45 GPa. The differences in median and quartile interval underscore in particular the strong softening of malignant bone with respect to the healthy bone. Furthermore, it should be emphasized that the Risedronate treatment leads to a significant recovering of the nanomechanical stiffness. To observe more finely these differences, spatial distribution of elastic modulus has been recorded (**Figure 5**).

In **Figure 5**, for each nanoindentation data spaced of 300 nm, the elastic modulus is represented by a colour value. The three types of bone present heterogeneity in the spatial elastic modulus repartition (**Figure 5(b)**,

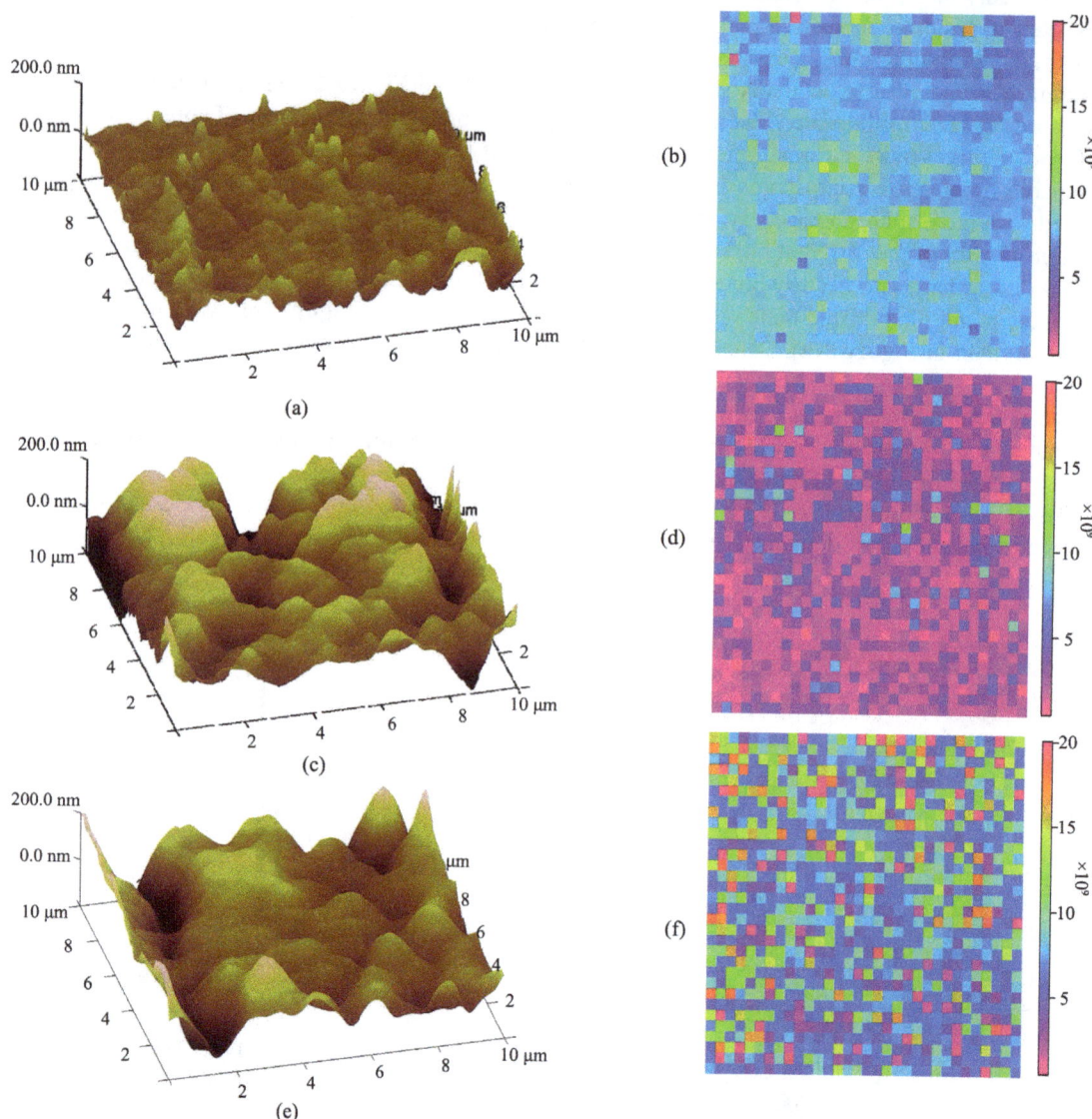

Figure 5. Ultrastructure and nanomechanical spatial heterogeneity of bone stiffness. (a) An a.c. intermittent contact-mode AFM height image (10×10 µm^2) viewed perpendicular to the naive healthy bone axis with a RMS of 21.0 nm; (b) 2D contour map (10×10 µm^2) of elastic modulus calculated from nanoindentation data (c) An a.c. intermittent contact-mode AFM height image (10×10 µm^2) viewed perpendicular to the vehicle bone with metastasis axis with a RMS of 55.3 nm. (d) 2D contour map (10×10 µm^2) of elastic modulus calculated from nanoindentation data; (e) An a.c. intermittent contact-mode AFM height image (10×10 µm^2) viewed perpendicular to the Risedronate treated bone axis with a RMS of 50.1 nm; (f) 2D contour map (10×10 µm^2) of elastic modulus calculated from nanoindentation data.

Figure 5(d), **Figure 5(f)**), but the bone with metastasis and Risedronate treatment is particular heterogeneous as revealed by the wide range of values of E present in the 10×10 μm^2 sample (**Figure 5(f)**).

In summary, bone is constantly being broken down and renewed. It is living tissue that needs exercise to gain strength. Bone metastasis occurs when cancer cells from the primary tumor relocate to the bone. Metastatic bone disease develops as a result of the many interactions between tumor cells and bone cells. This leads to disturbance of normal bone metabolism, with the increased osteoclast activity seen in most, if not all, tumor types providing a rational target for treatment.

Like many natural materials, bone is mechanically heterogeneous with spatial distributions in the shape, size and composition of its constituent. It is expected that nanomechanical heterogeneity influence elasticity, damage, fracture and remodelling of bone. We report here the first demonstration of the nanoscale stiffness distribution in bone metastasis before and after treatment of animals with Risedronate, a drug which is currently used for the treatment of bone metastases in patients with advanced cancers. This concept is generally applicable to a broad class of natural materials because nanomechanical heterogeneity expected to be ubiquitously presented.

2. Methods

2.1. Specific Drug for Treatment

Bisphosphonate Risedronate [2-(3-pyridinyl)1-hydroxyethylidene-bisphosphonic acid] was obtained from Procter and Gamble Pharmaceuticals (Mason, OH, USA). The drug was dissolved in water and stored at 4°C.

2.2. Mouse Model of Breast Cancer Bone Metastasis

All procedures involving mice including their housing and care, the method by which they were killed, and all experimental protocols were conducted in accordance with a code of practice established by the ethical committee in Lyon (France). This study was monitored on a routine basis by the attending veterinarian according to ensure continued compliance with the proposed protocols. Four-week-old female Balb/c athymic (nu/nu) mice were purchased from Charles River (St. Germain sur l'Arbresle, France). The bone metastasis experiments in mice were conducted as previously described, using B02 cells, a subpopulation of the human MDA-MB-231 breast cancer cell line that was selected for the high efficiency with which it metastasizes to bone after intravenous inoculation [22]. B02 cells (5×10^5 cells in 100 μL phosphate-buffered saline) were injected into the tail vein of anesthetized (130 mg/kg ketamin and 8.8 mg/kg xylazin) mice on day 0. Based on an average body weight of 20 g for 4-wk-old mice, risedronate (150 μg/kg body weight) was given daily to animals by subcutaneous injection in 100 μL PBS (vehicle). Control mice received a daily treatment with the vehicle only. On day 32 after tumor cell inoculation, radiographs of anesthesized animals were taken with the use of MIN-R2000 film (Kodak) in an MX-20 cabinet X-ray system (Faxitron X-ray Corporation). Osteolytic lesions are recognized on radiographs as demarcated radiolucent lesions in the bone. The area of the osteolytic lesions was measured using a Visiolab 2000 computerized image analysis system (Explora Nova, La Rochelle, France) and the extent of bone destruction per animal was expressed in mm^2, as described previously [22]. Anesthetized animals were killed by cervical dislocation following radiography at day 32.

2.3. Bone Histology and Histomorphometry

Bone histology and histomorphometric analysis of bone tissue sections were performed as previously described [22]. Vehicle- and bisphosphonate-treated tumor-bearing animals were killed at day 32, and both hind limbs from each animal were dissected, fixed in 80% (vol/vol) alcohol, dehydrated, and embedded in methylmethacrylate. A microtome (Polycut E, Reichert-Jung, Heidelberg, Germany) was used to cut 7 μm thick sections of undecalcified long bones, and the sections were stained with Goldner's trichrome. Histologic and histomorphometric analyses were performed on Goldner-stained longitudinal medial sections of tibial metaphysis with the use of a computerized image analysis system (Visiolab 2000). Histomorphometric measurements [bone volume (BV)/tissue volume (TV) and tumor volume (TV)/soft tissue volume (STV) ratios] were performed in a standard zone of the tibial metaphysis, situated at 0.5 mm from the growth plate, including cortical and trabecular bone. The BV/TV ratio represents the percentage of bone tissue. The TV/STV ratio represents the percentage of tumor tissue.

2.4. Sample Preparation and Characterization

2.4.1. Sample Preparation

Condylar tibia metaphysis were carefully sectioned under PBS (Gibco) flow by means of a sawing machine equipped with a diamond disk (Isomet, Buehler, Evanston, IL) into transversal sections about 2 mm thick from the proximal top (200 rpm speed).

Distal surfaces of sections were first set on AFM disks (Ted Pella Inc., Redding, CA) with wax (Kerr) and polished under distilled water flow by means of gradually disks fixed on polishing machine (Escil, Chassieu, France): G 1200, G 4000 silicon carbide and 3 µm, 0.1 µm microabrasive diamond polyester disks (40 rpm speed). Samples were stored in PBS solution until imaging and measuring.

The Tapping Mode™ AFM imaging at room humidity and temperature was used with Multimode AFM (Veeco, Santa Barbara, CA) on hydrated samples with a silicon probe MPP11100 (Veeco, Santa Barbara, CA) with tip radius about 10 nm and a resonant constant of 230 kHz. The roughness was determined by Root Mean Squared (RMS) approach on a 10×10 µm^2 AFM images.

2.4.2. Nanoindentation

Nanoindentation experiments were analyzed in ambient conditions using the PicoForce microscope (Veeco, Santa Barbara, CA) and NW-DT-NCHR cantilevers (42 N·m^{-1} of spring stiffness, 100 nm of tip radius, Nano World, Schaffhausen, Switzerland). Stiffness (dynamic elastic modulus, E) measurements of bone samples were performed as described previously [18]. Tai *et al.* verified that loading/unloading rates between 0.5 - 5 µm·s^{-1} did not lead to statistically representative differences in calculated bone moduli [18]. Therefore, we carried out displacement-controlled nanoindentation by loading at a rate of 1 µm·s^{-1} up to a trigger force of 5 µN followed by unloading at the same rate. To derive the elastic modulus, E, we use the Hertzian relation, derived for paraboloidal tips, between applied force, F, and resulting indentation, δ, namely

$$F = \frac{4}{3}\frac{E}{1-v^2}R^{1/2}\delta^{3/2}$$

where v is the Poisson ratio and R is the tip radius. The Poisson ratio for cortical bone is 0.325 [23]. Values of the elastic modulus were derived from the experimental force vs. indentation curve over the indentation domain ranging from 20% up to 95% of the maximum indentation corresponding to the maximum applied force of 5 µN. A minimum of 100 nm inter-indentation spacing was considered to be sufficiently large for minimal interference with adjacent residual in elastically deformed zone as well as remaining stresses.

2.5. Statistical Analysis

The three histograms shown in **Figure 4(b)** have been compared using One-Way ANOVA on ranks (*i.e.* Kruskal-Wallis test; Sigmastat, Systat Software, Chicago, IL). They differ significantly ($P < 0.001$). Consequently, a pairwise comparison was undertaken (using Dunn's method) which showed that the three pairs of groups differed significantly ($P < 0.05$).

2.6. Histological and Electron Microscopic Analysis

The samples were fixed in Karnovsky fixative, postfixed with 1% osmium tetroxide in 0.1 M cacodylate buffer for 1 h at 4°C, dehydrated through graded alcohol and embedded in Epon 812. Semi-thin sections were cut at 7 µm and stained with toluidine blue, and histologically analysed by light microscopy. Ultrathin sections were cut at 70 nm and contrasted with uranyl acetate and lead citrate, and examined with a Morgagni 268 electron microscope.

Acknowledgements

This work was supported by the NanoOSCAR ANR project from the "Agence Natiionale la Recherche", the "Fondation Avenir", the "Ligue contre le Cancer du Haut-Rhin, Région Alsace" and "Cancéropôle du Grand Est". LK and QW thank the "Faculté de Chirurgie Dentaire" of Strasbourg. Histology was made by Dr N. Messadeq.

References

[1] Boissier, S., Magnetto, S., Frappart, L., Cuzin, B., Ebetino, F.H., Delmas, P.D. and Clezardin, P. (1997) Bisphosphonates Inhibit Prostate and Breast Carcinoma Cell Adhesion to Unmineralized and Mineralized Bone Extracellular Matrices. *Cancer Research*, **57**, 3890-3894.

[2] Coleman, R.E. (2008) Risks and Benefits of Bisphosphonates. *British Journal of Cancer*, **98**, 1736-1740. http://dx.doi.org/10.1038/sj.bjc.6604382

[3] Stresing, V., Daubine, F., Benzaid, I., Monkkonen, H. and Clezardin, P. (2007) Bisphosphonates in Cancer Therapy. *Cancer Letters*, **257**, 16-35. http://dx.doi.org/10.1016/j.canlet.2007.07.007

[4] Fantner, G.E., Hassenkam, T., Kindt, J.H., Weaver, J.C., Birkedal, H., Pechenik, L., Cutroni, J.A., Cidade, G.A., Stucky, G.D., Morse, D.E. and Hansma, P.K. (2005) Sacrificial Bonds and Hidden Length Dissipate Energy as Mineralized Fibrils Separate during Bone Fracture. *Nature Materials*, **4**, 612-616. http://dx.doi.org/10.1038/nmat1428

[5] Gao, H., Ji, B., Jager, I.L., Arzt, E. and Fratzl, P. (2003) Materials Become Insensitive to Flaws at Nanoscale: Lessons from Nature. *Proceedings of the National Academy of Sciences of the United States of America*, **100**, 5597-5600. http://dx.doi.org/10.1073/pnas.0631609100

[6] Gupta, H.S., Wagermaier, W., Zickler, G.A., Raz-Ben Aroush, D., Funari, S.S., Roschger, P., Wagner, H.D. and Fratzl, P. (2005) Nanoscale Deformation Mechanisms in Bone. *Nano Letters*, **5**, 2108-2111. http://dx.doi.org/10.1021/nl051584b

[7] Tai, K., Ulm, F.J. and Ortiz, C. (2006) Nanogranular Origins of the Strength of Bone. *Nano Letters*, **6**, 2520-2525. http://dx.doi.org/10.1021/nl061877k

[8] Morgan, E.F., Bayraktar, H.H. and Keaveny, T.M. (2003) Trabecular Bone Modulus-Density Relationships Depend on Anatomic Site. *Journal of Biomechanics*, **36**, 897-904. http://dx.doi.org/10.1016/S0021-9290(03)00071-X

[9] Pope, M.H. and Outwater, J.O. (1974) Mechanical Properties of Bone as a Function of Position and Orientation. *Journal of Biomechanics*, **7**, 61-66. http://dx.doi.org/10.1016/0021-9290(74)90070-0

[10] Gupta, H.S., Stachewicz, U., Wagermaier, W., Roschger, P., Wagner, H.D. and Fratzl, P. (2006) Mechanical Modulation at the Lamellar Level in Osteonal Bone. *Journal of Materials Research*, **21**, 1913-1921. http://dx.doi.org/10.1557/jmr.2006.0234

[11] Rho, J.Y., Roy, M.E., 2nd, Tsui, T.Y. and Pharr, G.M. (1999) Elastic Properties of Microstructural Components of Human Bone Tissue as Measured by Nanoindentation. *Journal of Biomedical Materials Research*, **45**, 48-54. http://dx.doi.org/10.1002/(SICI)1097-4636(199904)45:1<48::AID-JBM7>3.0.CO;2-5

[12] Martin, R.B. and Burr, D.B. (1989) Structure, Function and Adaptation of Compact Bone. Raven Press, New York.

[13] Balooch, G., Balooch, M., Nalla, R.K., Schilling, S., Filvaroff, E.H., Marshall, G.W., Marshall, S.J., Ritchie, R.O., Derynck, R. and Alliston, T. (2005) TGF-Beta Regulates the Mechanical Properties and Composition of Bone Matrix. *Proceedings of the National Academy of Sciences of the United States of America*, **102**, 18813-18818. http://dx.doi.org/10.1073/pnas.0507417102

[14] Jaasma, M.J., Bayraktar, H.H., Niebur, G.L. and Keaveny, T.M. (2002) Biomechanical Effects of Intraspecimen Variations in Tissue Modulus for Trabecular Bone. *Journal of Biomechanics*, **35**, 237-246. http://dx.doi.org/10.1016/S0021-9290(01)00193-2

[15] Peterlik, H., Roschger, P., Klaushofer, K. and Fratzl, P. (2006) From Brittle to Ductile Fracture of Bone. *Nature Materials*, **5**, 52-55. http://dx.doi.org/10.1016/S0021-9290(01)00193-2

[16] Phelps, J.B., Hubbard, G.B., Wang, X. and Agrawal, C.M. (2000) Microstructural Heterogeneity and the Fracture Toughness of Bone. *Journal of Biomedical Materials Research*, **51**, 735-741. http://dx.doi.org/10.1002/1097-4636(20000915)51:4<735::AID-JBM23>3.0.CO;2-G

[17] Currey, J. (2005) Structural Heterogeneity in Bone: Good or Bad? *Journal of Musculoskeletal & Neuronal Interactions*, **5**, 317.

[18] Tai, K., Dao, M., Suresh, S., Palazoglu, A. and Ortiz, C. (2007) Nanoscale Heterogeneity Promotes Energy Dissipation in Bone. *Nature Materials*, **6**, 454-462. http://dx.doi.org/10.1038/nmat1911

[19] Engler, A.J., Richert, L., Wong, J.Y., Picart, C. and Discher, D.E. (2004) Surface Probe Measurements of the Elasticity of Sectioned Tissue, Thin Gels and Polyelectrolyte Multilayer Films: Correlations between Substrate Stiffness and Cell Adhesion. *Surface Science*, **570**, 142-154. http://dx.doi.org/10.1016/j.susc.2004.06.179

[20] Ehrlich, P.J. and Lanyon, L.E. (2002) Mechanical Strain and Bone Cell Function: A Review. *Osteoporosis International*, **13**, 688-700. http://dx.doi.org/10.1007/s001980200095

[21] You, L., Cowin, S.C., Schaffler, M.B. and Weinbaum, S. (2001) A Model for Strain Amplification in the Actin Cy-

toskeleton of Osteocytes Due to Fluid Drag on Pericellular Matrix. *Journal of Biomechanics*, **34**, 1375-1386.
http://dx.doi.org/10.1016/S0021-9290(01)00107-5

[22] Peyruchaud, O., Serre, C.M., NicAmhlaoibh, R., Fournier, P. and Clezardin, P. (2003) Angiostatin Inhibits Bone Metastasis Formation in Nude Mice through a Direct Anti-Osteoclastic Activity. *Journal of Biomechanics*, **278**, 45826-45832

[23] Cowin, S.C. (1999) Bone Poroelasticity. *Journal of Biomechanics*, **32**, 217-238.
http://dx.doi.org/10.1016/S0021-9290(98)00161-4

13

Energy Efficient Manufacturing of Nanocellulose by Chemo- and Bio-Mechanical Processes: A Review

Ashok K. Bharimalla[1], Suresh P. Deshmukh[2], Prashant G. Patil[1],
Nadanathangam Vigneshwaran[1*]

[1]ICAR-Central Institute for Research on Cotton Technology, Mumbai, India
[2]Department of General Engineering, Institute of Chemical Technology, Mumbai, India
Email: *Vigneshwaran.N@icar.gov.in, *nvw75@yahoo.com

Abstract

Nanocellulose is a new-age material derived from cellulosic biomass and has large specific surface area, high modulus and highly hydrophilic in nature. It comprises of two structural forms *viz.*, nanofibrillated cellulose (NFC) and nanocrystalline cellulose (NCC). This review provides a critical overview of the recent methods of bio- and chemo-mechanical processes for production of nanocellulose, their energy requirements and their functional properties. More than a dozen of pilot plants/commercial plants are under operation mostly in the developed countries, trying to exploit the potential of nanocellulose as reinforcing agent in paper, films, concrete, rubber, polymer films and so on. The utilization of nanocellulose is restricted mainly due to initial investment involved, high production cost and lack of toxicological information. This review focuses on the current trend and exploration of energy efficient and environment-friendly mechanical methods using pretreatments (both chemical and biological), their feasibility in scaling up and the future scope for expansion of nanocellulose application in diverse fields without impacting the environment. In addition, a nanocellulose quality index is derived to act as a guide for application based screening of nanocellulose production protocols.

Keywords

Biodegradable, Energy Conservation, Mechanical Process, Nanocellulose, Pretreatment

1. Introduction

In the last two decades, several reviews have been published on nanocellulose [1]-[5], preparation by TEMPO

---

*Corresponding author.

(2,2,6,6-tetramethylpiperidine-1-oxyl radical)-mediated oxidation [6], various mechanical processes [7] and their potential applications in composites [8]-[10] and liquid crystal displays [11]. Due to applications envisaged in various areas and their future growth predictions, book chapters [12] [13] books [14] [15] have been exclusively dedicated for the science and technology on nanocellulose. In addition, nanocellulose patents' trends are also well documented recently [16] [17]. Till now, the huge energy requirement for production of nanocellulose hampers their entry into the wider commercial market. Hence, in this review, our main focus is to discuss the advances in manufacturing processes for production of nanocellulose by chemo- and bio-mechanical means; in which the efficacy of chemical and biological pretreatments and their impact on subsequent mechanical treatments are discussed in detail.

2. Cellulose

Cellulose is a natural biopolymer made up of linear chain of several hundred to over ten thousand β (1→4) linked D-glucose, having the formula $(C_6H_{10}O_5)n$ and stabilized by intermolecular hydrogen bonds. The French chemist Anselme Payen in 1838, described fibrous solid material present in various plant tissues and determined its molecular formula to be $C_6H_{10}O_5$ by elemental analysis [18]; and, the term "cellulose" was first used in 1939 in a report of the French academy on the work of Payen. It is a crystalline structural polysaccharide and the most abundant form of living terrestrial biomass available on Earth. Formed by the repeated connection of D-glucose building blocks, the highly functionalized, linear stiff-chain homopolymer is characterized by its hydrophilicity, chirality, biodegradability, broad chemical modifying capacity, and its formation of versatile semi crystalline fiber morphologies [19]. This review focused on new frontiers, including environmentally friendly cellulose fiber technologies, bacterial cellulose biomaterials, and *in vitro* syntheses of cellulose together with future aims, strategies, and perspectives of cellulose research and its applications. The polymorphs of cellulose (I, II, III and IV) and their preparation protocols are discussed in detail in a book chapter in 2007 [20]. As per this reference, though cellulose has cellobiose as monomer, the ideal shape for cellulose does not have a two-fold structure and that a range of shapes should occur.

The present three kinds of hydroxyl groups within an anhydroglucose unit in a cellulose molecule exhibit different polarities, which contribute to formation of various kinds of inter- and intra-molecular hydrogen bonds among secondary "-OH" at the C-2, secondary "-OH" at the C-3 and primary "-OH" at the C-6 position. In addition, all the hydroxyl groups are bonded to a glucopyranose ring equatorially. This causes appearance of hydrophilic site parallel to the ring plane. On the contrary, the "-CH" groups are bonded to a glucopyranose ring axially, causing hydrophobic site perpendicular to the ring. These effects lead to formation of hydrogen bonds in parallel direction to a glucopyranose ring, and to van der Waals interaction perpendicular to the ring [21].

Cellulose occurs in almost the purest form in cotton fibers, while in wood and various parts of plants, it is found in combination with other materials, mainly lignin and hemicelluloses. Cotton fibers are of great interest as they lack lignin, which minimizes number of processes during manufacturing of nanocellulose. Cellulose is also produced by bacteria, algae, fungi and tunicates. **Figure 1** shows the schematic representation of structural arrangement of cellulose microfibrils in cottonfibers.

3. Nanocellulose

Nanocellulose represents a new family of nanomaterials that appear to have very broad applications in a variety of materials related domains where physical characteristics such as strength, weight, rheology, optical properties and the like can be affected in a very positive manner (TAPPI, 2011). Acronyms commonly used to denote nanocellulose include cellulose nanocrystals (CNC), nanocrystalline cellulose (NCC), cellulose nanoparticles (CNP), microfibrillated cellulose (MFC) and nanofibrillated cellulose (NFC). A recommendation on the preliminary terminology framework for nanocellulose was presented at an initial TAPPI's workshop held in Arlington on 9th June 2011 as given in **Figure 2**. In this classification, cellulose nanofibrils and cellulose microfibril are classified separately that we feel is confusing since it is very much difficult to differentiate these two materials by virtue of their overlapping in properties. While few researchers classify bacterial cellulose as a type of nanocellulose, it is beyond the scope of this review and hence, not included here. Hence, in this review, nanocellulose is classified only as two groups *viz.*, Nanocrystalline cellulose (NCC) having low aspect ratio (<100) and Nanofibrillated cellulose (NFC) having high aspect ratio (>100) and taken up for further discussion.

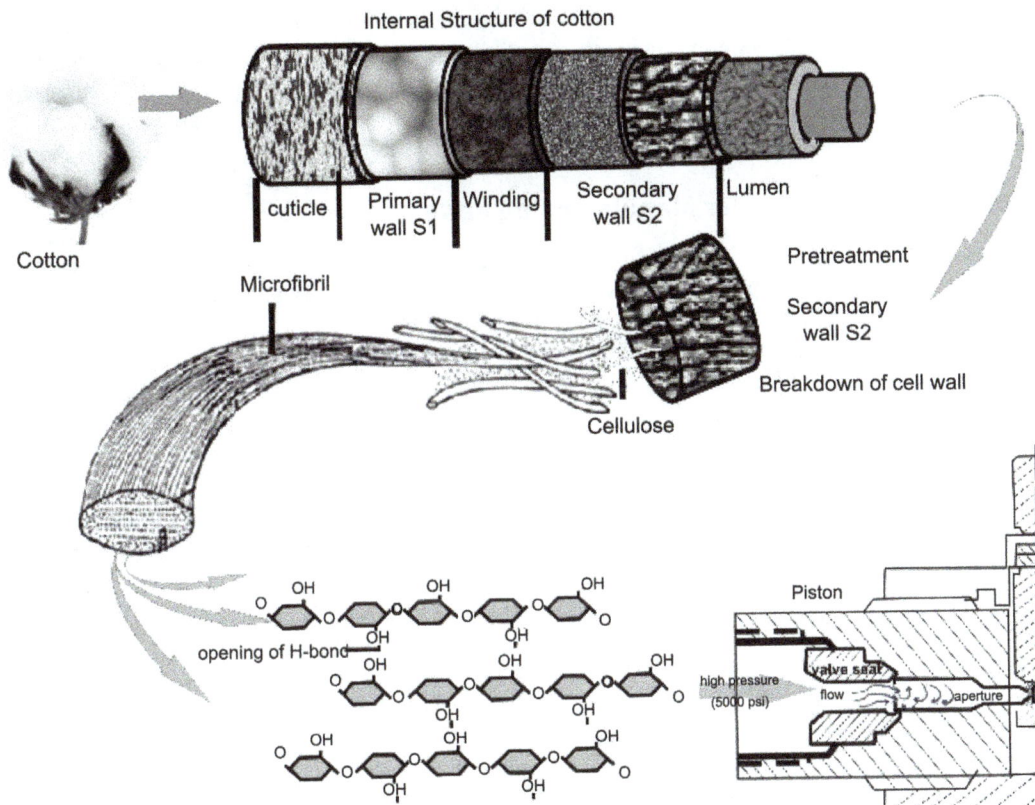

Figure 1. Schematic representation of structural arrangement of cellulose microfibrils in cottonfibers.

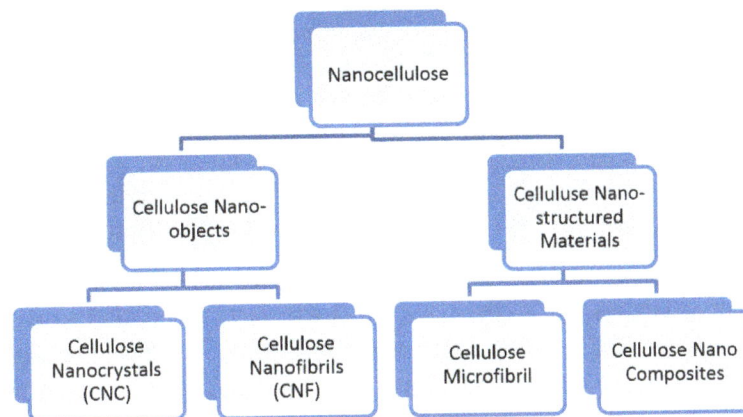

Figure 2. Naming hierarchy of nanocellulose as per the TAPPI's workshop held in Arlington on 9th June 2011.

3.1. Structural Properties of Nanocellulose

The nanocellulose is the smallest structural part of the cellulosic biomass of various organisms. While NFC is generally produced by mechanical process, NCC is by concentrated acid hydrolysis. Hence, the NFC is more in amorphous nature when compared to NCC. These basic structural differences lead to different types of application for NFC and NCC.

The determination of stiffness by theoretical and experimental means has shown that native cellulose (cellulose I) has a Young's modulus of 167.5 GPa and that of regenerated cellulose (cellulose II) is 162.1 GPa [22]. Using atomic force microscopy (AFM), various researchers evaluated the elastic modulus of both NCC and NFC. Self-standing TEMPO oxidized cellulose nanofibres are transparent and flexible, with high tensile strengths of 200 - 300 MPa and elastic moduli of 6 - 7 GPa [6]. The high strength of nanofibrillar cellulose combined with its potential economic advantages offers the opportunity to make lighter and strong materials with greater durability. Indeed, because of these properties nanofibres have attracted a lot of research effort in different disciplines and continues to be a subject of its utility in everyday materials such as paints, packaging, cosmetic bases, pigments, food modifiers, sensor applications, biomedical sciences and composites.

3.2. Energy Requirement for Manufacturing of Nanocellulose

For nanocellulose production, the inter-fibrillar hydrogen bonding (of cellulose) energy has to be overcome. Since, more than one type of hydrogen bond is present, a range of values need to be considered to quantify the hydrogen bond strength; and, it ranges between 19 and 21 MJ/kg mol [23]. Traditionally, mechanical homogenization process is used in large-scale production of nanocellulose that leads to huge energy consumption to the tune of 20,000 to 30,000 kWh per tonne [8]; and the required energy is 4 - 5 times that of the stored energy of that cellulosic biomass [24]. In one of the earliest patent [25], NFC was produced by passing a liquid suspension of cellulose through small diameter orifice in which the suspension is subjected to a pressure drop of at least 3000 psi and a high velocity shearing action followed by a high velocity decelerating impact, with repeated passes (11 numbers) till formation stable suspension. The production of cellulose nanofibrils from a bleached eucalyptus pulp (2% consistency) using a commercial stone grinder (SuperMassColliderTM) required the energy in the order of 5 - 30 kWh per kg [26]. To reduce the energy consumption, various types of pretreatments (chemical and biological) are being evaluated that results in drastic reduction of energy consumption less than 1000 kWh per tonne.

Energy requirement by mechanical pretreatment to produce MFC films with maximum obtainable properties for each processing method, the total energy required was approximately 9180 kJ/kg for the microfluidizer with pretreatment, 9090 kJ/kg for the grinder with pretreatment, and 5580 kJ/kg for the grinder without pretreatment and 31,520 kJ/kg for the homogenizer with pretreatment. Here, the pretreatment was a mechanical beating process (processed in a valley beater) [27].

3.3. Reduction in Energy Consumption by Chemical Pretreatment

The literature review on the effect of various chemical pretreatments on energy conservation or requirement for nanocellulose production is listed in the **Table 1**. Depending on the process and their analyses, the energy conservation/requirement are listed in the references. The energy calculated focused only on the actual energy required for mechanical comminution process without including the energy required during chemical pretreatments. Hence, a fair comparison of energy conservation is not possible.

Within the framework of EU-FP7 INNOBITE project, the Spanish applied research institute Tecnalia Research & Innovation and Ecopulp Finland Oy, a SME devoted to the production of shape moulded pulps out of waste paper, are developing a new method for converting waste paper into a new value-added material: newspaper-based nanocellulose [33]. The target is reducing the energy needed for the paper fibrillation process, for which a specialty pulp is generated via oxidation reactions. This chemical treatment converts the hydroxyl groups at the native cellulose fibres into carboxylate groups, thus creating anionic charges that will subsequently turn into repulsion forces within the internal structure of the fibres. Combining Ecopulp's industrial facilities and Tecnalia's know-how, such a specialty cellulose pulp has already been produced in mid-scale. So far, the energy requirement for complete paper fibrillation has been decreased to values of 2.1 kWh per kilo of dry matter that equals the range of other extensively applied industrial treatments such as pulp refining.

3.4. Reduction in Energy Consumption by Biological Pretreatment

While chemical pretreatment results in substantial reduction of energy consumption, there erupts the problem of waste disposal and release of chemical pollutants into the environment. Also, depending on the nature of the chemical treatment, the surface chemistry of cellulose gets modified. To circumvent these problems, biological pretreatment shall be an ideal means of reducing energy consumption without creating chemical effluents.

The high hemicellulose content of the pulp decreases the cell wall cohesion of the fibers, making cell wall delamination easier. But, this alone was not sufficient to avoid blocking of the orifice in the homogenizer and to reduce energy consumption. Hence, small additions of the mono-component endoglucanases enzyme promoted cell wall delamination and prevented the blocking of homogenizer [34].

Table 1. Effect of chemical pretreatments on energy conservation/requirement during nanocellulose production.

No.	Pretreatments	Effect of pretreatments	References	Energy conservation requirement	Process & remarks
1	(a) Ozone at a charge level of at least about 0.1 wt/wt%, based on the dry weight of the cellulosic material for generating free radicals in the slurry (b) Cellulase enzyme at a concentration from about 0.1 to about 10 lbs/ton based on the dry weight of the cellulosic material A combination of both (a) and (b) mentioned above	Partial depolymerization of cellulose	[28]	>2% energy conserved	Comminution processing
2	Organic (morpholine, piperidine or mixtures) or inorganic (inorganic halide, an inorganic hydroxide, or mixtures thereof) swelling agent or a mixture thereof	Swelling of cellulose	[29]	1400 kWh/t energy required 500 kWh/t energy required	Comminution processing 80% nanocellulose (defined as having an average diameter of less than 30 nm) 45% of the material having average diameters less than 30 nm
3	Carboxymethylation	Increases the anionic charges due to the formation of carboxyl groups in the surface	[30]	5500 kWh/t without pretreatment (per pass through)- Minimum of 4 passes 2200 kWh/t with pretreatment (per pass through)-4 passes 30000 kW/t (total) without pretreatment	Microfluidization process Ultra-fine friction grinding
4	Carboxymethylation OR by irreversibly attaching CMC onto cellulose fibres	Increases the anionic charges due to the formation of carboxyl groups in the surface	[31]	500 - 2300 kWh/t	Microfluidization process
5	Acid like sulfur dioxide, sulfurous acid, sulfur trioxide, sulfuric acid, lignosulfonic acid & their combinations or enzyme	Hydrolysis	[32]	<1000 kWh/t	Mechanical treatment

The bio-mechanical process overcomes the high energy requirement to a certain extent; however, the use of cellulose hydrolyzing enzymes also has a negative impact on the molecular weight and the chain length of the isolated nanocellulose [35]. Hence, pretreatment with the fungus secreting hydrogen bond-specific enzymes was tested to produce nanocellulose with mechanical strength marginally higher to that of those isolated via a conventional mechanical process since the bio-pretreatment produced nanocellulose had higher aspect ratio.

In spite of various protocols claiming energy reduction due to enzyme/biological pretreatments, energy calculations are not available for comparison of actual energy saving. In our lab, we are working with the cellulase enzyme pretreatments of various cellulosic biomasses (cotton linter pulp, soft wood pulp, hard wood pulp and bagasse) before proceeding for mechanical treatments by refiner and microfluidization for production of nanocellulose (both NCC and NFC). In case of NCC production from cellulose powder, a minimum of 10 passes are required in the high pressure homogenizer OR microfluidizer to reach the desired size range below 100 nm. But, in case of enzyme (cellulase) pretreatment, with 5 passes itself, we could achieve the desired size range; and hence, a 50% reduction in energy consumption can be claimed. But, the problem faced was the time & process required for the removal of enzyme after pretreatment; and the same needs to be quantified in terms of energy requirement. In case of NFC production from cellulosic pulp, it is not possible to treat with high pressure homogenizer OR microfluidizer without enzyme pretreatment. If a mechanical pretreatment like beating/refining is imparted to make it amenable for homogenization/microfluidization, then the energy required in case of both mechanical pretreatment and enzyme pretreatment were same. The energy requirement in enzyme pretreatment (including heating and stirring) was high enough to make them equivalent to that of mechanical pretreatment.

3.5. Comparison of Energy Consumption in Chemical and Biological Pretreatments

Figure 3 shows the comparison of three different processes for manufacturing nanocellulose (mechanical, chemo-mechanical and bio-mechanical) in terms of productivity, energy efficiency and ease of handling in radar chart. All three protocols display their own strength and weaknesses. The mechanical process shows its strength in productivity, chemo-mechanical process in ease of handling and bio-mechanical process in energy efficiency. In weaknesses, only bio-mechanical process suffers both in terms of productivity and ease of handling while mechanical process suffers in energy efficiency. For selection among these three different protocols during commercial production, apart from the above said three parameters, quality of the nanocellulose needs to be considered based on the required area of application. We have formulated a "Nanocellulose Quality Index" to choose the nanocellulose for desired application (**Figure 4**). The two influential (on final product) quality para-

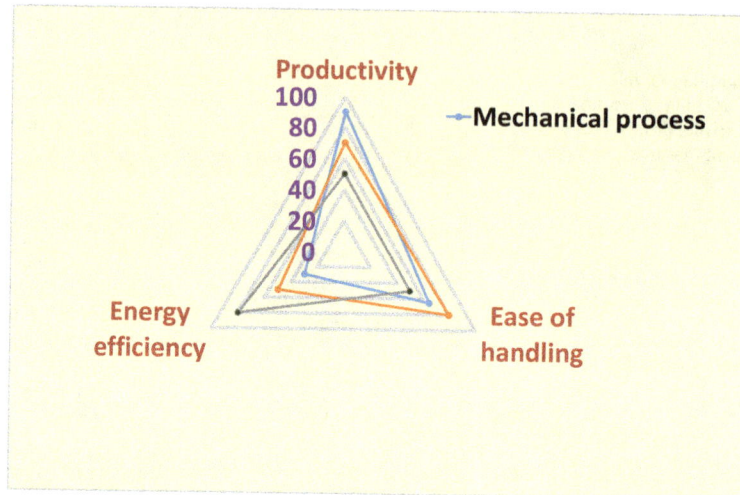

Figure 3. Comparison of three different processes for manufacturing nanocellulose in terms of productivity, energy efficience and ease of handling. Scale from 0 to 100 represents benefit/advantage in terms of percentage.

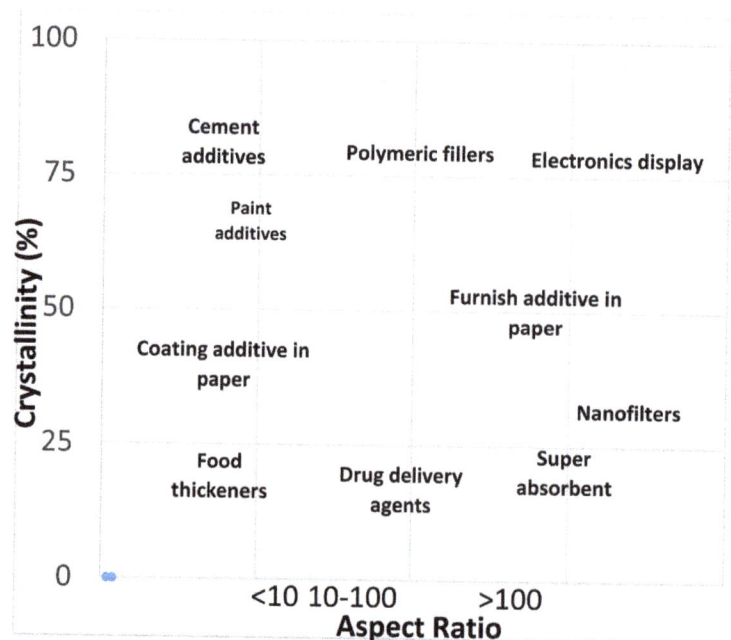

Figure 4. Nanocellulose Quality Index based on aspect ratio and crystallinity for various applications.

meters, aspect ratio and crystallinity are included in arriving at the index. While crystallinity improves strength and modifies rheological attributes, aspect ratio acts towards web formation in films, displays and improves strength. Based on their combined values, their suitability of applications can be determined as suggested in the **Figure 4**.

4. Outlook

In spite of various reports claiming reduction in energy consumption in nanocellulose manufacturing due to pretreatments, a holistic approach is required to include the energy equivalents of various axillaries and other processes involved during such pretreatments. Energy consumption in terms of chemicals/enzyme equivalence, time required for pretreatments, quality of output subjected to pretreatments needs to be evaluated and used for comparison. Such a measure is possible in pilot plants/commercial plants of nanocellulose production. The suggested nanocellulose quality index will act as a guide for application based screening of nanocellulose production protocols.

Acknowledgements

The authors are thankful to Dr. S. K. Chattopadhyay and Dr. Sujata Saxena of ICAR-Central Institute for Research on Cotton Technology, Mumbai, India for their kind suggestions and support for this work. This work was financially supported by the National Agricultural Innovation Project, Indian Council of Agricultural Research through its sub-project entitled "ZTM-BPD Unit of CIRCOT".

References

[1]   Lavoine, N., Desloges, I., Dufresne, A. and Bras, J. (2012) Microfibrillated Cellulose—Its Barrier Properties and Applications in Cellulosic Materials: A Review. *Carbohydrate Polymers*, **90**, 735-764. http://dx.doi.org/10.1016/j.carbpol.2012.05.026

[2]   Giri, J. and Adhikari, R. (2013) A Brief Review on Extraction of Nanocellulose and Its Application. *BIBECHANA*, **9**, 7.

[3]   Rebouillat, S. and Pla, F. (2013) State of the Art Manufacturing and Engineering of Nanocellulose: A Review of Available Data and Industrial Applications. *Journal of Biomaterials and Nanobiotechnology*, **4**, 24. http://dx.doi.org/10.4236/jbnb.2013.42022

[4]   Dufresne, A. (2013) Nanocellulose: A New Ageless Bionanomaterial. *Materials Today*, **16**, 220-227. http://dx.doi.org/10.1016/j.mattod.2013.06.004

[5]   Klemm, D., Kramer, F., Moritz, S., Lindstrom, T., Ankerfors, M., Gray, D., *et al.* (2011) Nanocelluloses: A New Family of Nature-Based Materials. *Angewandte Chemie International Edition*, **50**, 5438-5466. http://dx.doi.org/10.1002/anie.201001273

[6]   Isogai, A., Saito, T. and Fukuzumi, H. (2011) TEMPO-Oxidized Cellulose Nanofibers. *Nanoscale*, **3**, 71-85. http://dx.doi.org/10.1039/C0NR00583E

[7]   Abdul Khalil, H.P.S., Davoudpour, Y., Islam, M.N., Mustapha, A., Sudesh, K., Dungani, R., *et al.* (2014) Production and Modification of Nanofibrillated Cellulose Using Various Mechanical Processes: A Review. *Carbohydrate Polymers*, **99**, 649-665. http://dx.doi.org/10.1016/j.carbpol.2013.08.069

[8]   Siró, I. and Plackett, D. (2010) Microfibrillated Cellulose and New Nanocomposite Materials: A Review. *Cellulose*, **17**, 459-494. http://dx.doi.org/10.1007/s10570-010-9405-y

[9]   Abdul Khalil, H.P.S., Bhat, A.H. and Ireana Yusra, A.F. (2012) Green Composites from Sustainable Cellulose Nanofibrils: A Review. *Carbohydrate Polymers*, **87**, 963-979. http://dx.doi.org/10.1016/j.carbpol.2011.08.078

[10]   Azizi Samir, M.A., Alloin, F. and Dufresne, A. (2005) Review of Recent Research into Cellulosic Whiskers, Their Properties and Their Application in Nanocomposite Field. *Biomacromolecules*, **6**, 612-626. http://dx.doi.org/10.1021/bm0493685

[11]   Lagerwall, J.P.F., Schutz, C., Salajkova, M., Noh, J., Park, J.H., Scalia, G., *et al.* (2014) Cellulose Nanocrystal-Based Materials: From Liquid Crystal Self-Assembly and Glass Formation to Multifunctional Thin Films. *NPG Asia Materials*, **6**, e80. http://dx.doi.org/10.1038/am.2013.69

[12]   Aspler, J., Bouchard, J., Hamad, W., Berry, R., Beck, S., Drolet, F., *et al.* (2013) Review of Nanocellulosic Products and Their Applications. In: Dufresne, A., Thomas, S. and Pothen, L.A., Eds., *Biopolymer Nanocomposites: Processing, Properties, and Applications*, John Wiley & Sons, Inc., Hoboken, 461-508.

http://dx.doi.org/10.1002/9781118609958.ch20

[13] Spence, K., Habibi, Y. and Dufresne, A. (2011) Nanocellulose-Based Composites. In: Kalia, S., Kaith, B.S. and Kaur, I., Eds., *Cellulose Fibers: Bio- and Nano-Polymer Composites*, Springer, Berlin, 179-213. http://dx.doi.org/10.1007/978-3-642-17370-7_7

[14] Dufresne, A. (2012) Nanocellulose: From Nature to High Performance Tailored Materials. De Gruyter, Berlin. http://dx.doi.org/10.1515/9783110254600

[15] Surhone, L.M., Tennoe, M.T. and Henssonow, S.F. (2011) Nanocellulose. Betascript Publishing, Beau-Bassin.

[16] Charreau, H., Foresti, M.L. and Vazquez, A. (2013) Nanocellulose Patents Trends: A Comprehensive Review on Patents on Cellulose Nanocrystals, Microfibrillated and Bacterial Cellulose. *Recent Patents on Nanotechnology*, 7, 56-80. http://dx.doi.org/10.2174/187221013804484854

[17] Duran, N., Lemes, A.P. and Seabra, A.B. (2012) Review of Cellulose Nanocrystals Patents: Preparation, Composites and General Applications. *Recent Patents on Nanotechnology*, 6, 16-28. http://dx.doi.org/10.2174/187221012798109255

[18] Payen, A. (1838) Mémoire sur la composition du tissu propre des plantes et du ligneux. (Memoir on the composition of the tissue of plants and of woody [material]). *Comptes Rendus Hebdomadaires des Séances de l'Académie des Sciences*, 7, 7.

[19] Klemm, D., Heublein, B., Fink, H.P. and Bohn, A. (2005) Cellulose: Fascinating Biopolymer and Sustainable Raw Material. *Angewandte Chemie International Edition*, 44, 3358-3393. http://dx.doi.org/10.1002/anie.200460587

[20] French, A. and Johnson, G. (2007) Cellulose Shapes. In: Brown Jr., R.M. and Saxena, I., Eds., *Cellulose: Molecular and Structural Biology*, Springer, Dordrecht, 257-284. http://dx.doi.org/10.1007/978-1-4020-5380-1_15

[21] Kondo, T. (2007) Nematic Ordered Cellulose: Its Structure and Properties. In: Brown Jr., R.M. and Saxena, I., Eds., *Cellulose: Molecular and Structural Biology*, Springer, Dordrecht, 285-305. http://dx.doi.org/10.1007/978-1-4020-5380-1_16

[22] Tashiro, K. and Kobayashi, M. (1991) Theoretical Evaluation of Three-Dimensional Elastic Constants of Native and Regenerated Celluloses: Role of Hydrogen Bonds. *Polymer*, 32, 1516-1526. http://dx.doi.org/10.1016/0032-3861(91)90435-L

[23] Nissan, A.H., Byrd, V.L., Batten, G.L. and Ogden, R.W. (1985) Paper as an H-Bond Dominated Solid in the Elastic and Plastic Regimes. *Tappi Journal*, 68, 118-124.

[24] Zhu, J.Y., Sabo, R. and Luo, X. (2011) Integrated Production of Nano-Fibrillated Cellulose and Cellulosic Biofuel (Ethanol) by Enzymatic Fractionation of Wood Fibers. *Green Chemistry*, 13, 1339-1344. http://dx.doi.org/10.1039/c1gc15103g

[25] Turbak, A.F., Snyder, F.W. and Sandberg, K.R. (1983) Microfibrillated Cellulose. Patent No. US 4374702 A.

[26] Wang, Q.Q., Zhu, J.Y., Gleisner, R., Kuster, T.A., Baxa, U. and McNeil, S.E. (2012) Morphological Development of Cellulose Fibrils of a Bleached Eucalyptus Pulp by Mechanical Fibrillation. *Cellulose*, 19, 1631-1643. http://dx.doi.org/10.1007/s10570-012-9745-x

[27] Spence, K., Venditti, R., Rojas, O., Habibi, Y. and Pawlak, J. (2011) A Comparative Study of Energy Consumption and Physical Properties of Microfibrillated Cellulose Produced by Different Processing Methods. *Cellulose*, 18, 1097-1111. http://dx.doi.org/10.1007/s10570-011-9533-z

[28] Bilodeau, M.A. and Paradis, M.A. (2013) Energy Efficient Process for Preparing Nanocellulose Fibers. Patent No. EP 2861799 A1.

[29] Graveson, I. (2014) Low Energy Method for the Preparation of Non-Derivatized Nanocellulose. Patent No. WO 2014009517 A1.

[30] Taipale, T., Österberg, M., Nykänen, A., Ruokolainen, J. and Laine, J. (2010) Effect of Microfibrillated Cellulose and Fines on the Drainage of Kraft Pulp Suspension and Paper Strength. *Cellulose*, 17, 1005-1020. http://dx.doi.org/10.1007/s10570-010-9431-9

[31] Ankerfors, M. (2012) Microfibrillated Cellulose: Energy Efficient Preparation Techniques and Key Properties. Licentiate Thesis, KTH Royal Institute of Technology, Stockholm, 49.

[32] Nelson, K., Retsina, T., Pylkkanen, V. and O'Connor, R. (2014) Processes and Apparatus for Producing Nanocellulose, and Compositions and Products Produced Therefrom. Patent No. US 20140154757 A1.

[33] Tejado, A. (2013) Review Series 1—Biorefinery and Microfibrillated Cellulose. 1*st INNOBITE Workshop*, INNOBITE-WP8-DEL-D813-CIMV-20140314-v02doc.

[34] Pääkkö, M., Ankerfors, M., Kosonen, H., Nykänen, A., Ahola, S., Österberg, M., *et al.* (2007) Enzymatic Hydrolysis Combined with Mechanical Shearing and High-Pressure Homogenization for Nanoscale Cellulose Fibrils and Strong

Gels. *Biomacromolecules*, **8**, 1934-1941. http://dx.doi.org/10.1021/bm061215p

[35] Janardhnan, S. and Sain, M. (2011) Targeted Disruption of Hydroxyl Chemistry and Crystallinity in Natural Fibers for the Isolation of Cellulose Nano-Fibers via Enzymatic Treatment. *BioResources*, **6**, 1242-1250.

14

The Effect of Plasticizers on Mechanical Properties and Water Vapor Permeability of Gelatin-Based Edible Films Containing Clay Nanoparticles

Mahsa Rezaei[1], Ali Motamedzadegan[1,2]*

[1]Faculty of Food Industries, Ayatollah Amoli Branch, Islamic Azad University, Amol, Iran
[2]Department of Food Science, Sari University of Agricultural Sciences and Natural Resources, Sari, Iran
Email: *mmohse@yahoo.com

Abstract

The effects of glycerol and sorbitol as two plasticizers on mechanical properties, water vapor permeability, thermal properties, color and capability of heat sealing of gelatin films (of phytophagous fish, bovine gelatin with high gel-forming ability, and bovine gelatin with low gel-forming ability) containing clay nanoparticles were studied in this research. For this purpose, $6 \times 2 \times 3$ factorial experiments using the completely randomized design and comparison of the means at 95% confidence level ($\alpha = 0.05$) were performed. Higher concentrations of plasticizers increased percentage elongation to the breaking point. When glycerol concentration was raised to over 20%, flexibility of the layers improved but their water vapor permeability increased. The minimum passage of water vapor was that of fish-skin gelatin films containing clay nanoparticles and 30% sorbitol, and the maximum that of bovine gelatin films with high gel-forming ability which contained nanoparticles but no plasticizers ($p < 0.05$). There were no significant differences with respect to color in the various treatments ($p > 0.05$). All samples had heat sealing capability, and fish-skin gelatin films containing clay nanoparticles had better heat sealing capability compared with the other samples so that fish-skin gelatin films containing clay nanoparticles with 25% glycerol and 5% sorbitol had the highest flexibility and tensile strength, and remained attached to where they were heat sealed. Electron microscope images showed that films without plasticizers had uniform surfaces, but that samples containing glycerol at concentrations of over 0.20 g/g gelatin exhibited cavities between gelatin chains and that water vapor permeability in gelatin films containing clay nanoparticles.

---

*Corresponding author.

Keywords

Gelatin, Clay Nanoparticles, Plasticizer, Mechanical Properties, Water Vapor Permeability

---

1. Introduction

More than five billion tons of waste from packaging materials is produced annually in the world, 30% of which are plastic compounds. Pollution with synthesized plastics, which is called white pollution, forms a major part of environmental pollution in industrial countries, and also in developing countries like Iran that have weak plastic recovery systems. Concern for environmental pollution caused by synthesized plastics has attracted researchers to studying the possibility of using natural biodegradable polymers in the production of packaging materials [1]. Use of edible coatings and films for increasing the shelf life of food materials has been common from very old times. For example, covering fresh oranges and lemons with wax to delay the drying out of their peels was tested in the 12th and 13th centuries in China. These coatings substantially reduce evaporation of water from food materials and prevent exchange of respiratory gases and, thereby, decrease fermentation. Natural polymers are biodegradable in nature and are converted into natural products such as CO_2, water, ethane, and biomass during a composting process [2] [3].

Manufacture of earthen vessels was a considerable progress in food storage. About 4000 years ago, inhabitants of present day Pakistan used earthen vessels to store food. In 530 AD, Iranians made use of earthen vessels with lids to send water and food materials to conquered lands in Egypt, and reused the emptied vessels. In brief, making vessels for preserving and storing food materials was of interest from the Equator to the Poles and people tried to perfect these vessels. Considering what was said above, although the art of packaging food materials was as old as human history, most researchers believed that the packaging industry was founded in 1840 when the Frenchman Operetta developed the art of preserving food materials in glass containers with lids. Since then, we have been witnessing the ever-increasing flourishing of this industry [4].

Although, initially, the purpose of food packaging is to store seasonal products and facilitate the transfer of food materials from one place to another, nowadays the advertising on food packages has given food packaging a wider meaning. Based on sources related to this field of science and technology, packaging can be defined as follows:

1. Packaging is a system that reduces the required time to prepare merchandise for transportation, storage, and retail sale;

2. Packaging is a concept that ensures reliable delivery of merchandise in good condition and with minimum cost to the end consumer;

3. Packaging is an economic-technical operation that minimizes delivery costs while increasing sales and, hence, improves profitability.

Preparation and production of packaging materials with the goal of preserving and improving quality, increasing shelf life, and protecting food materials against various microbial infections and chemical decays have been one of the concerns and research subject of researchers active in the food industry in recent years. Production of plastic packaging materials such as polyolefins, polyesters, polyamides, etc. has expanded in recent years due to the abundance and low price of the raw materials required for their production. On the other hand, the increase in environmental pollution and the non-biodegradability of such materials has attracted the interest of many researchers in producing biodegradable packaging materials that have the capability to preserve food materials. That is why many attempts have been made to produce packaging materials of natural origin (proteins, fats, and carbohydrates) in the form of films or coatings, the various types of which will be described in the following sections. The type and concentration of plasticizers influence film properties through affecting the interactions between protein molecules. Plasticizers are added to decrease the rigidity and increase the flexibility of films. In the case of certain proteins, some plasticizers help us to obtain desirable mechanical properties while only slightly affecting their inhibitory features.

Many studies have been carried out on the technology of biodegradable and edible films and coatings, but the effects of plasticizers on gelatin films containing clay nanoparticles have not been thoroughly studied yet. This research intended to study the capability of edible films that are based on gelatin, contained clay nanoparticles, and to which glycerol and sorbitol are added as plasticizers.

2. Materials and Methods

2.1. Extraction of Fish Gelatin

Commercial gelatin type 1 (bovine gelatin with high gel forming ability and of 390 Bloom grams), commercial gelatin type 2 (bovine gelatin with low gel forming ability and of 170 Bloom grams), and glycerol and sorbitol were bought from Merck Company.

2.2. Preparing Nanocomposite Plasticizer/Fish Gelatin/Clay Nanoparticle Films

Nanocomposite films were made using the solution casting method [5]. The Nanoclay solution was prepared by mixing clay nanoparticles (5% gelatin by weight/FPH) with 50 ml of water stirred at room temperature for 24 hours, followed by complete homogenization of the solution for 10 minutes using ultrasonic devices. The gelatin/plasticizer solution for making the films was prepared at optimal concentration using the method described below.

The solvent evaporation method was used to prepare the films. During the first stage of the experiment, 3 grams of gelatin were mixed in 100 ml of water containing clay nanoparticles [6] together with a 0, 10, 15, 20, 25%, and 30% by weight combination of gelatin, sorbitol, and glycerol as shown in **Table 1**. This table shows treatments used in this research based on type of gelatin containing clay nanoparticles. The mixture was then poured into 8.4 cm long and 4.8 cm wide perforated plexi containers and dried in an oven for 24 hours. After drying, the films were removed from the containers and stored at 25°C and 50% RH in a desiccator containing saturated magnesium nitrate solution until testing time [7].

2.3. Measuring Film Thickness

A micrometer with an accuracy of 0.01 mm was used to measure film thickness, and the mean thickness of at least 10 randomly selected spots on the film was reported [7].

2.4. Mechanical Properties

Following the D882ASTM standard and using a Santam STM-5 texture analyzer at the Laboratory of Material Analysis of the Food Sciences and Technology Department at Sari Agricultural Sciences and Natural Resources University, the three parameters of tensile strength, elongation to the breaking point, and Young's modulus were calculated. The samples were conditioned at 25°C and RH of 50% ± 3% in a desiccator containing a saturated solution of magnesium nitrate for at least two days. The films were cut in the form of rectangular bands 4 cm long and 1 cm wide, and installed between the two jaws of the device at a specific initial distance. The jaws moved away from each other at the rate of 10 mm/min until the film was torn. Young's modulus, stress, and strain were measured at the breaking point. The films were tested immediately after they were removed from the desiccator [7].

2.5. Water Vapor Permeability

Water vapor flow rate was calculated according to the ASTM E96 standard. The films were sealed to cup mouths with paraffin and fixed by o-rings that were placed around them. The cups, which had a uniform depth

Table 1. Treatments used in this research.

Type of gelatin containing clay nanoparticles	Percentage glycerol by weight: percentage sorbitol to gelatin by weight (gram per gram of gelatin)							
	A	B	C	D	E	F	G	H
Bovine gelatin with high Bloom values	0:0	30:0	25:5	20:10	15:10	10:10	5:10	0:10
Bovine gelatin with low Bloom values	0:0	30:0	25:0	20:0	15:0	10:0	5:0	0:0
Fish skin	0:0	30:0	25:0	20:0	15:0	10:0	5:0	0:0

and internal diameter of 17.9 mm and were filled with a saturated solution of magnesium nitrate, were put in a desiccator with RH of 50% $\pm$ 3%. The whole apparatus was kept at 25°C. The relative humidity gradient on the two sides of the films was 50:0. The cups were weighed at specified intervals using a balance with an accuracy of 0.0001 gram. Water vapor flow rate was calculated using the following relation [7]:

$$WVP = WVTR_X \times \left[P0 \left(RH1 - RH2 \right) \right]. \tag{1}$$

In the above relation, x is the mean film thickness, P_o (kPa) the net water vapor pressure (which is 3.159 at 25°C), (RH1-RH2) the relative humidity gradient employed in the test, and WVTR (g/h·m^2) the water vapor flow rate.

2.6. Color and Turbidity

The image processing method was used to measure color and turbidity. Digital photos must first be taken to evaluate the L*, a*, and b* parameters using software. To do this, a 50 × 50 × 60 cm box (length, width, and height, respectively) was made and its inner walls were painted white for the lamp light to be reflected from all directions onto the sample. A 60W LED light bulb (T > 5000 K, Cixin) was used to provide light, and a digital camera to take photos of the samples. The camera was placed at a distance of 25 cm from the sample and its resolution was set to 1600 × 1200 pixels. The self-timer was set to a 10-second delay so that the box door could be closed when photos were taken. The digital photos were transferred to a computer and analyzed using Photoshop 8. To do this, a circle with the diameter of 100 pixels was selected, the Average Filter Blur option was used to calculate the average colors of the pixels in the selected section, and the Information window was utilized to extract the values of L*, a*, and b*. The extent of deviation of the color of the samples was calculated using the following relation:

$$Whiteness = 100 - \left[\left(100 - L^* \right)^2 + a^{*2} b^{*2} \right]^{1/2} \tag{2}$$

$$\Delta E = \left[\left(\Delta L^{*2} \right) + \left(\Delta a^{*2} \right) + \left(\Delta b^{*2} \right) \right]^{1/2}. \tag{3}$$

2.7. Glass Transition Temperature Test (T$_g$)

The method introduced by Rivero et al. for differential scanning calorimetry analysis was used, with some modifications, to measure the glass transition temperature. The equipment used in differential scanning calorimetry analysis measures energy that is exchanged in the form of heat (at constant pressure) in a physical or chemical process. The samples weighing 6 - 7 mg were completely fixed on a pan (of stainless steel or aluminum) and an empty pan was used as reference. The samples were analyzed in the temperature range of 50°C - 200°C, and the rate of temperature increase was 10°C/min [8].

T_m: melting temperature (°C).
ΔH: enthalpy (j/g dry basis).
T_g: glass transition temperature (°C).

2.8. Film Structure

A scanning electron microscope was used to determine film structure at the Nanoelectronics Laboratory in the School of Electrical and Computer Engineering of Tehran University. Sections with specified dimensions were taken from each sample, fixed on special pins, and placed in the gold deposition instrument. The completely gold-covered sections were then transferred into the SEM and images were taken at 10,000×, 30,000×, and 60,000× magnifications.

2.9. Heat Sealing

The prepared films, that were stored in a desiccator containing a saturated magnesium nitrate solution at RH of 50°C $\pm$ 3°C, were taken out and each two thin film layers from every treatment were placed on top of each on a distance of two centimeters. They were sealed together by a heat-sealing machine at the same time and temperature, and were then exposed to tensile force applied by a Santam STM-5 texture analyzer to measure the strength of the heat sealing, and the elongation to the breaking point, the tensile strength of the spot where the films were heat sealed together.

2.10. Solubility in Water

The water solubility of the edible films was determined using the Gontard *et al.* method [9]. The films were cut and weighed. Percentage water solubility was the percentage of the dry matter of the films that dissolved after they were immersed in water for 24 hours. The percentage of the original raw material in each film was determined after the film was kept at 100°C for 24 hours. The films that were cut into 2 by 2 centimeter pieces were immersed in 50 ml of water that contained a small amount of sodium azide (0.02% w/v) to prevent growth of microorganisms. They were stirred intermittently for 24 hours at 20°C, and the films that did not dissolve were removed and dried as much as possible (at 100°C for 24 hours) to determine the weight of the dry matter.

Percentage solubility was calculated using relation 4:

$$\text{Percentage solubility} = \left(M_d - M_s \right) / M_d \times 100 . \tag{4}$$

In the above relation, M_d was the original dry weight of the film and M_s the dry matter weight of the undissolved film.

2.11. Microbial Test

The antimicrobial activity of the films was measured using the agar diffusion method. Food pathogens such as E. coli and *L. monocytogenes* were used as test organisms. They were cultured under sterile conditions on TSB and BHI Broth culture media at 37°C for 16 hours. The old liquid culture media (0.1 ml) of each bacterial species was transferred (under sterile conditions) to Eppendorfs containing 0.9 ml of sterile water and diluted to twice their original volumes. One ml of the diluted liquid medium was added to the TSB and BHI (100 ml) agar media before plating and a drill for making wells in agar was used to make wells with the diameter of 4.5 mm. The sample liquids (80 µl) were added to the wells and put in an incubator at 37°C for 24 hours. The inhibitory region was determined by measuring the diameter of the zone of inhibition around the wells [5].

2.12. Antioxidant Activity

The method introduced by Benzie and Strain [10] was used to measure the reduction potential of iron ions in the samples. This method is based on increased absorption at 595 nm due to the formation of the tripyridyl triazine complex (TPTZ) Fe (II in the presence of reducing agents at 37°C. Gelatin and FPH dissolved in distilled water, while film samples dissolved in 0.5M acetic acid. Absorption was read after 30 minutes on a spectrophotometer, and the standard $FeSO_4.7H_2O$ curve was obtained, which relates $FeSO_4.7H_2O$ concentration to absorption at 595 nm. Results were expressed in µM of FeSO4.7H20 equivalent per gram of the sample.

3. Results and Discussion

3.1. Mechanical Properties

Figure 1 shows the mechanical behavior of gelatin films containing clay nanoparticles without plasticizers. The films exhibited high tensile strength and low percentages of elongation to the breaking point, which indicated their brittle property. Cao *et al.* [7] reported similar behavior in bovine gelatin films with high or low Bloom values containing clay nanoparticles. All three types of gelatin films containing clay nanoparticles behaved similarly, but Bovine gelatin films with high Bloom values containing clay nanoparticles had greater tensile strength and required higher tensile force. Bovine gelatin films with low Bloom values containing clay nanoparticles and fish-skin gelatin films containing clay nanoparticles exhibited identical behaviors while, at identical stress, fish-skin gelatin films containing clay nanoparticles had higher elongation to the breaking point compared to bovine gelatin films with low Bloom values that contained clay nanoparticles.

Figure 2 indicates the mechanical behavior of fish-skin gelatin films containing clay nanoparticles and 25% glycerol and sorbitol by weight. Gelatin films containing clay nanoparticles and sorbitol had greater tensile strength compared to those that had glycerol instead of sorbitol, which could be due to increased fluidity of lateral branches and because of matrix lubrication when glycerol was added to these films. It seems glycerol wea-

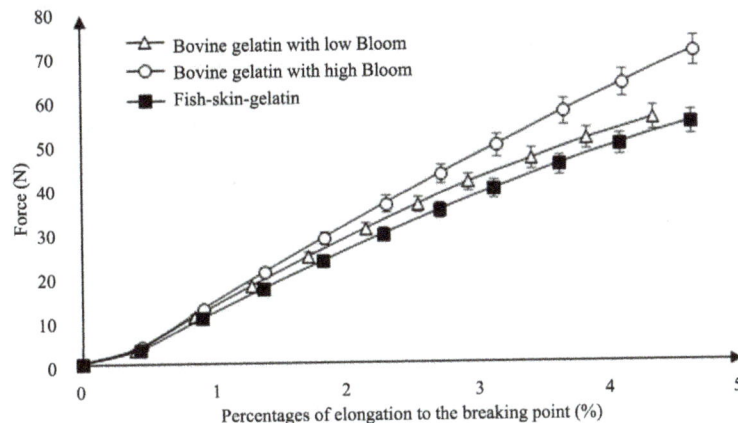

Figure 1. Mechanical behavior of gelatin films containing clay nanoparticles without plasticizer.

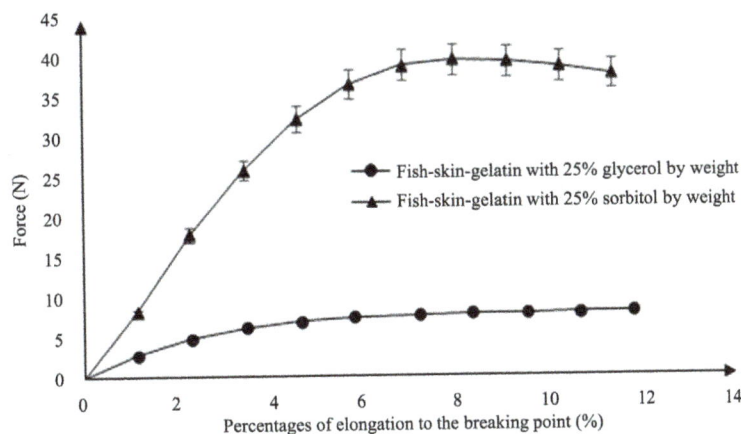

Figure 2. Mechanical behavior of fish-skin gelatin containing clay nanoparticles with 25% glycerol and sorbitol by weight.

kens film structure leading to reductions in tensile strength and to substantial increases in the percentage elongation to the breaking point.

Figure 3 and **Figure 4** show the effects of glycerol and sorbitol on mechanical behavior such as tensile strength, percentage elongation to the breaking point, and Young's modulus in gelatin films containing clay nanoparticles. Increases in the concentration of the plasticizers improved percentage elongation to the breaking point except when glycerol percentage by weight was raised from 25% to 30% in fish-skin gelatin films that contained clay nanoparticles, which could be due to increased fluidity in film structure. It appears glycerol at the concentration of 25% by weight greatly weakened gelatin structure, which led to reductions in tensile strength and Young's modulus in all of the samples. It seems glycerol increased percentage elongation to the breaking point and reduced tensile strength and Young's modulus in films treated by it through weakening gelatin structure by weakening the bonds through coming in between the lateral branches. After a few days, reductions were observed in film flexibility, probably because the plasticizers had moved to the surface of the films, following which the films had hardened. Young's modulus is the ratio of stress to strain in the linear portion of the curve, and can be obtained by measuring the slope of the stress-strain curve in its linear portion. Young's modulus is the required force to break the bonds in the samples, and it is related to the strength of the related material

Adding sorbitol to bovine gelatin films with high or low Bloom value and containing clay nanoparticles reduced tensile strength and Young's modulus. In films made from bovine gelatin with high Bloom values that contained clay nanoparticles, considerable reductions in tensile strength were observed when sorbitol was added at the concentration of 10% by weight, but no significant changes in tensile strength were detected when its

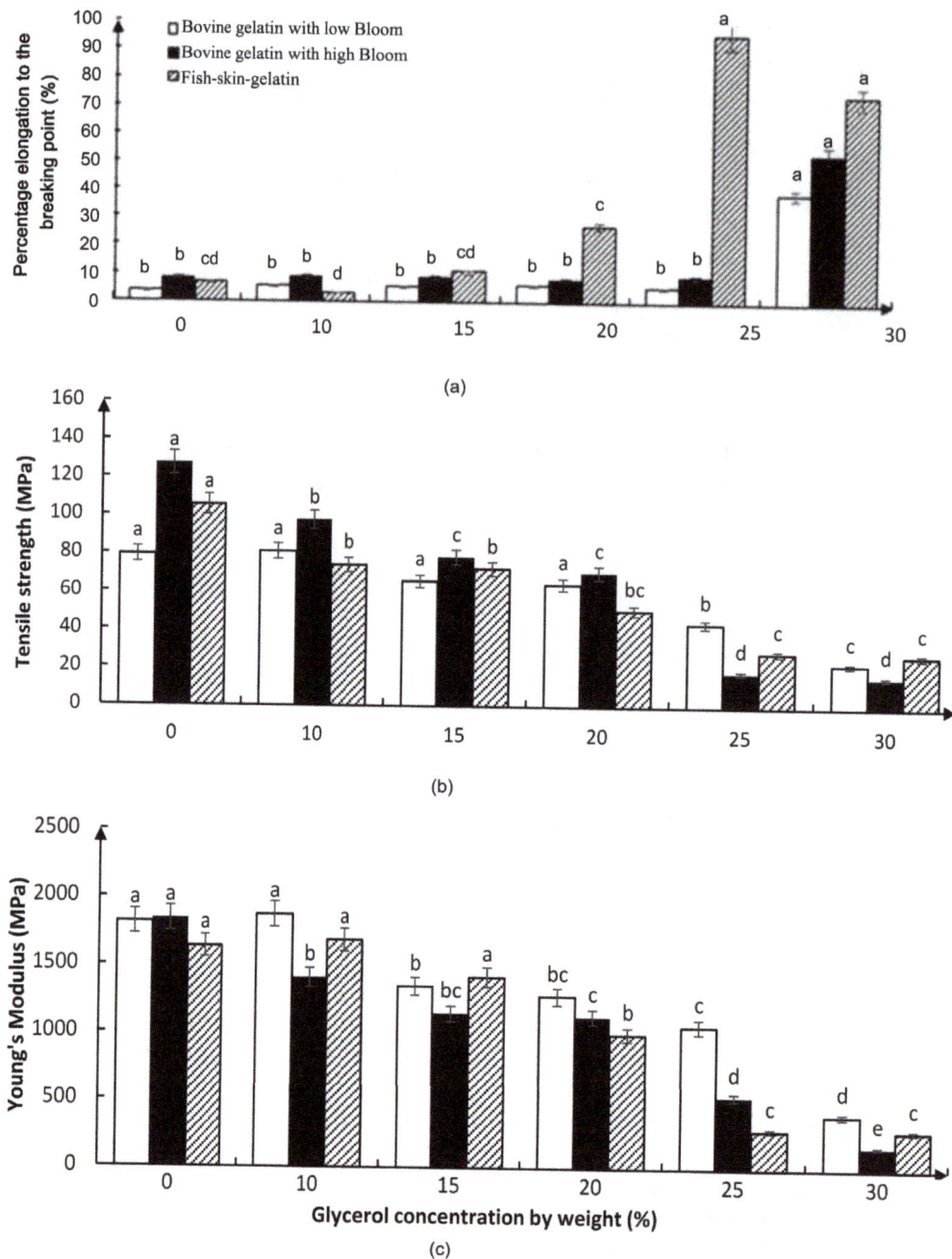

(a)

(b)

(c)

Figure 3. Results related to the effects of glycerol on (a) percentage elongation to the breaking point; (b) tensile strength; (c) Young's modulus in the three types of gelatin films containing clay nanoparticles. Different letters suggest averages of the studied property in each of the tests were significantly different ($p < 0.05$).

concentration was raised from 10% up to 30% by weight. Moreover, no significant changes were noticed in Young's modulus in these samples when sorbitol concentration increased from 10% up to 25% by weight. When sorbitol concentration was raised from 25% to 30% by weight, young's modulus improved. It seems glycerol was a better plasticizer than sorbitol with respect to mechanical properties because it increased percentage elongation to the breaking point besides imparting suitable tensile strength to the films. Films with high ultimate tensile strength and low percentage increase to the breaking point are brittle, and those with low ultimate tensile strength and high percentage elongation to the breaking point are weak.

Figure 5 presents the combined effects of sorbitol and glycerol on mechanical properties of gelatin films with

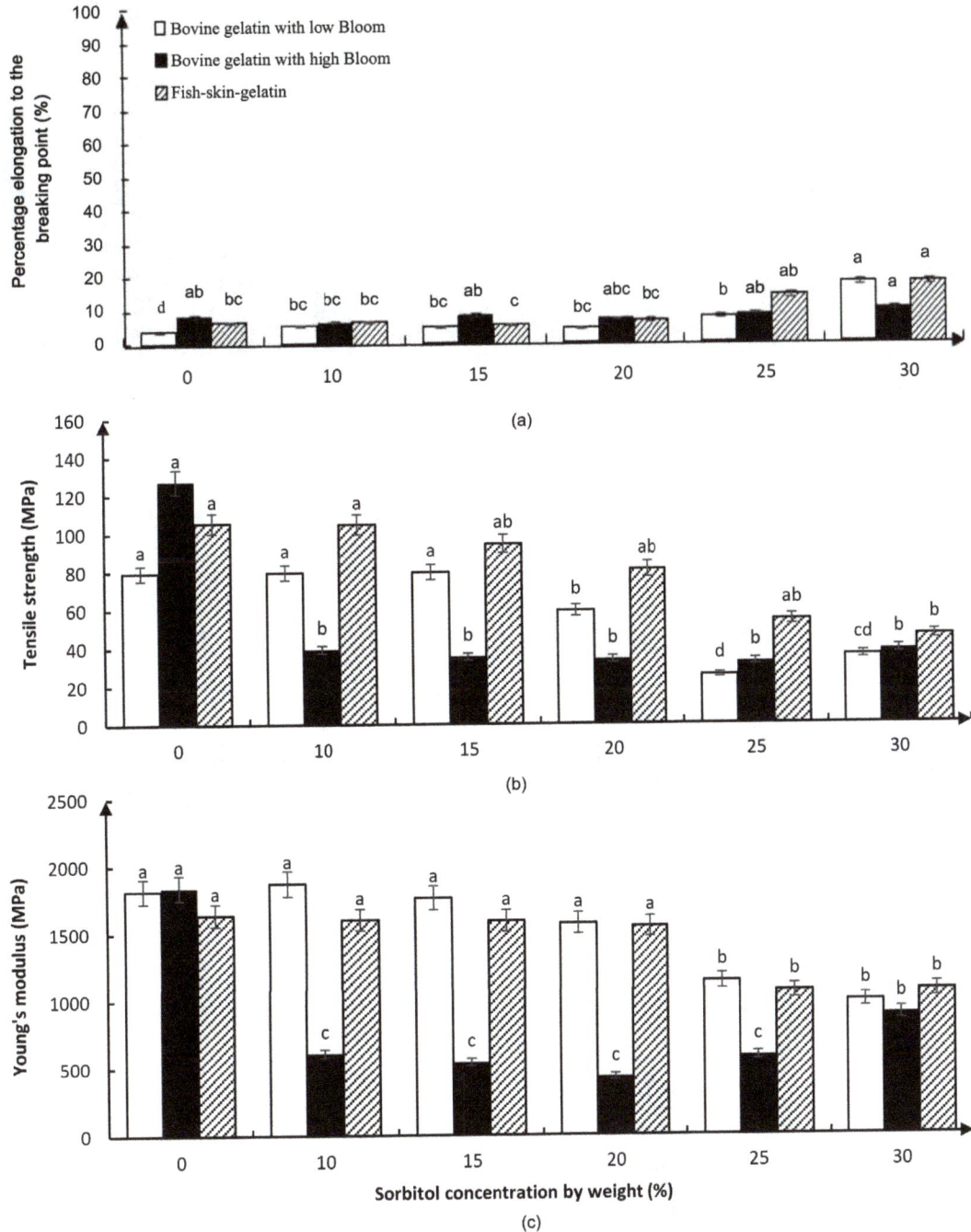

Figure 4. Results related to the effects of sorbitol on (a) percentage elongation to the breaking point (b) tensile strength, and (c) Young's modulus in the three types of gelatin films containing clay nanoparticles. Different letters in each section suggest the averages of the studied property in each of the tests were significantly different (p < 0.05).

clay nanoparticles. In all three types of the studied gelatin films, films without plasticizers had the maximum tensile strength and Young's modulus (p < 0.05). With increases in plasticizer concentration, tensile strength and Young's modulus declined and percentage elongation to the breaking point improved. In bovine gelatin films with high or low Bloom values that contained clay nanoparticles, tensile strength and Young's modulus increased when plasticizers were added at concentrations higher than 0.20 g sorbitol/g gelatin (p < 0.05). This could be due to water repellency in films containing sorbitol. These results conform to those found by Ghasem-

(a)

(b)

(c)

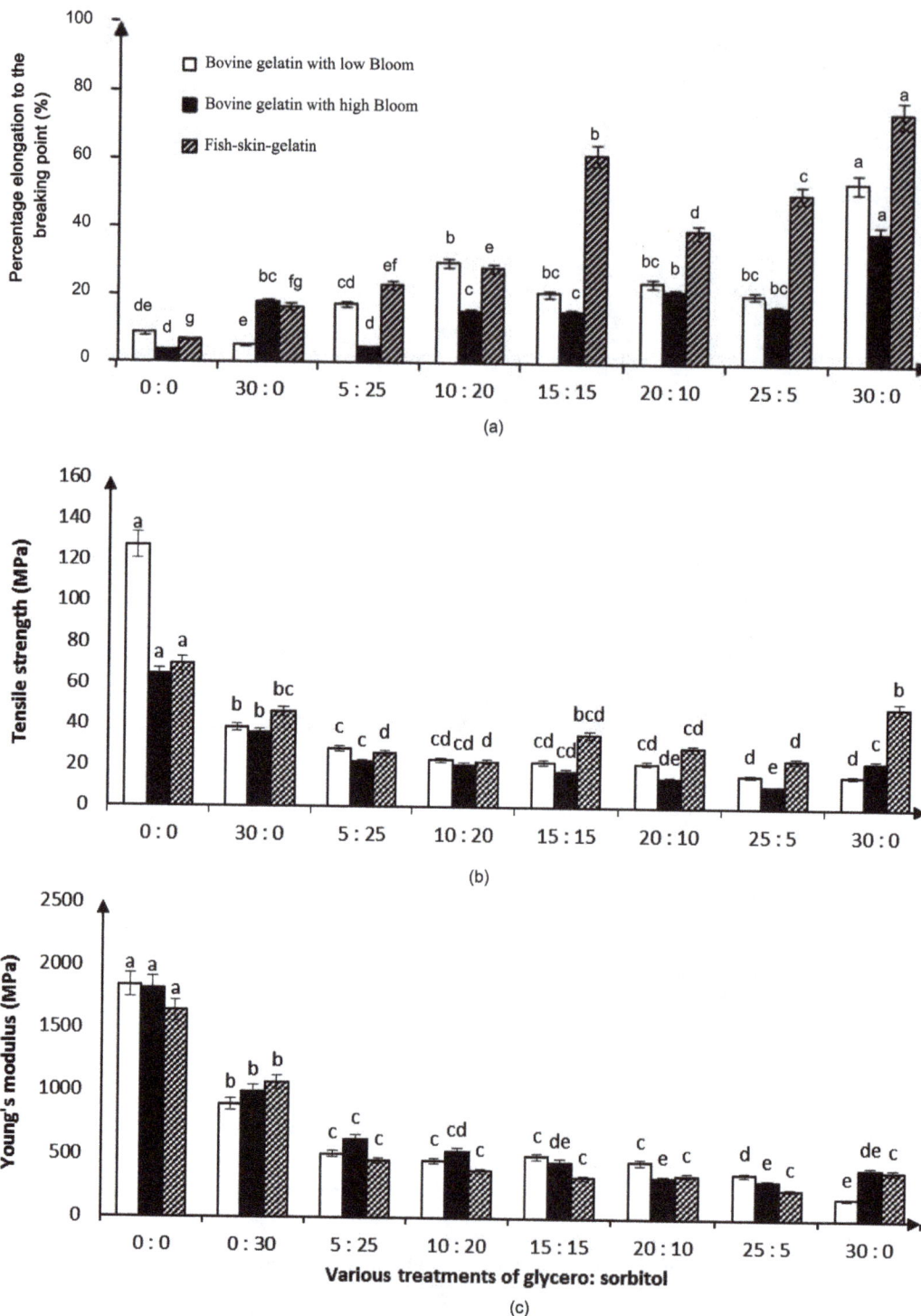

Figure 5. Results related to the various treatments of glycerol: sorbitol on (a) percentage elongation to the breaking point; (b) tensile strength; and (c) Young's modulus in three types of gelatin films containing clay nanoparticles. Different letters in each section suggest the averages of the studied property in each of the tests were significantly different (p < 0.05).

loo *et al.* [11] and by Oses *et al.* [12]. Films with higher concentrations of sorbitol were less flexible and had greater strength. Increases in glycerol concentration reduced tensile strength and Young's modulus and increased percentage elongation to the breaking point so that films containing glycerol at 0.30 g/g gelatin had the maximum percentage elongation to the breaking point ($p < 0.05$). These results agree with those reported by Kilburn *et al.* [13]. Plasticizers come between polymer chains, weaken bonds between polymer molecules, increase the free space between molecules and, finally, cause motility of molecular chains, increase flexibility, and reduce tensile strength [14].

3.2. Water Vapor Permeability

Table 2 shows the effects of glycerol and sorbitol on water vapor permeability of gelatin films. All samples of gelatin films containing clay nanoparticles but no plasticizers exhibited high water vapor permeability. This phenomenon could be due to the absence of lateral branches and the presence of empty spaces in the structure of the films. Water vapor permeability is influenced by factors such as the hydrophilic or hydrophobic nature of the

Table 2. Results of measuring the water vapor flow rate.

Type of film	Thickness (micrometer)	Glycerol concentration (gram per gram of gelatin)	Sorbitol concentration (gram per gram of gelatin)	Water vapor permeability $\times$ 10 (gram/millimeter per kilopascal/hour/square meter
Bovine gelatin with high Bloom values containing clay nanoparticles	60	0.00	0.00	1.89 a
	70	0.10	0.00	1.83 a
	70	0.15	0.00	1.51 b
	70	0.20	0.00	1.28
	60	0.25	0.00	1.30 c
	70	0.30	0.00	1.78 a
	60	0.00	0.00	1.89 a
	70	0.00	0.10	1.40 b
	70	0.00	0.15	1.32 b
	70	0.00	0.20	0.86 c
	70	0.00	0.25	0.65 c
	70	0.00	0.30	0.89 c
Bovine gelatin with low Bloom values containing clay nanoparticles	50	0.00	0.00	1.42 a
	50	0.15	0.00	1.22 b
	50	0.15	0.00	1.10 b
	50	0.20	0.00	1.05 b
	60	0.25	0.00	1.21 b
	60	0.30	0.00	1.70 a
	50	0.00	0.00	1.42 a
	50	0.00	0.10	1.03 b
	50	0.00	0.15	1.00 b
	50	0.00	0.20	0.45 c
	60	0.00	0.25	0.41 c
	50	0.00	0.30	0.34 c
Fish skin gelatin containing clay nanoparticles	60	0.00	0.00	1.86 a
	50	0.10	0.00	0.76 c
	70	0.15	0.00	0.52 d
	70	0.20	0.00	0.85 c
	70	0.25	0.00	1.32 b
	70	0.30	0.00	1.29 b
	60	0.00	0.00	1.86 a
	60	0.00	0.10	0.92 b
	60	0.00	0.15	0.57 c
	70	0.00	0.20	0.62 c
	70	0.00	0.25	0.12 d
	70	0.00	0.30	0.20 d

Different letters in each section suggest the averages of the studied property in each of the related tests were significantly different ($p < 0.05$).

material, and the presence or absence of cracks and fissures in the structure of the material and in its three-dimensional structure [15]. Sorbitol is a powder with six carbon atoms in its molecular structure, but glycerol is a liquid with three carbon atoms in each molecule. Adding these plasticizers to all of the samples reduced their water vapor permeability, which could be due to the close proximity of the lateral branches at low plasticizer concentrations. However, when glycerol concentration was raised to over 20% by weight and sorbitol to over 25% by weight, water vapor permeability improved due to increased fluidity, and probably because of increased spaces between the lateral branches resulting from the improved fluidity. Comparison of films containing plasticizers revealed that those with glycerol had the maximum and those with sorbitol the minimum water vapor permeability, probably because glycerol is more hydrophilic than sorbitol. The minimum water vapor permeability was observed in the fish-skin gelatin containing clay nanoparticles and sorbitol at 25% by weight.

Table 3 shows the measured rates of water vapor flow, which were higher in gelatin films containing clay nanoparticles that had glycerol compared to those that included sorbitol instead of glycerol ($p < 0.05$). Films containing low concentrations of glycerol were brittle but their flexibility improved and their water vapor permeability increased when glycerol concentration was raised to over 20%. In films without plasticizers, the brittleness and lower flexibility of the films, the joining together of the branches in gelatin structure, and the creation of empty spaces caused the formation of pores (microscopic break) facilitating water vapor permeability. Therefore, it seems water vapor passes from the spots where the capillary layers are broken and not through pores on the surface of the films, while the reverse happens in films containing plasticizers.

Permeability is influenced by the hydrophilic or hydrophobic nature of the material, the presence of cracks and fissures, spatial inhibition, and the structure of the films [15]. The effects of glycerol on the mechanical properties and inhibition of water vapor by protein films were studied by Hochstetter *et al.* [16], Lee *et al.* [17], and Hernandez *et al.* [18]. These researchers stated that increasing the concentration of glycerol improved water vapor permeability and absence of plasticizers in films was the cause of their brittleness.

Table 3. Results of water vapor flow rate.

Film type	Treatment Glycerol: sorbitol	Thickness (micrometer)	Glycerol concentration (g/g gelatin)	Sorbitol concentration (g/g gelatin)	Water vapor permeability × 10 (g/mm /kPa/h/m^2)
Bovine gelatin with high Bloom values containing clay nanoparticles	0:0	60	0.00	0.00	1.95 a
	30:0	70	0.00	0.30	0,92 def
	25:5	70	0.05	0.25	0.63 f
	20:10	70	0.10	0.20	0.75 ef
	15:15	70	0.15	0.15	1.14 cde
	20:20	70	0.20	0.10	1.22cd
	5:25	70	0.25	0.05	1.51 bc
	0:30	80	0.30	0.00	1.84 ab
Bovine gelatin with low Bloom values containing clay nanoparticles	0:0	60	0.00	0.00	1.21 ab
	30:0	60	0.00	0.30	0.35 d
	25;5	70	0.05	0.25	0.64 cd
	20:10	70	0.10	0.20	0.79 bcd
	15:15	70	0.15	0.15	1.06 bc
	20:20	70	0.20	0.10	1.25 ab
	5:25	70	0.25	0.05	1.61 a
	0:30	80	0.30	0.00	1.68 a
Fish-skin gelatin containing clay nanoparticles	0:0	80	0.00	0.00	1.86 a
	30:0	70	0.00	0.30	0.21 f
	25:5	60	0.05	0.25	0.54 e
	20:10	70	0.10	0.20	0.81 d
	15:15	70	0.15	0.15	0.91 d
	20:20	70	0.20	0.10	1.31 b
	5:25	60	0.25	0.05	1.09 c
	0:30	70	0.30	0.00	1.30

Different letters in each section suggest the averages of the studied property in each of the tests were significantly different ($p < 0.05$).

In our research, the minimum water vapor permeability was that of the fish-skin gelatin film containing clay nanoparticles and sorbitol at 0.30 g/g gelatin, and the maximum that of the bovine gelatin films with high Bloom values containing clay nanoparticles but no plasticizers (p < 0.05).

3.3. Color and Turbidity

Using a box under controlled conditions (light, position of the camera, distance between the camera and the sample, camera angle, the sample and the light source) for taking images, and employing Photoshop software, the L*, a*, and b* parameters were evaluated. Results indicated light intensity increased in bovine gelatin films with high Bloom values when sorbitol concentration was raised compared to glycerol, while in bovine gelatin films with low Bloom values and containing clay nanoparticles, and in fish-skin gelatin films with clay nanoparticles, light intensity declined. No significant differences were observed in the colors of the treatments (p < 0.05).

3.4. Glass Transition Temperature

Two heat absorption phenomena were observed in all of the samples. First, the amorphous portion of gelatin, that is in the glass state at ambient temperature [6] [19] [20], changed from the glass to the plastic state (that was observed as a step change in the specific heat). Following that, a heat absorption peak related to the melting of the crystalline portion was observed. These results suggest gelatin has a semi-crystalline structure. Marshall et al. [19], Pinhouse et al. [20] and Sobral et al. [6] previously reported similar results. In the region of glass transition, a minor heat absorption phenomenon was observed resulting from the thermal history of the polymer and the physical scheduling phenomenon, and was thoroughly studied in gelatin [21] [22]. **Table 4** presents glass transition temperatures and melting temperatures of the three gelatin films containing clay nanoparticles. The glass transition test of bovine gelatin with high and low Bloom values containing clay nanoparticles have been investigated herein.

Fish-skin gelatin films containing clay nanoparticles with sorbitol at 0.25 g/g gelatin together with glycerol at 0.05 g/g gelatin had the highest glass transition temperature, and bovine gelatin films with low Bloom values containing clay nanoparticles and sorbitol at 0.25 g/g gelatin together with glycerol at 0.05 g/g gelatin had the maximum melting temperature. This shows the role played by sorbitol in increasing melting temperature and glass transition temperature, probably because sorbitol fills the empty spaces in gelatin structure and increases film stability, which leads to reduced water vapor permeability resulting from increases in sorbitol concentration.

3.5. Film Structure

Differences in the morphology of gelatin films containing clay nanoparticles were studied by taking images using a scanning electron microscope. As shown in **Figure 6**, three different morphologic types were observed among the samples. In films without plasticizers, the surfaces were uniform at 1000× magnification, which showed the uniformity and homogeneity of structure in these films (**Figure 6**). Of course, this was observed in all of the samples but, in samples containing glycerol at concentrations over 0.20 g/g gelatin, glycerol increased fluidity between the lateral branches in the structure of gelatin. This created empty spaces between gelatin branches leading to higher percentages of elongation to the breaking point and to higher water vapor permeability in gelatin films containing clay nanoparticles. In samples that contained sorbitol at concentrations over 0.20

Table 4. Results of the glass transition test.

Film type	Treatment (Glycerol: sorbitol)	Glass transfer temperature (°C)	Melting point (°C)
Bovine gelatin with high Bloom values containing clay nanoparticles	25:5	64.00	97.16
Bovine gelatin with high Bloom values containing clay nanoparticles	5:25	46.50	78.33
Bovine gelatin with low Bloom values containing clay nanoparticles	25:5	54.33	100.33
Bovine gelatin with low Bloom values containing clay nanoparticles	5:25	46.33	76.50
Fish-skin gelatin containing clay nanoparticles	25:5	74.66	97.33
Fish-skin gelatin containing clay nanoparticles	5:25	52.00	81.33

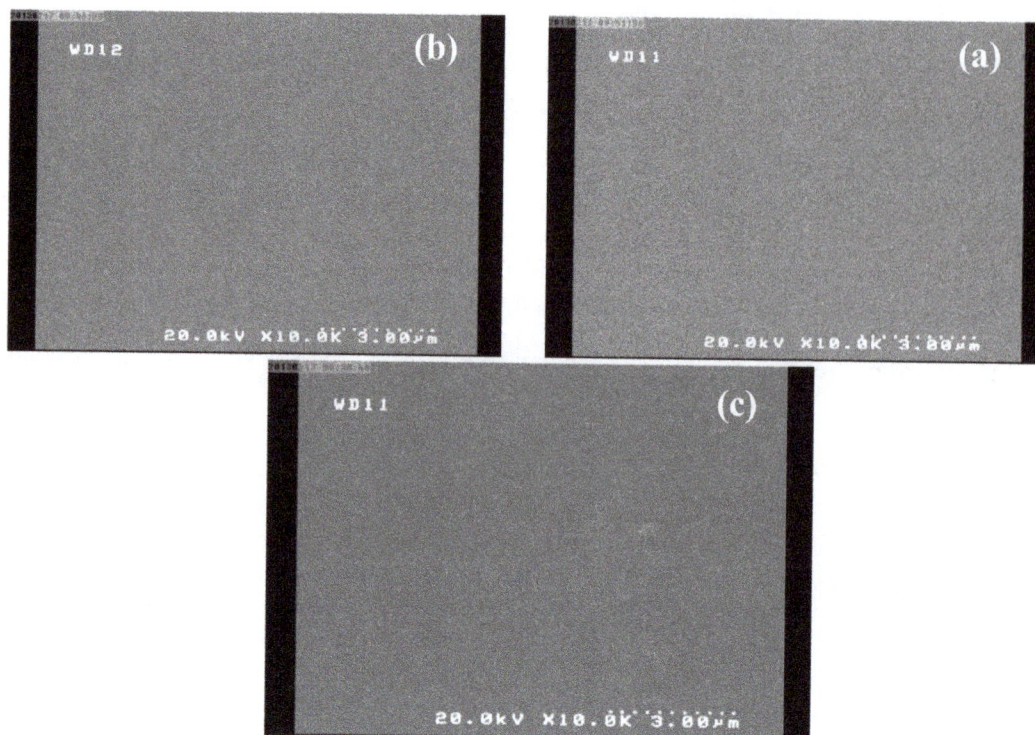

Figure 6. Images taken with the scanning electron microscope at 10000× magnification: (a) bovine gelatin with high Bloom containing clay nanoparticles; (b) Bovine gelatin with low Bloom containing clay nanoparticles; (c) Fish-skin gelatin containing clay nanoparticles.

g/g gelatin, sorbitol crystals positioned themselves in the empty spaces between gelatin branches, which led to greater strength and reduced water vapor permeability in the samples, particularly in films produced from fish skin (**Figure 7**).

3.6. Heat Sealing

Table 5 lists results of the heat sealing test of the samples. Based on the tensile strength test of the spot the gelatin films containing clay nanoparticles were heat sealed together in this research, all of the samples could be heat sealed. Among the heat sealed films, fish-skin gelatin films containing clay nanoparticles and glycerol at 0.30 g/g gelatin had the maximum percentage of elongation to the breaking point (107.48%) and also the highest tensile strength (24.88 megapascal) ($p < 0.05$). Moreover, among the treatments optimized by a combination of the two plasticizers, fish-skin gelatin films containing clay nanoparticles and glycerol at 25% and sorbitol at 5% had the maximum elasticity and tensile strength. The heat-sealed films did not separate from each other at spots they were heat-sealed but were torn at other places (which showed the high heat sealability of these films). This phenomenon may be caused the melting of the gelatin structure due to heating and the re-entwining resulting from the applied heat. Moreover, based on results of the differential scanning calorimetry analysis, the mentioned samples melted at relatively lower temperatures compared to the other samples.

4. Conclusion

Gelatin films containing clay nanoparticles are brittle by nature, large quantities of water vapor pass through them, and they have low glass transition temperatures and low melting points. Glycerol alone improves the mechanical properties of gelatin films containing clay nanoparticles, and sorbitol improves water vapor permeability of these films. The mutual effects of these two plasticizers result in films with suitable mechanical properties, water vapor permeability and glass transition temperature, and raise their melting points. Fish-skin gelatin films containing clay nanoparticles and sorbitol at 25% and glycerol at 5% exhibited the best properties because of their low water vapor permeability, suitable mechanical properties, appropriate color, heat-sealing capability,

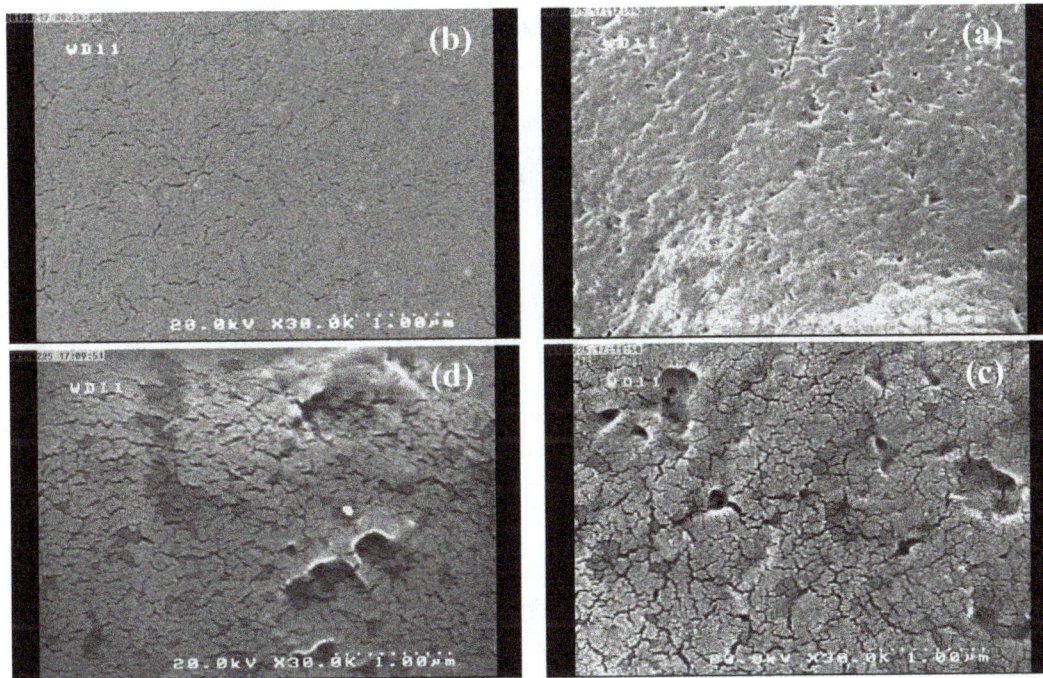

Figure 7. Images taken with an electron microscope: (a) The treatment of glycerol:sorbitol (25:5) on bovine gelatin with high Bloom containing clay nanoparticles at 10,000× magnification; (b) the treatment of glycerol:sorbitol (25:5) on fish-skin gelatin containing clay nanoparticles at 30,000× magnification; (c) the treatment glycerol:sorbitol (25:5) on bovine gelatin with high Bloom containing clay nanoparticles at 30,000× magnification; (d) the treatment glycerol:sorbitol (5:25) on bovine gelatin with low Bloom containing clay nanoparticles at 3000× magnification.

Table 5. Results of the heat-sealing test.

Film type	Treatment Glycerol:sorbitol	Thickness of the heat-sealing spot (micrometer)	Elongation to the breaking point (percentage)	Tensile strength of the heat-sealing spot (megapascal)
Bovine gelatin with high Bloom values containing clay nanoparticles	0:0	60	1.96 c	19.13 a
	30:0	80	6.45 bc	14.97 ab
	25:5	60	13.55 bc	15.47 ab
	20:10	70	26.66 b	13.79 ab
	15:15	70	19.05 bc	12.29 ab
	20:20	70	57.87 a	10.69 bc
	5:25	60	26.63 b	9.84 bc
	0:30	80	27.98 b	6.44 c
Bovine gelatin with low Bloom values containing clay nanoparticles	0:0	60	0.98 d	9.35 abc
	30:0	60	4.51 cd	13.40 a
	25:5	60	5.00 cd	13.20 a
	20:10	70	7.06 bcd	9.21 abc
	15:15	70	10.32 bcd	11.56 ab
	20:20	70	26.10 a	8.72 bc
	5:25	70	16.20 abc	3.42 d
	0:30	60	19.81 ab	5.58 cd
Fish-skin gelatin with clay nanoparticles	0:0	60	1.91 c	15.94 bc
	30:0	70	11.97 c	9.56 bc
	25:5	70	16.35 c	8.52 c
	20:10	60	22.09 c	14.66 bc
	15:15	60	25.70 c	11.71 bc
	20:20	60	55.75 b	17.29 b
	5:25	60	64.29 b	17.23 b
	0:30	60	107.48 a	24.88 a

Different letters in each section suggest the averages of the studied property in each of the tests were significantly different ($p < 0.05$).

and high strength at the heat-sealed spots. Morphological study showed that films without plasticizers had uniform surfaces, but that samples containing glycerol at concentrations of over 0.20 g/g gelatin exhibited cavities between gelatin chains, and that water vapor permeability in gelatin films containing clay nanoparticles. The results herein proved that the effect of plasticizers on mechanical properties and water vapor permeability of gelatin-based edible films containing clay nanoparticles is obvious and deserves further study.

References

[1] Liu, Z., Ge, X., Lu, Y., Dong, S., Zhao, Y. and Zeng, M. (2012) Effects of Chitosan Molecular Weight and Degree of Deacetylation on the Properties of Gelatine-Based Films. *Food Hydrocolloids*, **26**, 311-317. http://dx.doi.org/10.1016/j.foodhyd.2011.06.008

[2] Motamedzadegan, A., Davarniam, B., Asadi, G., Abedian, A. and Ovissipour, M. (2011) Optimization of Enzymatic hydrolysis of Yellowfin Tuna *Thunnus albacares* Viscera Using Neutrase. *International Aquatic Research*, **2**, 173-181.

[3] Shahiri Tabarestani, H., Sedaghat, N., Jahanshahi, M., Motamedzadegan, A. and Mohebbi, M. (2015) Physicochemical and Rheological Properties of White-Cheek Shark (*Carcharhinus dussumieri*) Skin Gelatin. *International Journal of Food Properties*. http://dx.doi.org/10.1080/10942912.2015.1050595

[4] Motamedzadegan, A., Ebdali, S. and Regenstein, J.M. (2013) Gelatin: Production, Applications and Health Implications, Halal and Kosher Regulations and Gelatin Production. Chap. 12, Nova Publishers.

[5] Kanmani, P. and Rhim, J.W. (2014) Physicochemical Properties of Gelatin/Silver Nanoparticle Antimicrobial Composite Films. *Food Chemistry*, **148**, 162-169. http://dx.doi.org/10.1016/j.foodchem.2013.10.047

[6] Sobral, P.D.A., Menegalli, F., Hubinger, M. and Roques, M. (2001) Mechanical, Water Vapor Barrier and Thermal Properties of Gelatin Based Edible Films. *Food Hydrocolloids*, **15**, 423-432. http://dx.doi.org/10.1016/S0268-005X(01)00061-3

[7] Cao, N., Yang, X. and Fu, Y. (2009) Effects of Various Plasticizers on Mechanical and Water Vapor Barrier Properties of Gelatin Films. *Food Hydrocolloids*, **23**, 729-735. http://dx.doi.org/10.1016/j.foodhyd.2008.07.017

[8] Rivero, S., García, M. and Pinotti, A. (2010) Correlations between Structural, Barrier, Thermal and Mechanical Properties of Plasticized Gelatin Films. *Innovative Food Science & Emerging Technologies*, **11**, 369-375. http://dx.doi.org/10.1016/j.ifset.2009.07.005

[9] Gontard, N., Duchez, C., Cuq, J.L. and Guilbert, S. (1994) Edible Composite Films of Wheat Gluten and Lipids: Water Vapour Permeability and Other Physical Properties. *International Journal of Food Science & Technology*, **29**, 39-50. http://dx.doi.org/10.1111/j.1365-2621.1994.tb02045.x

[10] Benzie, I.F. and Strain, J. (1996) The Ferric Reducing Ability of Plasma (FRAP) as a Measure of "Antioxidant Power": the FRAP Assay. *Analytical biochemistry*, **239**, 70-76. http://dx.doi.org/10.1006/abio.1996.0292

[11] Ghasemlou, M., Khodaiyan, F. and Oromiehie, A. (2011) Physical, Mechanical, Barrier, and Thermal Properties of Polyol-Plasticized Biodegradable Edible Film Made from Kefiran. *Carbohydrate Polymers*, **84**, 477-483. http://dx.doi.org/10.1016/j.carbpol.2010.12.010

[12] Osés, J., Fernández-Pan, I., Mendoza, M. and Maté, J.I. (2009) Stability of the Mechanical Properties of Edible Films Based on Whey Protein Isolate during Storage at Different Relative Humidity. *Food Hydrocolloids*, **23**, 125-131. http://dx.doi.org/10.1016/j.foodhyd.2007.12.003

[13] Kilburn, D., Claude, J., Schweizer, T., Alam, A. and Ubbink, J. (2005) Carbohydrate Polymers in Amorphous States: An Integrated Thermodynamic and Nanostructural Investigation. *Biomacromolecules*, **6**, 864-879. http://dx.doi.org/10.1021/bm049355r

[14] Sousa, A.M., Sereno, A.M., Hilliou, L. and Gonçalves, M.P. (2010) Biodegradable Agar Extracted from Gracilaria Vermiculophylla: Film Properties and Application to Edible Coating. *Materials Science Forum*, **636-637**, 739-744. http://dx.doi.org/10.4028/www.scientific.net/MSF.636-637.739

[15] Wang, Q. and Padua, G.W. (2005) Properties of Zein Films Coated with Drying Oils. *Journal of Agricultural and Food Chemistry*, **53**, 3444-3448. http://dx.doi.org/10.1021/jf047994n

[16] Hochstetter, A., Talja, R.A., Helén, H.J., Hyvönen, L. and Jouppila, K. (2006) Properties of Gluten-Based Sheet Produced by Twin-Screw Extruder. *LWT-Food Science and Technology*, **39**, 893-901. http://dx.doi.org/10.1016/j.lwt.2005.06.013

[17] Lee, S., Lee, M. and Song, K. (2005) Effect of Gamma-Irradiation on the Physicochemical Properties of Gluten Films. *Food Chemistry*, **92**, 621-625. http://dx.doi.org/10.1016/j.foodchem.2004.08.023

[18] Hernández-Muñoz, P., López-Rubio, A., del-Valle, V., Almenar, E. and Gavara, R. (2004) Mechanical and Water Barrier Properties of Glutenin Films Influenced by Storage Time. *Journal of Agricultural and Food Chemistry*, **52**, 79-83. http://dx.doi.org/10.1021/jf034763s

[19] Marshall, A. and Petrie, S. (1980) Thermal Transitions in Gelatin and Aqueous Gelatin Solutions. *Journal of Photographic Science*, **28**, 128-134.

[20] Pinhas, M.F., Blanshard, J., Derbyshire, W. and Mitchell, J. (1996) The Effect of Water on the Physicochemical and Mechanical Properties of Gelatin. *Journal of Thermal Analysis*, **47**, 1499-1511. http://dx.doi.org/10.1007/BF01992842

[21] Badii, F., MacNaughtan, W. and Farhat, I. (2005) Enthalpy Relaxation of Gelatin in the Glassy State. *International Journal of Biological Macromolecules*, **36**, 263-269. http://dx.doi.org/10.1016/j.ijbiomac.2005.06.008

[22] Badii, F., Martinet, C., Mitchell, J. and Farhat, I. (2006) Enthalpy and Mechanical Relaxation of Glassy Gelatin Films. *Food Hydrocolloids*, **20**, 879-884. http://dx.doi.org/10.1016/j.foodhyd.2005.08.010

15

Method of Formation of Waveguides on the Basis of Any Polymeric Materials

Valentin Tsvetkov[1]*, Sergei Pasechnik[1], Aleksei Dronov[2], Jacob Y. L. Ho[3], Vladimir Chigrinov[3], Hoi-Sei Kwok[3]

[1]Problem Laboratory of Molecular Acoustics, Moscow State University of Instrument Engineering and Computer Sciences, Moscow, Russia
[2]National Research University of Electronic Technology, Moscow, Zelenograd, Russia
[3]State Key Laboratory on Advanced Displays and Optoelectronics Technologies, Clear Water Bay, Kowloon, Hong Kong, China
Email: *tsvetkov_v_a@mail.ru

Abstract

This paper describes a new type of polymeric waveguides which has the core, cladding medium and active nodes made from the same material. Part of the polymer is removed in cladding medium by formation of nanopores. The pores can be filled with liquid crystals (LC) in order to create an active composite medium needed for electrically controlled nodes formation.

Keywords

Polymeric Materials, Nanopores, Waveguides, Optical

1. Introduction

Recently, a large number of works was devoted to polymeric materials applicable for optoelectronic waveguide devices, including waveguide films formation technology descriptions and their optical characteristics.

The polymers show some advantages over commonly used crystalline materials such as Si, Ge or multicomponent semiconducting compounds. In particular, they provide a simple and relatively cheap technology applicable for production of big size units. For example, thin polymers films are well suited for formation of the plane waveguides of micrometer thickness [1]-[5]. In most cases, the polymer waveguides are formed by trivial methods extrusion, or by spin-coating process [4]. It is necessary in these cases to provide the lower value of a refractive index (RI) in the cladding part of a waveguide with the value of the RI in the central part (the core of a

---

*Corresponding author.

waveguide). Usually, this aim is achieved by formation of outer (cladding) layers of a waveguide from materials such as polymethyl methacrylate (PMMA), or perfluorinated polyimide (PFPI) with relatively low values of the RI. Such decision creates additional technological difficulties. In particular, input and output parts of the waveguides with the active nodes have an optical and structural coupling problem.

2. Materials and Methods

To avoid these difficulties, it is desirable to use the same material for both the core and cladding waveguide parts. Moreover, in this case it is possible to obtain controlled cladding layer by using an electrically controllable materials, such as a LC's. It can be applied for elaboration and production of new waveguide units such as X, Y splitters, deflectors, wavelength selectors, etc. In this paper we describe a new method of waveguide formation based on usage of the same polymer for both core and cladding parts. It is realized by formation in the cladding part of the waveguide the system of opened or closed pores, which provide a reduction of the RI in cladding part. The degree of the reduction depends on both the pore size and pore density. The filling of pores with liquid crystals may provide changes of effective refractive index of a cladding via electric field, which makes possible to obtain the electrically controlled waveguide. Earlier we used similar polyethylene terephthalate films (PET) as the working medium for the displays [6] [7].

3. Experiments and Explanations

To verify the proposed method, we used a membrane series made by Institute of Nuclear Research in Dubna (Russia). Membrane thickness was 12 and 23 micrometers. The series included more than 20 membranes with pore diameters from 20 nanometers to 5 micrometers with different pore densities.

PET films were made by extrusion and therefore have significant optical anisotropy [8]. The published data and the results of our measurements show that the mentioned above materials have high losses of light of order of 7.3 dB/cm for $\lambda = 0.473$ μ; 4.0 dB/cm for $\lambda = 0.633$ μ and 1.44 dB/cm for $\lambda = 1.550$ μ.

The images shown in **Figures 1(c)-(f)** demonstrate that the light passes along the waveguide center (the waveguide core) surrounded by cladding medium permeated by the pores.

We have found that there is significant light scattering in the region of interface between porous and nonporous parts of the film at the pore's diameters comparable or even smaller than a wavelength of propagating light beam (see bright waveguide core framing). This effect was observed in the whole range of wavelength (450,···,1550 nm) used in experiments. It can be explained as an integral effect of scattering from a number of chaotically distributed cylindrical obstacles. Different aspects of light scattering from individual cylinders im-

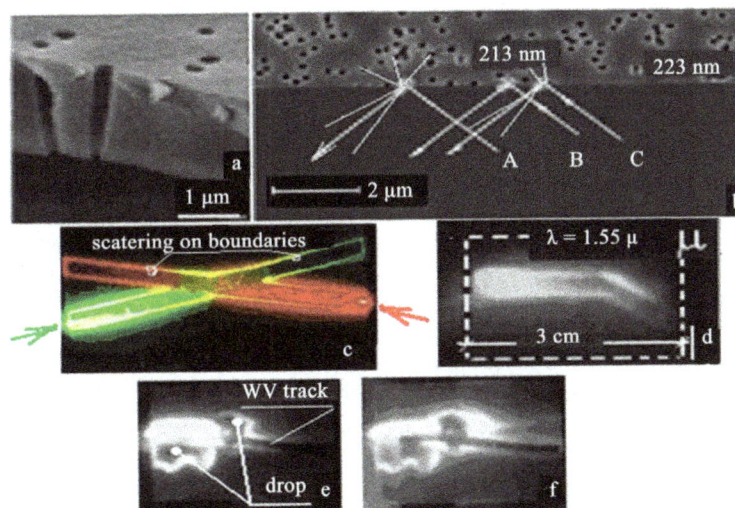

Figure 1. Cross-section of the PET film with pores (a); explanation of light beam reflectance from the interface of a pore less film and a film having randomly arranged pores (b); (c) (d) light wave propagation in some waveguide elements; (e) (f) wave tracks passing the drops.

mersed in isotropic media were considered previously [8]-[10]. In our case, polymer film can be anisotropic which prevent the direct application of previously obtained results. Moreover, usage of liquid crystals as anisotropic media inside anisotropic film made theoretical description of the problem to be quite complicated. So, we focus firstly on the qualitative explanation of the obtained results.

We believe that the observed scattering effect appears due to the pores random distribution in cladding part, resulting in a section between the core and the waveguide cladding medium looks like a "matte surface" (**Figure 1(b)**). To prove this, a drop of an isotropic liquid with a refractive index greater than the film's refractive index was placed at the interface. It resulted in evolution of the scattering region due to penetration of a liquid into pores, as shown in **Figure 1(e)** and **Figure 1(f)**. In particular, it can be seen that diffuse scattering disappears as the droplet spreads.

A similar scattering disappearance effect can be obtained by filling pores with LC and further application of a control voltage, which makes possible to form an electrically controlled node.

We believe that one of the ways to eliminate the parasitic scattering at the boundaries is the usage of regular or quasi-regular pore matrices formation. To check this thought, the membranes with a strictly regular arrangement of pores were produced (**Figure 2**).

For the first experiments, the polyimide films of 4 micrometer in thickness with the RI = 1.8 were used. Films were formed by spin coating the glass substrates of sizes 2×2 cm. The non-through pores of 2 μm in depth with the pores' diameter equal to 200 or 400 nm were obtained by interference lithography and ion-plasma etching methods. The distance between the pores was equal to 400 or 800 nm, respectively (**Figure 2**). In addition, to enhance the future technological production of waveguides, the successful attempt to transfer by replication the pores film topography from the glass substrate to the polymer substrate was made. **Figure 2(c)** shows that replica has the same spectral composition as the original sample.

The obtained difference between RIs of porous and non-porous parts (about 0.35) allows, in principle, to support a waveguide regime of propagating electromagnetic waves in wide wavelength range (from UV to THz).

Consequently, according to the proposed technology it is possible to use any other polymer film with different RI. It extends the possibility of application of different types of polymer, like TOPAS, widely used for production of THz waveguides [11].

Since both the pore's diameter and period of a pore distribution are smaller than the visible light wavelengths, a formation of the super-wave grating (SWG) regime is possible, which leads to a noticeable bright staining. This phenomenon can be used to build color displays without any polarizers or dyes, if switching on and off modes of the grating will be realized , for example, by filling the pores with electrically active medium, such as the LC's and application of a control voltage.

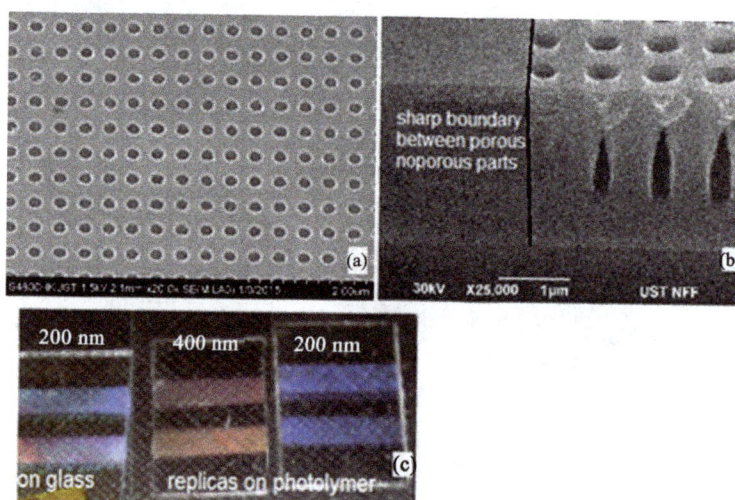

Figure 2. (a) Samples of waveguides structures: a-distribution of pores; (b) cross section of the sample; (c) views of WG deposited on glass substrate and its replicas on photopolymer.

At illumination of the edge of a substrate by a non-collimated laser beam of a diameter 3 - 4 mm, the input beam was partly reflected from the outer surface of the substrate and partly transmitted to the waveguide. The transmitted beam crossed the boundary between the porous and non-porous parts (**Figure 3**) at some small angle.

Afterwards, this beam was partly reflected from the interface in the accordance with geometrical optics laws, namely:

1. The angles of incidence and reflection are equal.

2. The reflected light is mainly polarized in the direction perpendicular to the waveguide plane. The boundary between the two parts of the waveguide shows no sign of scattering, contrary to the case of inherent scattering at the boundaries of the waveguide observed at a random arrangement of pores.

We observed, that the reflected beam could be totally deleted by placing onto waveguide surface a drop of an isotropic liquid with the RI equal to or greater than the polymer film RI. In this case the introduced beam was not reflected from the interface between the porous and nonporous regions of the film and penetrated into the porous part filled with a liquid, due to matching of the RIs. At the same time, the reflection occurred at the boundary between regions with filled and empty pores.

Obviously, that a similar effect can be induced by application of electric field to the pores filled with the LC's. Thus the active element can be created from the same polymer as the waveguide core.

The above described experiments confirm a possibility to control light beams propagating in the waveguide.

It makes possible to get electrically control boundary, which can be switched between modes "light reflects" and "light passes". With the set of such interfaces, it is possible to build a multi-step M × N channel router similar to that shown in **Figure 4**.

4. Conclusions

1. It was shown experimentally that usage of the proposed technology provided a production of any passive waveguide structure from any available polymer. The perspective of an electric control of light beams propagating in waveguides, when the pores will be filled in with electrically controlled material, such as LC, is considered. Light losses in such waveguides are determined only by their own material loss, because any manipulations will not have effect on core properties.

2. The proposed methods and technologies are universal and allow to create waveguide structures for different wavelengths because the difference of the core and cladding medium refractive indices is high (0.35), and unattainable by any other waveguide formation method. An important consequence of such difference is the ability

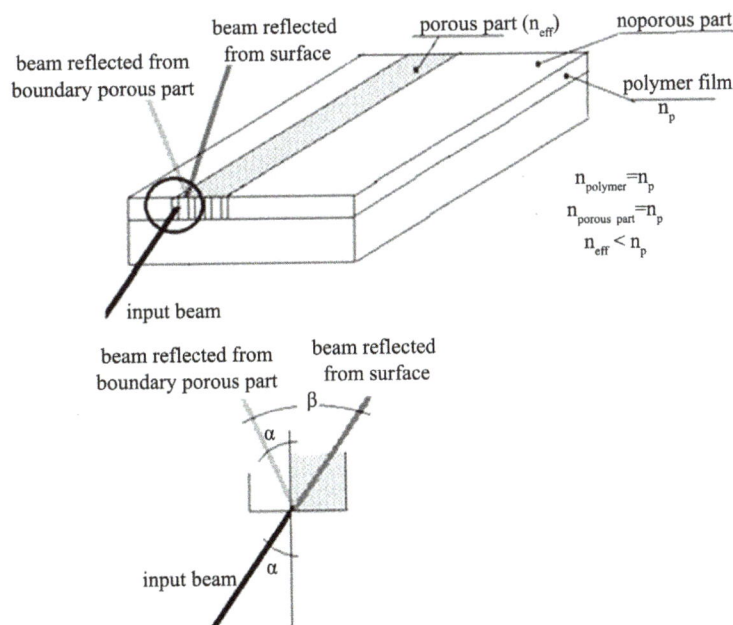

Figure 3. Path of the rays in the reflection from the interface.

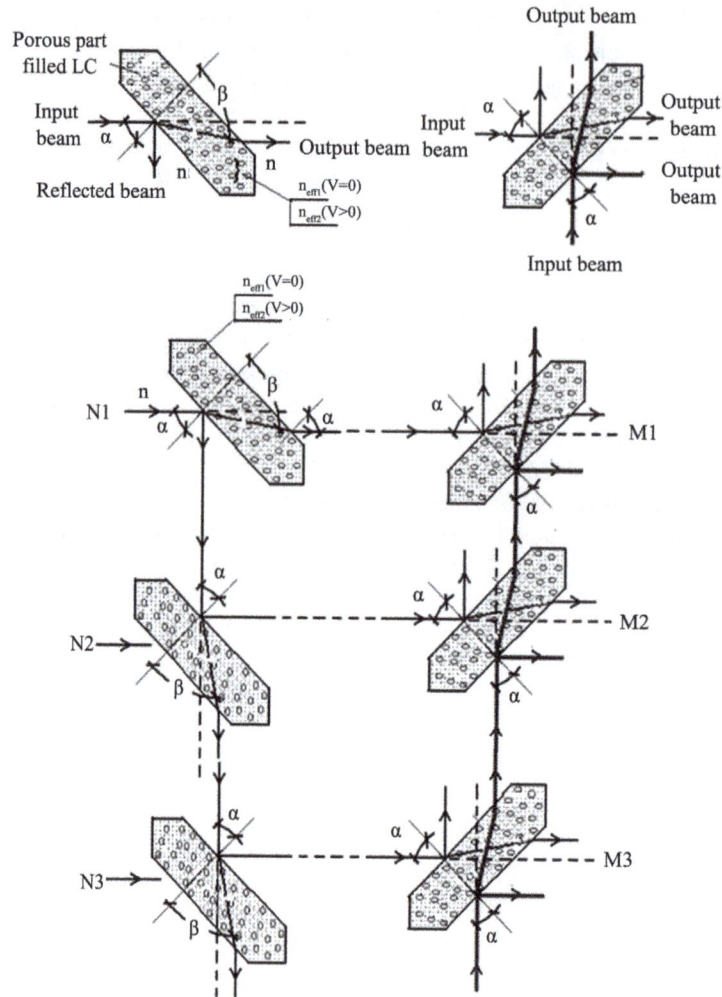

Figure 4. Router M × N canals.

to handle any wavelengths radiation from UV to THz.

Now the work is in the stage of progress.

Conflict of Interests

The authors declare that there is no conflict interests regarding the publication this paper.

Acknowledgements

This work was partially supported by the Ministry of Education and Science of the Russian Federation (Grant No. 3.1921.2014/K) and by the City University of Hong Kong under Grants SRG 7002475 and SRG 7002866.

References

[1]   Uddin, M.A. and Chan, H.P. (2008) Materials and Process Optimization in the Reliable Fabrication of Polymer Photonic Devices. *Journal of Optoelectronics and Advanced Materials*, **10**, 1-15.

[2]   Zhou, S.F., Reekie, L., Chan, H.P., Chow, Y.T., Chung, P.S. and Luk, K.M. (2010) Polymer Fiber Bragg Gratings for the THz Region. *Optics Express*, **8**, 11707-11712.

[3]   Low, A.L.Y., Singh, G.K., and Yong, Y.S. (2004) High-Index and Low Loss Polyethylene Terephthalate Optical Waveguides. *Optical and Quantum Electronics*, **36**, 997-1003.
http://dx.doi.org/10.1007/s11082-004-2039-2

[4] Iiyama, K., Ishida, T., Ono, Y., Maruyama, T. and Yamagishi, T. (2011) Fabrication and Characterization of Amorphous Polyethylene Terephthalate Optical Waveguides. *IEEE Photonics Technology Letters*, **23**, 275-277.

[5] Iiyama, K., Ishida, T., Sasaki, K., Kitamura, K. and Maruyama, T. (2011) Low Loss Amorphous Polyethylene Terephthalate (PET) Optical Waveguides. 16*th Opto-Electronics and Communications Conference*, *OECC* 2011, Taiwan, 4-7 July 2011, 301-302.

[6] Semerenko, D., Shmeliova, D., Pasechnik, D., Murauskii, A., Tsvetkov, V. and Chigrinov, V. (2010) Optically Controlled Transmission of Porous Polyethylene Terephthalate Films Filled with Nematic Liquid Crystal. *Optics Letters*, **35**, 2155-2157. http://dx.doi.org/10.1364/OL.35.002155

[7] Semerenko, D., Shmeliova, D., Pasechnik, S., Chigrinov, V., Murauskii, A. and Tsvetkov, V. (2010) The Investigation of Polarization Properties of the Porous PET Films Filled with LC. *International Conference on Advanced Infocomm Technology* (*ICAIT* 2010), Oral 3-3A-3, Hainan, China, 20-23 July 2010.

[8] Apel, P.Y., Blonskaya, I.V., Dmitriev, S.N., Orelovitch, O.L., Presz, A. and Sartowska, B.A. (2007) Fabrication of Nanopores in Polymer Foils with Surfactant-Controlled Longitudinal Profiles. *Nanotechnology*, **18**, Article ID: 305302. http://dx.doi.org/10.1088/0957-4484/18/30/305302

[9] Uemura, Y., Fujimura, M., Hashimoto, T. and Kawai, H. (1978) Application of Light scattering from Dielectric Cylinder Based upon Mie and Rayleigh—Gans—Born Theories to Polymer Systems. *Polymer Journal*, **10**, 341-351. http://dx.doi.org/10.1295/polymj.10.341

[10] Liou, K.-N. (1972) Electromagnetic Scattering by Arbitrary Oriented Ice Cylinders. *Applied Optics*, **11**, 667-672. http://dx.doi.org/10.1364/AO.11.000667

[11] Topas-Brochure. http://www.topas.com/sites/default/files/files/TOPAS_Brochure_E_2014_06(1).pdf

16

Eco-Friendly Production of Silver Nanoparticles from Peel of Tangerine for Degradation of Dye

Eman Alzahrani

Chemistry Department, Faculty of Science, Deanship of Scientific Research, Taif University, Taif, KSA
Email: em-s-z@hotmail.com

Abstract

Green chemistry methods for production of nanoparticles have many advantages, such as ease of use, which makes the methods desirable and economically viable. The aim of the present work was to green synthesise silver nanoparticles (SNPs) using aqueous tangerine peel extract in different ratios (2:1, 1:1, 1:2). The formed SNPs were characterised using ultraviolet-visible (UV-Vis) spectrophotometry, and transmission electron microscopy (TEM). The UV-Vis spectra showed that the highest absorbance was observed when the ratio of peel tangerine extract to silver nitrate solution was 1:2. The transmission electron micrographs showed the formation of poly dispersed nanoparticles. It was found that the average diameter of the nanoparticles was 30.29 ± 5.1 nm, 16.68 ± 5.7 nm, and 25.85 ± 8.4 nm, using a tangerine peel solution and silver nitrate solution ratio of 2:1, 1:1, and 1:2, respectively. The formed SNPs were evaluated as catalysts for methyl orange dye degradation, and the results confirmed that SNPs can speed up the degradation of the dye.

Keywords

Green Process, Synthesis Silver Nanoparticle, Tangerine Peel Extract, Degradation of Dye, Methyl Orange

1. Introduction

Nanomaterials have been receiving attention due to their unique physical and chemical properties compared with their larger-size counterparts [1]-[3]. An example of nanostructured materials are silver nanoparticles (SNPs), and they have been utilised in different fields such as physics, chemistry, biology, and medicine. Recently, silver nanoparticles have been found to exhibit interesting antibacterial activities, burn treatments, coating stainless steel metals, and sun creams [4]-[7].

The noble metal nanoparticles have been fabricated by chemical reduction with stabilising reagents (NaBH$_4$, citrate, or ascorbate) [8], thermal decomposition [9], photo reduction in reverse micelles [10], and radiation chemical reduction [11]. Many of these approaches are expensive, consume a lot of energy, result in low yields, and the chemicals used in their production are toxic and hazardous [12].

Recently, silver nanoparticles have been formed by biological approaches, such as using microorganisms [13], enzymes [14], and fungus [15]. The disadvantages of these approaches are they need special culture preparation and isolation techniques for synthesis of the nanoparticles [16] [17]. Silver nanoparticles have also been fabricated using plant extracts as reducing and capping agents. The main advantages of using plant extracts are that the process is simple, cheap, scaling-up, eco-friendly, and safe [18]-[21].

Shankar *et al.* [22] fabricated silver nanoparticles using neem leaf; however, the required time for a 90% reduction was about 4 hours. In this work, a simple and fast chemical reaction for the fabrication of silver nanoparticles using tangerine peel extract as a reducing and stabilising agent was investigated. To the best of our knowledge, this is the first paper examining the use of tangerine peel extract for large-scale production of SNPs. The fabricated nanoparticles were characterised using UV-Visible absorption spectroscopy, and transmission electron microscopy. In this paper, using SNPs as a catalyst for degradation of dye was studied. The chosen dye was methyl orange dye (MO) which is an anionic azo dye.

2. Experimental

2.1. Chemicals and Materials

Tangerine peel was collected from a local shop (Taif, KSA). Silver nitrate (AgNO$_3$) (99.8%) was purchased from Sigma-Aldrich (Poole, UK). The Whatman filter paper (pore size 25 μm and diam. 15 cm) was purchased from Sigma-Aldrich (Poole, UK). Distilled water was employed for preparing all the solutions and reagents. Methyl orange, sodium borohydride (NaBH$_4$) and hydrogen peroxide (H$_2$O$_2$) were purchased from Fisher Scientific (Loughborough, UK).

2.2. Instrumentation

Transmission electron microscopy (TEM) from JEOL Ltd. (Welwyn Garden City, UK), a water bath from Poly Science (Niles, Illinois, US), a UV-Vis spectrophotometer from Thermo Scientific™ GENESYS 10S (Toronto, Canada).

2.3. Synthesis of Silver Nanoparticles (SNPs)

20 g of tangerine peel was added to 100 mL deionised water. The mixture was boiled at 80°C for 5 min. The solution was filtered using Whatman paper then added to silver nitrate solution (2 mM) indifferent ratios (2:1, 1:1, 1:2) at room temperature (25°C ± 3°C) without stirring. Reduction of silver ions into silver nanoparticles commonly followed by colour change and the formation of SNPs can be visually observed [3]. Therefore, the change of colour was visually monitored to check for the formation of SNPs. The fabricated silver nanoparticles solutions were stored at 2°C and were utilised within one week.

2.4. Characterisation of the Synthesised Silver Nanoparticles

2.4.1. UV-Vis Spectroscopy

The prepared silver nanoparticle solutions were observed by ultraviolet-visible spectrophotometer. The bioreduction of the Ag$^+$ ions in solution was checked by periodic sampling of aqueous component and observing the UV-Vis spectrum at different time intervals. 100 μL of solution was diluted with 1 mL deionised water then the absorbance of the mixture was measured using a UV-Vis spectrophotometer and compared with 1 mL of distilled water as a blank over the range 350 - 800 nm operated at resolution of 1 nm.

2.4.2. TEM Analysis

The formation of SNPs was confirmed by using transmission electron microscopy (TEM). In addition, the size and the morphology of the SNPs were characterised by TEM. For this, 5 μL of the sample solution was put onto lacy carbon-coated 3 mm diameter copper grids. TEM images were acquired with a Gatan Ultrascan 4000 digital camera attached to a JEOL 2010 transmission electron microscope running at 20 kV. The size of the prepared SNPs was measured using image J software.

2.5. Catalytic Reduction of Dye Using SNPs

Using SNPs as a catalyst in dyedegradation was checked using methyl orange dye. This was performed by mixing 1 mL of methyl orange solution (3×10^{-4} M) with 1.5 mL distilled water, 1 mL of H_2O_2 solution (0.001 M) or 1 mL of $NaBH_4$ (0.001 M) solution, and 500 μL of SNPs. The progress of the reaction was monitored using UV-Vis spectrophotometer at a wavelength of 465 nm. For the uncatalysed reaction, 500 μL of SNPs was replaced by an equal amount of distilled water.

3. Results and Discussion

3.1. Visual Observation Study

The present study deals with the biosynthesis of silver nanoparticles (SNPs) using tangerine peel. The main advantages of using green processes are rapidity, low cost, low environmental impact, and a single-step procedure for biofabrication [12] [23]. In this study, different ratios of silver nitrate solution and aqueous extract of tangerine peel (2:1, 1:1, and 1:2) were utilised without any stirring and hence the use of energy was minimal. It is known that the reduction of silver ion (Ag^+) into silver (Ag) nanoparticles during exposure to the plant extracts commonly followed by colour change and the formation of SNPs was visually observed [3]; therefore, the colour of the mixtures were observed.

Figure 1 shows the colour change of the solutions, which were silver nitrate solution (1) and tangerine peel solution (2), to dark brown after synthesis of silver nanoparticles using different ratios of silver nitrate solution to tangerine peel solution, (3) 2:1, (4) 1:1, and (5) 1:2, at different times: 15 min (Figure 1(a)), 30 min (Figure 1(b)), and after 45 min (Figure 1(c)). It was observed that the colour of the mixture changed immediately after mixing the silver nitrate solution with the extract solution, which confirmed that aqueous silver ions can be reduced by aqueous extract of tangerine peel to form stable SNPs in water. The reason for the brown colour is due to the excitation of surface plasmon vibrations in the silver metal nanoparticles [3] [24].

3.2. UV-Visible Spectral Analysis of SNPs

It is known that the UV-Vis spectrophotometer can be utilised for checking the formation of metal nanoparticles in aqueous suspensions [25]. Therefore, the progress of silver nanoparticle formation was monitored spectrophotometrically as a function of time of reaction over the range 350 - 800 nm, and the peak wavelength (λ_{max})

Figure 1. Photographs showing colour changes of silver nitrate solution (2 mM) after adding of extract of tangerine peel solution after reaction time: (a) 15 min; (b) 30 min; (c) 45 min. (1) silver nitrate solution (2) extraction of tangerine peel solution, (3) ratio of silver nitrate solution to tangerine peel solution 2:1, (4) 1:1, and (5) 1:2.

and value of the absorbance (O.D.) were recorded. **Table 1** presents the wavelength of maximum absorption (λ_{max}) and value of the absorbance (O.D.) of the reaction medium at different times (15 min, 30 min, and 45 min). It was noted that absorbance increased with time, from 15 min to 45 min. Formation of SNPs happened within 45 min of reaction, making this study one of the fastest green processes to produce SNPs, compared with a previous work [26] which took approximately 2 h. After 45 min, there was no increase in the absorbance due to the depletion of the silver ions (the initial material).

The ratio of silver nitrate solution to tangerine peel extraction in the reaction mixture was investigated, as can be seen in **Table 1**. It was found that the highest absorbance was obtained when the ratio of peel tangerine extract to silver nitrate solution was 1:2, and the absorbance was in the range 0.416 - 1.181. This confirmed that the yield of the formed SNPs can be affected by metal-extract proportions, as was noted in a previous study [27]. In addition, it was found that the maximum absorption peaks were between 419 to 435 nm. It was observed throughout the reaction period that fabricated SNPs were dispersed in the aqueous solution and stable for more than 6 months, with little sign of aggregation.

3.3. TEM Study

The fabricated SNPs were characterised using TEM analysis. **Figure 2** shows the TEM electron microscopy micrographs showing the formation of nanoparticles. The TEM micrographs showed that all the fabricated SNPs using different ratios of silver nitrate solution and tangerine peel extract were spherical in shape and they were dispersed in aqueous medium.

It was found that the size of the formed SNPs depends on the ratio of extract tangerine peel to silver nitrate solution. The size of SNPs was 30.29 ± 5.1 when the ratio of extract tangerine peel to silver nitrate solution was 2:1 while if the ratio solution was 1:1, the size of nanoparticles was 16.68 ± 5.7. Moreover, it was found that the size of the prepared SNPs using silver nitrate solution and tangerine peel solution (2:1) was 25.85 ± 8.4.

3.4. Effect of Silver Nanoparticles in Catalysing Degradation of Methyl Orange

The rate of degradation of MO. dye by sodium borohydride ($NaBH_4$) or hydrogen peroxide (H_2O_2) was monitored (0 - 60 min) in the absence and presence of silver nanoparticles that were prepared using the ratio of extract of tangerine peel to silver nitrate solution 1:2 (The ratio that gave the highest absorbance). The effect of silver nanoparticles in catalysing degradation of the dye was monitored spectrophotometrically, as can be seen in **Figure 3**. It was found that the absorbance of the dye decreased rapidly when using SNPs compared to without using SNPs. These results confirmed that SNPs can be used to catalyse the degradation of organic pollutants.

Table 1. Absorption maxima (λ) and absorbance SNPs prepared from tangerine peel, showing effect of reaction time, and silver nitrate solution: Peel tangerine extract ratio.

| Sample | Time (min) | Ratio of peel tangerine extract to silver nitrate solution | | | | | |
| | | 2:1 | | 1:1 | | 1:2 | |
		λ_{max}	Absorbance	λ_{max}	Absorbance	λ_{max}	Absorbance
1	0	-	-	-	-	-	-
	15	428	0.255	431	0.341	428	0.451
	30	435	0.514	428	0.677	419	0.919
	45	435	0.692	431	0.754	435	1.088
2	0	-	-	-	-	-	-
	15	425	0.242	419	0.396	419	0.416
	30	433	0.479	431	0.690	430	0.949
	45	431	0.676	431	0.811	433	1.086
3	0	-	-	-	-	-	-
	15	422	0.273	431	0.416	431	0.436
	30	419	0.582	428	0.705	425	1.060
	45	435	0.715	419	0.819	422	1.181

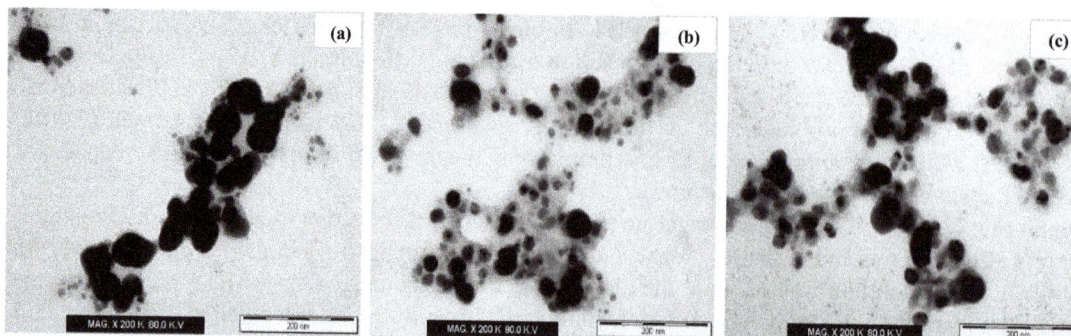

Figure 2. TEM images of the fabricated SNPs using different ratio of extract of tangerine peel solution to silver nitrate solution: (a) 20:10; (b) 10:10; (c) 10:20.

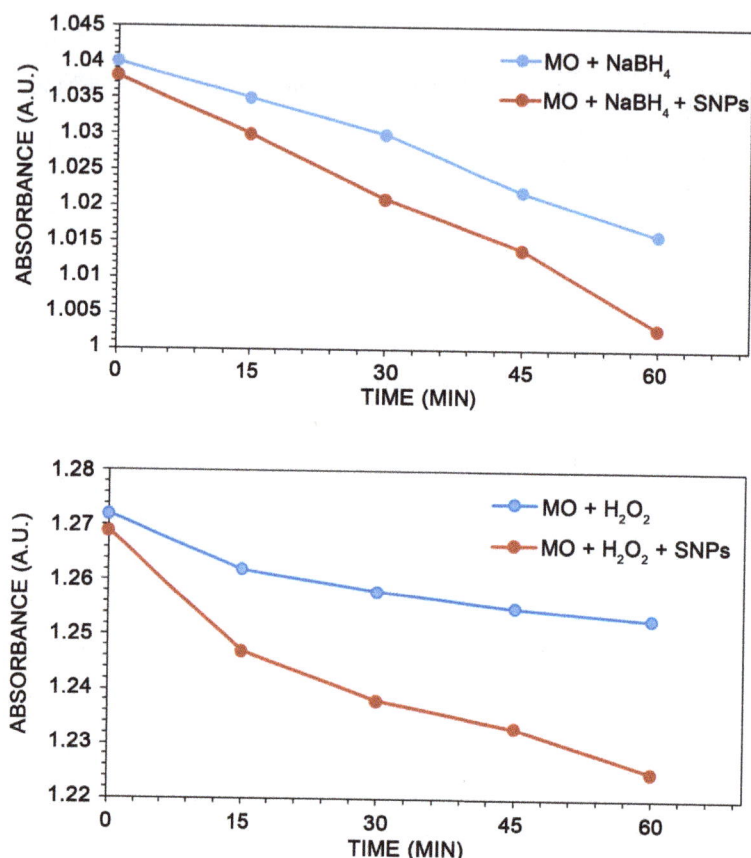

Figure 3. Degradation of the methyl orange solution (3×10^{-4} M) in the absence and presence of SNPs (500 μL) and 1 mL of NaBH$_4$ solution (0.001 M) or H$_2$O$_2$ solution (0.001 M).

4. Conclusion

The bioreduction of aqueous silver ions by the tangerine peel extract was demonstrated in the present study. It was found that this procedure is very simple, fast, eco-friendly, and can form quite stable SNPs in solution, confirming that tangerine peel can be a good source for preparation of silver nanoparticles, and is an alternative method for synthesis of metal nanoparticles. In addition, it was found that the ratio of tangerine peel extract solution to silver nitrate solution can affect the sizes and the yields of SNPs. Moreover, this study showed that SNPs can be utilised to catalyse degradation of dye. It would be interesting to utilise the fabricated SNPs for degradation of organic pollutants.

References

[1] Fayaz, A.M., *et al.* (2010) Biogenic Synthesis of Silver Nanoparticles and Their Synergistic Effect with Antibiotics: A Study against Gram-Positive and Gram-Negative Bacteria. *Nanomedicine: Nanotechnology, Biology and Medicine*, **6**, 103-109. http://dx.doi.org/10.1016/j.nano.2009.04.006

[2] Vijayakumar, M., *et al.* (2013) Biosynthesis, Characterisation and Anti-Bacterial Effect of Plant-Mediated Silver Nanoparticles Using *Artemisia nilagirica. Industrial Crops and Products*, **41**, 235-240. http://dx.doi.org/10.1016/j.indcrop.2012.04.017

[3] Alzahrani, E. and Welham, K. (2014) Optimization Preparation of the Biosynthesis of Silver Nanoparticles Using Watermelon and Study of Its Antibacterial Activity. *International Journal of Basic and Applied Sciences*, **3**, 392-400. http://dx.doi.org/10.14419/ijbas.v3i4.3358

[4] Song, J.Y. and Kim, B.S. (2009) Rapid Biological Synthesis of Silver Nanoparticles Using Plant Leaf Extracts. *Bioprocess and Biosystems Engineering*, **32**, 79-84. http://dx.doi.org/10.1007/s00449-008-0224-6

[5] Rai, M., Yadav, A. and Gade, A. (2009) Silver Nanoparticles as a New Generation of Antimicrobials. *Biotechnology Advances*, **27**, 76-83. http://dx.doi.org/10.1016/j.biotechadv.2008.09.002

[6] Kumar, A., *et al.* (2008) Silver-Nanoparticle-Embedded Antimicrobial Paints Based on Vegetable Oil. *Nature Materials*, **7**, 236-241. http://dx.doi.org/10.1038/nmat2099

[7] Furno, F., *et al.* (2004) Silver Nanoparticles and Polymeric Medical Devices: A New Approach to Prevention of Infection. *Journal of Antimicrobial Chemotherapy*, **54**, 1019-1024. http://dx.doi.org/10.1093/jac/dkh478

[8] Liz-Marzán, L.M. and Lado-Tourino, I. (1996) Reduction and Stabilization of Silver Nanoparticles in Ethanol by Nonionic Surfactants. *Langmuir*, **12**, 3585-3589. http://dx.doi.org/10.1021/la951501e

[9] Esumi, K., *et al.* (1990) Preparation and Characterization of Bimetallic Palladium-Copper Colloids by Thermal Decomposition of Their Acetate Compounds in Organic Solvents. *Chemistry of Materials*, **2**, 564-567. http://dx.doi.org/10.1021/cm00011a019

[10] Sun, Y., Atorngitjawat, P. and Meziani, M.J. (2001) Preparation of Silver Nanoparticles via Rapid Expansion of Water in Carbon Dioxide Microemulsion into Reductant Solution. *Langmuir*, **17**, 5707-5710. http://dx.doi.org/10.1021/la0103057

[11] Henglein, A. (1993) Physicochemical Properties of Small Metal Particles in Solution: "Microelectrode" Reactions, Chemisorption, Composite Metal Particles, and the Atom-to-Metal Transition. *The Journal of Physical Chemistry*, **97**, 5457-5471. http://dx.doi.org/10.1021/j100123a004

[12] Ponarulselvam, S., Panneerselvam, C., Murugan, K., Aarthi, N., Kalimuthu, K. and Thangamani, S. (2012) Synthesis of Silver Nanoparticles Using Leaves of *Catharanthus roseus* Linn. G. Don and Their Antiplasmodial Activities. *Asian Pacific Journal of Tropical Biomedicine*, **2**, 574-580. http://dx.doi.org/10.1016/S2221-1691(12)60100-2

[13] Konishi, Y., Ohno, K., Saitoh, N., Nomura, T., Nagamine, S., Hishida, H., *et al.* (2007) Bioreductive Deposition of Platinum Nanoparticles on the Bacterium *Shewanella algae. Journal of Biotechnology*, **128**, 648-653. http://dx.doi.org/10.1016/j.jbiotec.2006.11.014

[14] Willner, I., Baron, R. and Willner, B. (2006) Growing Metal Nanoparticles by Enzymes. *Advanced Materials*, **18**, 1109-1120. http://dx.doi.org/10.1002/adma.200501865

[15] Zhang, X.R., He, X.X., Wang, K.M. and Yang, X.H. (2011) Different Active Biomolecules Involved in Biosynthesis of Gold Nanoparticles by Three Fungus Species. *Journal of Biomedical Nanotechnology*, **7**, 245-254. http://dx.doi.org/10.1166/jbn.2011.1285

[16] Saxena, A., Tripathi, R. and Singh, R. (2010) Biological Synthesis of Silver Nanoparticles by Using Onion (*Allium cepa*) Extract and Their Antibacterial Activity. *Digest Journal of Nanomaterials and Biostructures*, **5**, 427-432.

[17] Jain, D., Daima, H.K., Kachhwala, S. and Kothari, S.L. (2009) Synthesis of Plant-Mediated Silver Nanoparticles Using Papaya Fruit Extract and Evaluation of Their Anti Microbial Activities. *Digest Journal of Nanomaterials and Biostructures*, **4**, 557-563.

[18] Azar, A.R.J. and Mohebbi, S. (2013) One-Pot Greener Synthesis of Silver Nanoparticles Using Tangerine Peel Extract: Large-Scale Production. *Micro & Nano Letters*, **8**, 813-815. http://dx.doi.org/10.1049/mnl.2013.0473

[19] Moulton, M.C., Braydich-Stolle, L.K., Nadagouda, M.N., Kunzelman, S., Hussain, S.M. and Varma, R.S. (2010) Synthesis, Characterization and Biocompatibility of "Green" Synthesized Silver Nanoparticles Using Tea Polyphenols. *Nanoscale*, **2**, 763-770. http://dx.doi.org/10.1039/c0nr00046a

[20] Dubey, M., Bhadauria, S. and Kushwah, B. (2009) Green Synthesis of Nanosilver Particles from Extract of *Eucalyptus hybrida* (Safeda) Leaf. *Digest Journal of Nanomaterials and Biostructures*, **4**, 537-543.

[21] Parashar, V., Parashar, R., Sharma, B. and Pandey, A.C. (2009) *Parthenium* Leaf Extract Mediated Synthesis of Silver Nanoparticles: A Novel Approach towards Weed Utilization. *Digest Journal of Nanomaterials and Biostructures*, **4**,

45-50.

[22] Shankar, S.S., Rai, A., Ahmad, A. and Sastry, M. (2004) Rapid Synthesis of Au, Ag, and Bimetallic Au Core-Ag Shell Nanoparticles Using Neem (*Azadirachta indica*) Leaf Broth. *Journal of Colloid and Interface Science*, **275**, 496-502. http://dx.doi.org/10.1016/j.jcis.2004.03.003

[23] Mittal, A.K., Chisti, Y. and Banerjee, U.C. (2013) Synthesis of Metallic Nanoparticles Using Plant Extracts. *Biotechnology Advances*, **31**, 346-356. http://dx.doi.org/10.1016/j.biotechadv.2013.01.003

[24] Xu, H. and Käll, M. (2002) Surface-Plasmon-Enhanced Optical Forces in Silver Nanoaggregates. *Physical Review Letters*, **89**, Article ID: 246802. http://dx.doi.org/10.1103/PhysRevLett.89.246802

[25] (2011) Biofabrication of Ag Nanoparticles Using *Moringa oleifera* Leaf Extract and Their Antimicrobial Activity. *Asian Pacific Journal of Tropical Biomedicine*, **1**, 439-442. http://www.ncbi.nlm.nih.gov/pmc/articles/PMC3614222/pdf/apjtb-01-06-439.pdf

[26] Kim, J.S., Kuk, E., Yu, K.N., Kim, J.-H., Park, S.J., Lee, H.J., *et al.* (2007) Antimicrobial Effects of Silver Nanoparticles. *Nanomedicine: Nanotechnology, Biology and Medicine*, **3**, 95-101. http://dx.doi.org/10.1016/j.nano.2006.12.001

[27] Ganaie, S.U., Abbasi, T., Anuradha, J. and Abbasi, S.A. (2014) Biomimetic Synthesis of Silver Nanoparticles Using the Amphibious Weed Ipomoea and Their Application in Pollution Control. *Journal of King Saud University-Science*, **26**, 222-229. http://dx.doi.org/10.1016/j.jksus.2014.02.004

17

Controlling Diameter, Length and Characterization of ZnO Nanorods by Simple Hydrothermal Method for Solar Cells

Ahmed H. Kurda[1], Yousif M. Hassan[1], Naser M. Ahmed[2]

[1]Physics Department, College of Science, University of Salahaddin, Erbil, Kurdistan of Iraq
[2]Nano-Optoelectronic Research & Technology Laboratory, School of Physics, Universiti Sains Malaysia, Penang, Malaysia
Email: ahmedkurda.69@gmail.com, yousif.60@Hotmail.com, naser@usm.my

Abstract

Zinc oxide (ZnO) nanorods have been synthesized by solution processing hydrothermal method in low temperature using the spin coating technique. Zinc acetate dehydrate, Zinc nitrate hexahydrate and hexamethylenetetramine were used as a starting material. The ZnO seed layer was first deposited by spin coated of ethanol zinc acetate dehydrate solution on a glass substrate. ZnO nanorods were grown on the ZnO seed layer from zinc nitrate hexahydrate and hexamethylenetetramine solution, and their diameters, lengths were controlled by precursor concentration and development time. From UV-Visible spectrometry the optical band gap energy of ZnO nanorods was calculated to be 3.3 eV. The results of X-Ray Diffraction (XRD) showed the highly oriented nature of ZnO nanorods the hardest (002) peak reflects that c-axis elongated nanorods are oriented normal to the glass substrate. The Field Emission Scanning Electron Microscope (FESEM) was employed to measure both of average diameter of ZnO nanorods, Energy Dispersive X-Ray (EDX) is used to identify the elemental present and to determine the element composition in the samples.

Keywords

Hydrothermal Method, Nanorods, Spin Coating, ZnO

1. Introduction

Zinc oxide (ZnO) is inexpensive n-type of semiconductor compound, which has shown promise for commercial applications in photovoltaic cells, [1] nanosensors, [2] [3] photocatalysise, [4] nanolasers [5] and light emitting diodes [6]. ZnO has a large band gap 3.37 eV, large excitonic binding energy 60 meV and high carrier mobility at room temperature. ZnO is composed of a hexagonal wurtzite crystal structure with unit cell a = 3.253 Å and c = 5.215 Å. Many reports have described syntheses ZnO nanorods through demonstrating various chemical processes that are simple and may also have industrial applications.

Nanostructured ZnO is fabricated using various thin film techniques as spray pyrolysis [7], sputtering [8], metal organic chemical vapor deposition [9] and hydrothermal method [10]. Hydrothermal method is widely adopted for the fabrication of transparent and conducting oxide due to its simplicity, safety, no needs costly vacuum system and hence cheap method for large area coating. The hydrothermal method also offers other advantages such as high surface area morphology at low crystallization temperature, the easy control of chemical components and fabrication of thin film at low cost for elucidating the structure and optical properties of ZnO nanorods.

In this work ZnO nanorods have been produced by thermal method using the solution, zinc acetate dehydrate in ethanol as a seed layer. The growth of ZnO nanorods, diameter and length are controlled by changing the solution concentration and immersion time in equimolar of zinc nitrate hexahydrate and hexamethylenetetramine (HMTA) in deionized water at a 90°C and their morphologies, preferential orientation and optical properties were examined in particular.

2. Experimental Work

The hydrothermal method synthesis and thin film process arrangement are presented schematically in **Figure 1** the glass substrate was cleaned with ethyl alcohol, urine, and acetone several times. The cleaned glass samples were further treated with UVO for 15 minutes to make rid of organic materials, the ZnO seed layer was first prepared as follows: A 5 mM ethanol solution of zinc acetate dehydrate (Zn(CH$_3$COO)$_2$·2H$_2$O, Aldrich, 98%)

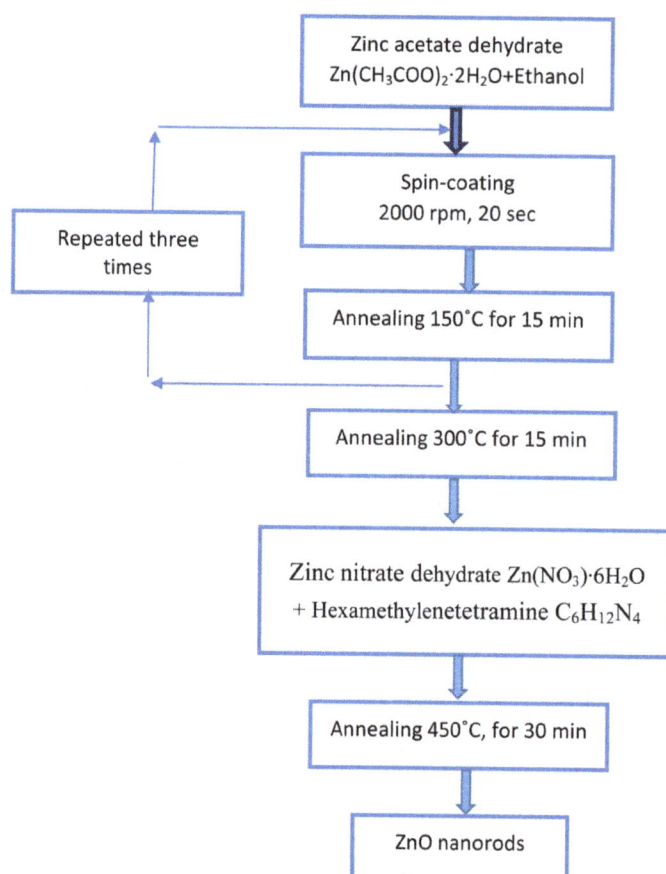

Figure 1. The flow chart showing the procedure for preparation ZnO nanorods.

was spin coated on the cleaned glass at a spinning speed of 2000 rpm for 20 s with a 10 s wait time, then annealing at 150°C for 15 min. The procedure was repeated three times, and finally the ZnO seed layer was annealed at 350°C for 15 minutes. The solution of growing ZnO nanorods was prepared by dissolving equimolar zinc nitrate hexahydrate (Zn(NO$_3$)·6H$_2$O, Aldrich 98%) and hexamethylenetetramine (HMTA) (C$_6$H$_{12}$N$_4$, Aldrich, 99%) in deionized (DI) water. The solution concentration was varied from 15 to 35 mM for controlling the ZnO nanorods. The ZnO seed layer deposited on glass was immersed in the solution, where the glass was face down, and the baker was kept at 90°C for 60, 90, 120, 150, 180 min. Alteration in the immersion time at a given concentration can control the length of the ZnO nanorods. The ZnO nanorods film was rinsed with Deionized water an ethyl alcohol several times. Ultimately, the film was annealed at 450°C for 30 minutes.

The average diameter and length of the ZnO nanorods were measured by using the field emission scanning electron microscope FESEM (Model: FEI Nova NanoSEM 450). The transmission spectra of the films were measured by a double beam UV/visible (UV-4100) spectrophotometer with a wave length rang 200 nm - 800 nm and the optical band gap was measured from the transmission spectra.

X-Ray Diffraction (XRD) was utilized for the physical construction of the ZnO thin films. XRD patterns were obtained with a (Model: PANalytical X'pert PRO MRD PW 3040) single scan diffractometer with CuKα (λ = 1.54050 Å) radiation and scanning range of 2θ set between 20° and 80°. The diameter and length of ZnO nanorods were measured using field emission scanning electron microscope FESEM (Model: FEI Nova NanoSEM 450). Energy Dispersive X-Ray Spectrometer (EDX) used for quantitative detection of elements in the prepared samples.

3. Results and Discussions

3.1. Optical Properties of ZnO Nanorods

Figure 2 shows the optical transmittance spectrum of nanocrystalline ZnO nanorods at 90°C for precursor concentration 35 mM from immersion time 180 minutes annealed at 450°C for 30 minutes using UV-Visible region from 200 nm - 800 nm. The transmittance is over 80% in the visible region from 400 nm to 800 nm for all the samples. Sharp absorption edge is located at 380 nm which is due to the fact that the ZnO is a direct ban gap semiconductor. The corresponding optical band gap of ZnO thin film is estimated by extrapolation of the linear relationship between $(\alpha hv)^2$ and hv according to Equation [11].

$$\alpha hv = A(hv - Eg)^{1/2} \qquad (1)$$

where α is the absorption coefficient, hv is the photon energy, Eg is the optical band gap and A is a constant. **Figure 3** depicts the plot of $(\alpha hv)^2$ versus photon energy hv. The value of the direct optical band gap Eg is calculated from the intercept of $(\alpha hv)^2$ vs hv curve had also been plotted. The presence of a single slop in the plot suggests that the ZnO nanorod has direct and allowed transition. The band gap value of ZnO nanorod is found to be 3.3 eV which is slightly smaller to bulk ZnO (3.37 eV).

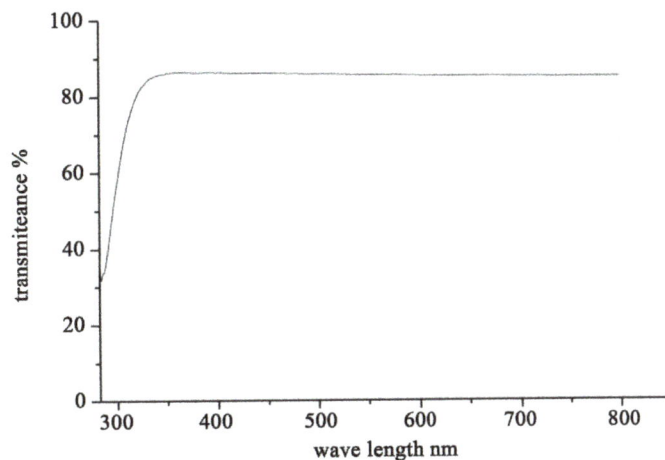

Figure 2. The transmittance spectrum of ZnO nanorods at 90°C for precursor concentration 35 mM from immersion time 180 min.

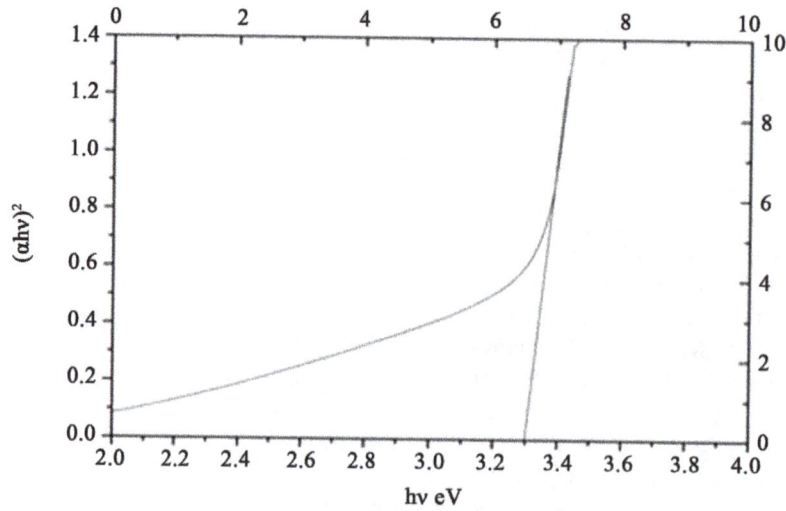

Figure 3. Plot of $(\alpha h v)^2$ vs photon energy $h v$ of ZnO nanorods.

This difference is due to the fact the values of band gap Eg depend on many factors, e.g. the granular structure, the nature and concentration of precursors, the structural defects and the crystal structure of the films. Moreover, departures from stoichiometry form lattice defects and impurity stats. Dengue Bao *et al.* [12] reported that the band gap difference between the thin film and crystal is due to the grain boundaries and the imperfection of the polycrystalline thin films. D. L. Zhange *et al.* [13] reported that this band gap difference between the film and bulk ZnO is due to the grain boundary, the stress and the interaction potentials between defects and host materials in the films.

3.2. Structural Analysis of ZnO Nanorods

Figure 4 depicts the X-Ray Diffraction (XRD) pattern of the crystal structure and orientation of the nanocrystalline ZnO nanorods deposited on glass substrate using spin coating at 2000 rpm, pre-heated at 150°C and annealed in air at 450°C. From the XRD pattern, one can clearly observe a diffraction peak at $2\theta = 34.426°$. Strong preferential growth is observed along c-axis, *i.e.* (002), suggesting that the prepared ZnO nanorods have the wurtizit structure.

The unit cell "*a*" and "*c*" of the crystalline ZnO nanorods with (002) orientation is calculated using the relation (2) and (3):

$$a = \sqrt{1/3}\,\lambda / \sin\theta \tag{2}$$

$$c = \lambda / \sin\theta \tag{3}$$

The values obtained for the unit cell $a = 3.007$ Å and $c = 5.21$ Å are in a good agreement with those reported in the JCPDS standard data (card no. 80 - 0074). The calculated parameters are given in **Table 1**.

From the XRD spectrum, grain size (D) of the film is calculated using debay scherrer formula [14].

$$D = k\lambda / \beta cos\theta \tag{4}$$

where k is a constant to be taken 0.49 [14] and, λ, β, and θ are the XRD wave length ($\lambda = 1.5406$ Å), full width at half maximum (FWHM) and Bragg angle respectively. By inserting the different values from **Table 2** in the Scherrer formula grain size of (002) oriented thin film is 44.12 nm which is same as reported in literature [15].

The dislocation density (δ), which represents the amount of defects in the crystal, is estimated from the following equation:

$$\delta = 1/D^2 \tag{5}$$

Strain (ε) of the thin film is determined from the following formula:

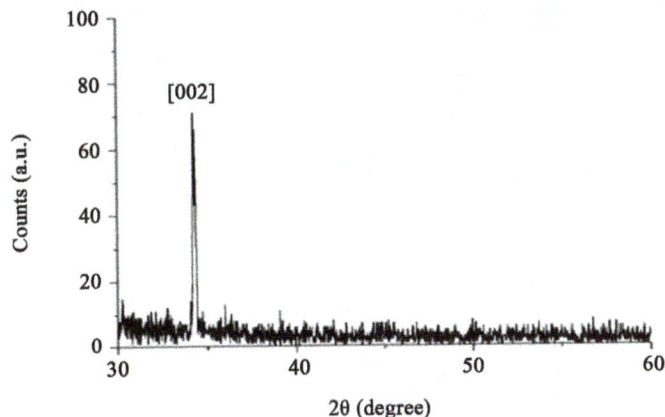

Figure 4. X-Ray Diffraction of the ZnO nanorods grown at 90°C for 180 min from the 35 mM precursor concentration.

Table 1. Lattice parameters of the ZnO nanorods.

	a (Å)		c (Å)	
Standard	Calculated	Standard	Calculated	
3.253	3.007	5.215	5.21	

Table 2. Structure parameters of the ZnO nanorods.

Plan	d (Å)	FWHM $(\beta)°$	$2\theta°$	D (nm)	$\delta \times 10^{-4}$ (nm)$^{-2}$	$\varepsilon \times 10^{-3}$
002	2.6055	0.1968	34.426	44.12	5.13	8.049

$$\varepsilon = \beta\cos\theta/4 \qquad (6)$$

The calculated structural parameters of the thin film are presented in **Table 2**.

3.3. Morphological Analysis of ZnO Nanorods

In **Figure 5**, the (FESEM) shows the average diameter (d) of the ZnO nanorods increases from (57, 64, 83, 120 and 230 nm) as the precursor concentration increase from 15, 20, 25, 30, and 35 mM, respectively, where the immersion time is fixed for 180 min at 90°C. Length of the grown ZnO nanorods is about 1 μm regardless of concentration, which indicates that changes in the precursor concentration at the fixed immersion time can affect only the diameter of the hexagonal ZnO nanorods. The rate of increase diameter of the ZnO nanorods is estimated to be approximately 34.4 nm/mM.

Length of ZnO nanorods can also be varied when the immersion time changes in the fixed concentration. **Figure 6** shows that the average length of the ZnO nanorods increases from (241, 459, 522, 820 nm and 1.2 μm) as the immersion time t increases from 60, 90, 120, 150 and 180 min, respectively, at the precursor concentration of 35 mM. Length of ZnO nanorods indicates that growth rate is 6.3 nm/min.

Figure 7 shows the (EDX) spectrum and atomic composition of the ZnO/glass (002) layers for precursor concentration 25 mM at immersion time 180 min. A description of the atomic composition of the elements in the layers is shown in percentages, as presented in the inset table in **Figure 7**, the concentration of these elements is indicated by the peaks, and clearly shows that the elements corresponding to the peaks comprised the layer. No contaminated element detected in the layers.

4. Conclusion

In this work, we have grown ZnO nanorods on glass substrates by solution processing hydrothermal method in low temperature using the spin coating technique. The structural, morphological and optical properties were in-

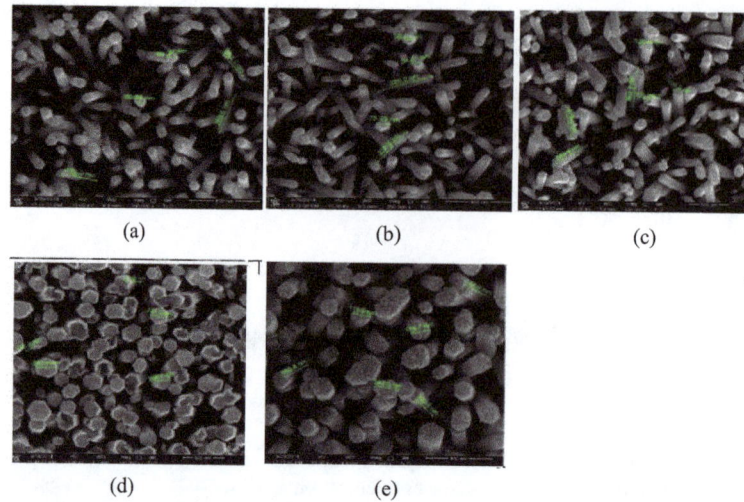

Figure 5. Surface FESEM images and diameter of hexagonal ZnO nanorods grown at 90°C various concentrations (a) 15; (b) 20; (c) 25; (d) 30; and (e) 35 mM precursor concentration for 180 min.

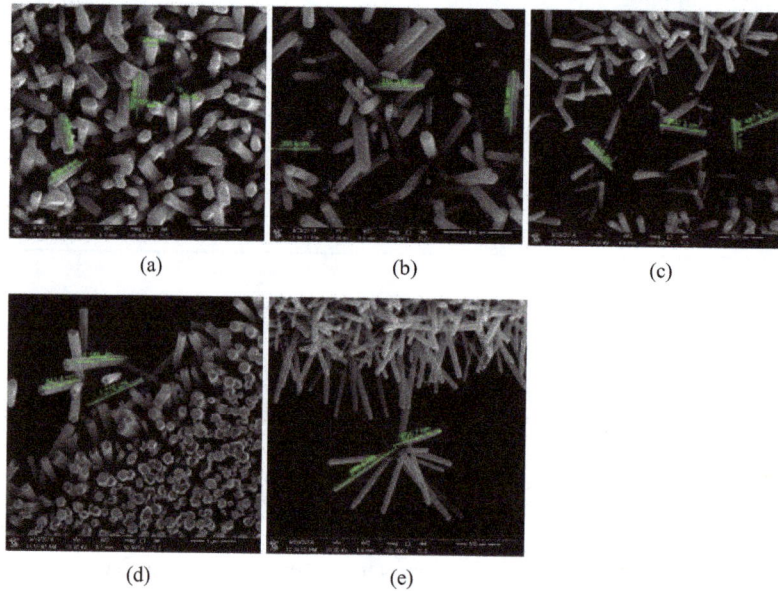

Figure 6. Surface FESEM images and length of hexagonal ZNO nanorods grown at 90°C for immersion time (a) 60; (b) 90; (c) 120; (d) 150; and (e) 180 min from precursor concentration 35 mM.

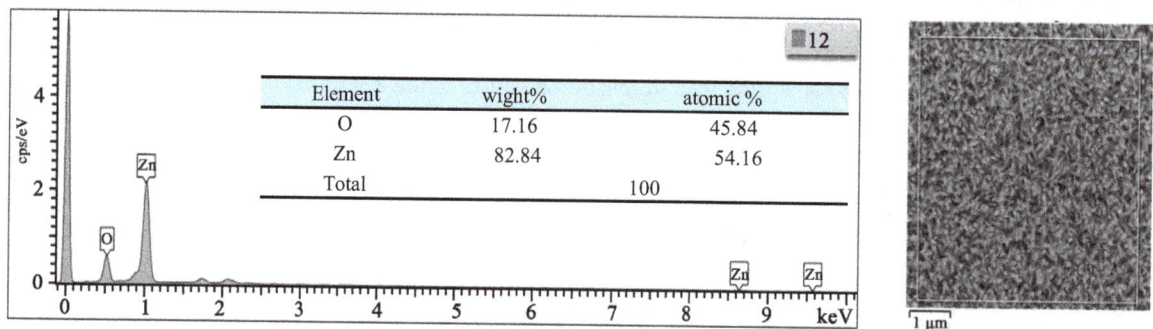

Element	wight%	atomic %
O	17.16	45.84
Zn	82.84	54.16
Total		100

Figure 7. The (EDX) of the ZnO nanorodes at 90°C for precursor concentration 35 mM from immersion time 180 min.

vestigated. The hydrothermal method is a relatively simple technique: there are many factors which affected the quality of the film. We have optimized different parameters to obtain a good crystalline structure of ZnO nanorods with intense and sharp peak. The optical transmittance is over 80% in the wave length range from 400 nm - 800 nm and the band energy band gap is found to be 3.300 eV. According to XRD results, the as deposited films exhibited a hexagonal wurtized structure with (002) preferential orientation after annealing at 400°C in air ambiance for 30 min. The XRD pattern consists of a single (002) peak which occurred due to ZnO crystals and grows along the c-axis. The grain size estimated to be 44.46 nm. The average diameter and average length of the ZnO nanorods obtained from the FESEM. The average diameter of ZnO nanorods, which are increasing from (57, 64, 83, 120 and 230 nm) as the precursor concentration increases at 90°C for immersion time 180 min, and the average length of ZnO nanorods increases from (241, 459, 522, 820 nm and 1.2 μm) when the immersion time was increased at 90°C for precursor concentration 35 mM. The (EDX) analyses of the samples clearly show that the sample prepared by above route has pure ZnO nanorod phases.

References

[1] Ku, C.H. and Wu, J.G. (2007) Electron Transport Properties in ZnO Nanowire Array/Nanoparticle Composite Dye-Sensitized Solar Cells. *Applied Physics Letters*, **91**, Article ID: 093117.

[2] Cheng, X.L., Zhao, H., Huo, L.H., Gao, S. and Zhao, J.G. (2004) ZnO Nanoparticute Thin Film, Preparation, Charectrarazation and Gas-Sensing Properties. *Sensors and Actuators B: Chemical*, **102**, 248-252. http://dx.doi.org/10.1016/j.snb.2004.04.080

[3] Kim, J.Y., Jeong, H. and Jang, D.J. (2001) Synthesis of a Graphene-Carbon Nanotube Composite and Its Electrochemical Sensing of Hydrogen Peroxide. *J. Nanoplast. Res.*, **13**, 6699-6706.

[4] Chacrabarty, S. and Dutta, B.K. (2004) Photocatalytical Degradation of Model Textile Dyes in Wastewater Using ZnO as Semiconductor Catalyst. *Journal of Hazardous Materials*, **112**, 269-278.

[5] Huang, M.H., Mao, S., Fieck, H., Yan, H., Wu, Y., Kind, H., Weber, E., Russo, R. and Yang, P. (2001) Room Temperature Ultraviolet Nanowirenano Lasers. *Science*, **292**, 1897-1899.

[6] Saito, N., Haneda, H., Sekiguchi, T., Ohashi, N., Sakaguchiandk, I. and Koumoto, K. (2002) Low Temperature of Light-Emitting Zinc Oxide Micropatterns Using Self-Assembled Monolayer. *Advanced Materials*, **146**, 418-421. http://dx.doi.org/10.1002/1521-4095(20020318)14:6<418::AID-ADMA418>3.0.CO;2-K

[7] Krunks, M. and Mellicov, E. (1995) Zinc Oxide Thin Films by the Spray Pyrolysis Method. *Thin Soled Filme*, **270**, 33-36.

[8] Nunes, P., Costa, D., Fortunate, E. and Martiens, R. (2002) Performenes Presented by Zinc Oxide Thin Films Deposited by R. F. Magnetron Sputtering. *Vacumm*, **64**, 293-297.

[9] Wu, J.-J. and Liu, S.-C. (2002) Low-Temperature Growth of Well-Aligned ZnO Nanarods by Chemical Vapor Deposition. *Advanced Materials*, **14**, 215-218. http://dx.doi.org/10.1002/1521-4095(20020205)14:3<215::AID-ADMA215>3.0.CO;2-J

[10] Ni, Y.H., Wei, X.W., Hong, J.M. and Ye, Y. (2005) Hydrothermal Synthesis and Optical Properties of ZnO Nanorods. *Materials Science and Engineering: B*, **121**, 42-47. http://dx.doi.org/10.1016/j.mseb.2005.02.065

[11] Caglar, M., Ilican, S. and Caglar, Y. (2009) Influence of Dopant Concentration on the Optical Properties of ZnO: In Films by Sol-Gel Method. *Thin Solid Films*, **517**, 5023-5028.

[12] Bao, D., Gu, H. and Kuang, A. (1998) Sol-Gel Derived C-Axis Oriented ZnO Thin Films. *Thin Solid Films*, **32**, 47-39.

[13] Zhang, D.L., Zhang, J.B., Wu, Q.M. and Miao, X.S. (2010) Microstructure, Morphology, and Ultraviolet Emission of zinc Oxidenanocrystallin Films by the Modified Successive Ionic Layer Adsorption and Reaction Method. *Journal of the American Ceramic Society*, **93**, 3284-3290.

[14] Khan, Z.R., Zulfequar, M. and Khan, M.S. (2010) Optical and Structural Properties of Thermally Evaporate Cadmium Sulfide Thin Films on Silicon (100) Wafer. *Materials Science and Engineering: B*, **174**, 145-149.

[15] Foo, K.L., Hashim, U., Muhammad, K. and Voon, Ch.H. (2014) Sol-Gel Synthesized Zinc Oxide Nanorods and Their Structural and Optical Investigation for Optoelectronic Application. *Nanoscale Research Letters*, **429**, 1-10.

18

A Cold Fusion-Casimir Energy Nano Reactor Proposal

Mohamed S. El Naschie

Department of Physics, University of Alexandria, Alexandria, Egypt
Email: Chaossf@aol.com

Abstract

Using a compact heap of Fullerene nano particle moduli of a nano matrix device we propose that by maximizing the Casimir forces between these particles as a desirable effect, we can achieve a gradual rather than a sudden implosion pressure. This we expect will result in a mini holographic universe from which energy can be extracted in a way constituting a nano energy reactor functioning effectively on a hybrid principle somewhere between a Casimir effect and a cold fusion process.

Keywords

Nano Reacter, Casimir Effect, Cold Fusion Process

1. Introduction

The physical E-infinity theory [1]-[4] which is based on transfinite set theory [4] [5] as well as substantial numbers of theorems and techniques borrowed from K-theory [7], E-infinity mathematical theory [6], sub-factors, knot theory, von Neumann continuous geometry [4] and in particular A. Conne's noncommutative geometry [8] holds that all forms of energy and matter represented by Einstein's iconic equation $E = mc^2$ are nothing but the zero point energy fluctuations of the real vacuum of spacetime [9]-[11]. In other words, in our dialectic philosophy we have a place for a non-materialistic matter described precisely by set-theoretical operations [5] [8] [10]. We habitually call this for us esoteric matter, spacetime but it can be called more mathematically, an infinite dimensional, hierarchal empty set [4]-[11]. The objective of the present paper is to show how we can extract energy from this spacetime by building a nano reactor that in a sense combines Casimir effect [10] [11] with cold fusion [12] [13]. We should stress from the outset that the literature upon which the present work is based is vast and our references [1]-[20] are the minimum required for a concise presentation. In addition we give in Appendix details of how to construct our Cantorian quantum spacetime. We draw the attention of the reader to the fact that our Banach-Tarski sphere decomposition (**Figure 1**) plays in the present theory a similar role to Schwinger's source [18] [21].

2. Preliminary Remarks

Let us start by recalling some facts about our Cantorian spacetime [4] [10] [11]. It is described by $\langle D \rangle = \langle -2, \phi^3 \rangle$ where $\phi = (\sqrt{5}-1)/2$ and $\langle -2 \rangle$ is the expectation topological dimension of the projection while $\langle \phi^3 \rangle$ is the expectation value of the Hausdorff dimension of the same projection. From the above analysis and in view of previously obtained results [1] [2] [11], we notice immediately that ϕ^3 is the cobordism of the quantum wave given by $D_Q(W) = (-1, \phi^2)$ as well as being the latent Casimir topological force of spacetime all apart of being the inverse of its Hausdorff dimension, namely [4] [10] [11]

Figure 1. Cantorian spacetime of E-infinity theory which is considered here to model our actual spacetime may be envisaged advantageously as in this artist impression. This is basically a two dimensional projection in which each of the larger balls (circles) are a zero set $(0;\phi)$ representing the quantum particle while the surface (circumference) represents the empty set $(-1, \phi^2)$ which in turn represents the quantum wave [1] [17]. This wave is then surrounded by an infinite hierarchy of smaller (fractal) spheres (surfaces) which may be seen as the emptier set $(-2, \phi^3)$, *i.e.* the surface of the empty set quantum wave. Remarkably the average set of all zero and empty sets is an expectation value equal $\langle -2; \phi^3 \rangle$. In other words $\langle -2; \phi^3 \rangle$ is our quantum spacetime which is the cobordism of the quantum wave which in turn is the cobordism of the quantum particle floating and propagating with the help of its wave in our Cantorian E-infinity spacetime [1] [2] [10] [11]. It is likewise remarkable that ϕ^3 is simultaneously equal to the topological Casimir force as well as the topological mass of the ordinary energy of spacetime. Thus all matter and energy manifestations in our cosmos are essentially a manifestation of the zero point energy of the vacuum of spacetime. To obtain Einstein maximal energy density we just need to find first the topological energy density by adding Kaluza-Klein $D = 5$ to ϕ^3 of the spacetime vacuum and find the fractal Kaluza-Klein dimension $5 + \phi^3$ then multiply this with the average Cantorian interval speed of light $c = \phi$ squared. The result is $(5+\phi^3)\phi^2 = 2$. Inserting in Newton's kinetic energy one finds $E(\text{Einstein}) = \frac{1}{2}m(v \to c)^2(2) = mc^2$ exactly as should be. The preceding explanation amounts to a paradigm shift in physics where the totally empty vacuum of spacetime is taken as fundamental and everything else is derivable from it. To prove this point was a dream of Serbian American inventor N. Tesla who died in 1943 as well as Soviet physicist A. Zakharof. In fact in his later years Nobel Laureate J. Schwinger was a champion of cold fusion [12] which comes very near to our present concept of a Casimir-nano energy reactor [10] [11]. We also stress that we are making tacit use of the Banach-Tarski decomposition theorem as a Schwinger-like source [18] [21].

$$D(H) = 1/\phi^3 = 4 + \phi^3 = 4 + \cfrac{1}{4 + \cfrac{1}{4 + \cdots}} = 4.23606799 \simeq 4 \qquad (1)$$

Furthermore, it cannot pass unnoticed that ϕ^3 is the topological mass of the ordinary part of the spacetime topological energy $E_T(O) = (\phi^3)(\phi^2)/2 = \phi^5/2$, which leads to the ordinary energy density $E(O) = mc^2(\phi^5/2) \cong mc^2/22$ [10] [11]. Similarly dark energy is clearly the part of the topological energy of space associated with the surface of the zero set quantum particle, *i.e.* the empty set quantum wave. It is given as the product of the topological dark mass 5 which is the dimensions of Kaluza-Klein spacetime additive topological volume with the square of the topological Cantorian interval speed of light C_T (Topological) = ϕ. That means $E_T(D) = 5\phi^5/2$ which leads to $E(D) = (5\phi^2/2)mc^2 \cong mc^2(21/22)$ [10] [11]. After all this E-infinity and set theoretical manipulation, all that we really want to show and use is the fact that Casimir local effect and dark energy global effect are essentially the forces which produce and fuse atomic and sub-atomic particles together. A. Zakharov described this intuitively as elementary particles floating in a spacetime spanned by these particles [13] which in turn are made of the fractal spheres spacetime [14]-[16] as imagined in **Figure 1**. This could also be viewed as a spacetime underpinning a transfinite version of 'Hooft's cellular automata quantum theory [20].

3. Fission, Fusion and Casimir Gradual Implosion

What we just said in the preceding section may strike one at first glance as difficult to grasp since fusion, as we know it from a fusion bomb, *i.e.* hydrogen bomb requires a huge amount of heat which can be provided in practice only via a fission bomb, *i.e.* an atom bomb [14]. From such a perspective cold fusion would seem like an unrealistic dream that can take place in nature only on the sun and similar hot stars or in a plasma [12] [13]. However we are proposing to manipulate the situation on the far smaller energy of nano measure spacetime, which is far less "hot" than the Planck or the Compton measure of elementary particles [9]. Thus we are dealing in effect with the Casimir energy part with a twist. To understand what we mean we go back to the fission weapon [14]. In this case ignition takes place automatically at a critical mass [14]. That way one uses either the gun method where two sub-critical masses are shot together or the second well known method, which is the implosion method when a shockwave changes the density of a non-critical mass suddenly making it critical [14]. Here we propose using nano particles to cause a gradual controlled low speed implosion of a heap of suitable nano particles to maximise the Casimir effect leading to a semi cold fusion on the level of the resolution of a quantum field theory of nano particles so that large amounts of the so produced far higher than usual Casimir energy which will escape to the surface of the heap can be extracted [10]-[13]. In other words, we are creating mini universes resembling moduli of a mini multiverse [9] giving us in effect a cold fusion Casimir nano energy reactor [10]-[13]. In this regard two well known theoretical effects will affect the real physical situation. The first is the physics and mathematics of surfaces and the 96% Dvoretzky measure concentration [15]. The second is the artificial high dimensionality of fractal sphere packing [16]. In what follows we would like to expand the preceding highly condensed ideas.

4. Phase One: Preliminary Design of a Cold Fusion Casimir Nano Energy Reactor

Let us leave the mathematical modelling of our reactor aside for now and ask ourselves how we will go on to build our reactor guided by the insight we gained so far. In practical terms it will all boil down to filling a small spherical shell with a huge amount of nano particles of various size and mixing it randomly and experiment with it to find an optimal density design. We expect to extract heat from this model by joining many such spaces in a fractal tree or network resembling a spacetime matrix which is, in the end analysis, our nano reactor [10]-[13]. In the most simplistic of terms we identify the things which we have to normally eliminate from a nanotechnological devices and just do the opposite here, *i.e.* maximizes what we normal minimize, namely the Casimir effect.

5. The Total Maximal Energy Locked in Our Fractal Kaluza-Klein Spacetime

To understand the theoretical scheme behind our design we need to understand the following. The entire topological energy density of the universe is the product of a fractal Klein-Kaluza spacetime dimension $5 + \phi^3$ and the topological dimension of the empty set ϕ^2 [17]

$$E(\text{topological}) = (5 + \phi^3)\phi^2.$$

(2)

Inserting in Newton's formula one finds [17]

$$E(\text{Einstein}) = \frac{1}{2}m(v \to c)^2(5 + \phi^3)\phi^2 = mc^2(2/2) = mc^2.$$

(3)

On the other hand the topological mass of dark energy $m = 5$ and the topological mass of ordinary energy $m = \phi^3$ could be either $m = 5$ or $m = \phi^3$ by union, namely $m = 5 + \phi^3$ and with $c = \phi$. This gives $E(\text{Einstein topological}) = (5 + \phi^3)\phi^2$ and $E(\text{Einstein}) = \frac{1}{2}(2)mc^2 = mc^2$ as anticipated [17]. In other words, in all the time since 1915, Einstein's beauty harboured in E the quintessence of the quantum theory dualism of wave and particle [17]. To end this section let us show how to arrive to $D = 5 + \phi^3$ from monadic quantum particle zero set and quantum wave empty set [10] [17]. Let us see first how the two dimensional projection of spacetime is constructed from the union of infinitely many sets for positive topological dimensions for $n = 0$ to $n =$ plus infinity. This is a simple summable infinite series

$$\sum_{n=0}^{+\infty} \phi^n = 1/(1 - \phi) = (1/\phi)^2 = 2 + \phi.$$

(4)

For the ground state vacuum we have to do the same for $n = 0$ to $n =$ minus infinity. This is then

$$\sum_{n=0}^{-\infty} (1/\phi)^n = 1/(1 - \phi) = (1/\phi)^2 = 2 + \phi.$$

(5)

The total sum is thus our fractal Klein-Kaluza dimensionality

$$D(k - k) = (2)(2 + \phi) = 5 + \phi^3.$$

(6)

From this master expression successive reinterpretation leads to various fundamental results in an incredibly simple way. First taking on rational approximation $2 + \phi \to 2.5$ we find that $2(2 + \phi)$ goes to the classical $D = 5$ Kaluza-Klein theory. Second we counted the unit border $\phi^0 = (1/\phi)^0 = 1$ twice as should be from one view point, however from another point of view it should be only once relative to an observer living in positive dimension so that we have effectively a bosonic space with $D = 4 + \phi^3$. Finally taking only the fractal set and disregarding the unit set, we have $5 + \phi^3 - 2 = 3 + \phi^3$ which upon rationalization, gives us our familiar $D = 3$ classical dimension.

6. Why E-Infinity and the Reasonable Effectiveness of Highly Structured Theories

Any reflective mind, whether one who occasionally uses and contributes to physical E-infinity theory or one from its inner wheels, must at a certain point ask himself why is this theory so simple mathematically and yet miraculously effective physically [1] [2] [4]. There may be more than one answer and more than one view point to explain this remarkable fact. As far as the present author is concerned, for the time being and awaiting a may be better explanation in the future, the following rationale may be partially satisfactory to a point. First E-infinity theory is rooted in the powerful topological global method even though it gives numerical exact results [4]. However these exact solutions and the associated exact numeric are basically due to the involved highly structured nature of the theory. Our second point is that there may be at least two main aspects of the preceding conclusion, namely the golden mean highly structured rings [6] upon which our computation is completely founded. In addition an extension of this well known aspect of E-infinity mathematical theory, the physical E-infinity Cantorian spacetime theory is itself high structured conceptually. This highly structured physical conception is what allows us to move from one physical aspect to another with almost a miraculous ease and simplicity. On the most fundamental level ϕ represents the Hausdorff dimension of the zero set [4] [11]. Going up the ladder one step higher to physics, this zero set is the model for the pre-quantum particle. Similarly the empty set is the pre-quantum wave and has a Hausdorff dimension equal ϕ^2. Now the Casimir plates gap enclose between them a quasi empty set in the limit while the virtual photons outside resemble pre-quantum particles. That is why we have a topological Casimir pressure equal to the difference between the empty set ϕ^2 and the zero set ϕ giving us $1 - \phi^2 = \phi^3$. On the other hand to be either a particle or a wave is simply $\phi + \phi^2 = 1$ which is the maximal possible probability belonging to the unit set because ϕ is also the topological probability of finding a Cantor point in a random Mauldin-Williams uniformly random Cantor set. The dimension of a manifold supporting such topological probability is a quotient space whose dimension is the ratio of $\phi + \phi^2$ to $(\phi)(\phi^2)$ which means $1/\phi^3 = 4 + \phi^3$, the well known expectation value of our E-infinity Cantorian spacetime which was developed based on transfinite set theory and Finkelstein's quantum sets [19]. The reader can now imagine how this web of expected and unexpected relations connecting initially seemingly unrelated objects arises as a highly structured and coherent theory in the golden mean resolves many problems in high energy physics and cosmology in the incredibly simple way we described earlier on. We should emphasize the role of the marvellous mathematical device invented by Felix Hausdorff [4] [9]. It is a generalization of the topological dimension to transfinite point set but it is equally a measure of disorder complexity and entropy. Consequently it plays a role in measuring the energy density as well as the mass which made it possible to find a new interpretation of Einstein's $E = mc^2$ [17]. To put it in a single honest sentence, the physical E-infinity theory and its direct connection to K-theory, the mathematical E-infinity theory as well as n-category theory is far cleverer than the present author who is supposed to have started it. I do not think one can say more than that.

7. Conclusion

A spacetime zero point vacuum nano reactor is outlined conceptually based on quantum set theory [19] and E-infinity theory [4] [6] [11]. The crucial step in the actual design is the realization that spacetime itself is virtually an infinite reservoir of cosmic energy and that the experimentally confirmed Casimir effect is the local manifestation of fractal Cantorian spacetime fluctuation ϕ^3 contained in its Hausdorff core dimension $D = 4 + \phi^3$ where $\phi = (\sqrt{5} - 1)/2$ and $D = 4$ are the Einstein spacetime topological dimension of special and general relativity. On the other hand dark energy, which derives from the observed accelerated cosmic expansion, is nothing but the global accumulation of the local Casimir effect [10] [11]. As in the electric Faraday cage, the information of a black hole and electrostatics is all surface rather than bulk bounded phenomena [9], the theoretical reason is deeply related to the measure concentration predicted for high-dimensional Banach-like manifolds of Dvoretzky's celebrated theorem [15]. Consequently our nano reactor design is based on a moduli unit mini holographic universe assembled into a matrix in a way to produce artificially high dimensionality and extracting energy from its surface. The optimal high dimensionality is $5 + \phi^3$ where five is the dimensionality of Kaluza-Klein spacetime and ϕ^3 is the universal fluctuation of spacetime. The inside of each moduli is basically a self similar matrix created by fractal packing of quasi spherical nano particles. In this way the Casimir effect will be maximized creating slow imploding pressure which increases the Casimir force further still and makes a cold fusion environment possible enhancing our reactor and making it *de facto* a hybrid Casimir-cold fusion-dark energy nano reactor [10]-[12].

References

[1] He, J.-H. (2014) A Tutorial Review on Fractal Spacetime and Fractional Calculus. *International Journal of Theoretical Physics*, **53**, 3698-3718. http://dx.doi.org/10.1007/s10773-014-2123-8

[2] Auffray, J.-P. (2014) E-Infinity Dualities, Discontinuous Spacetimes, Xonic Quantum Physics and the Decisive Experiment. *Journal of Modern Physics*, **5**, 1427-1436. http://dx.doi.org/10.4236/jmp.2014.515144

[3] Nottale, L. (1996) Scale Relativity and Fractal Spacetime: Application to Quantum Physics, Cosmology and Chaotic Systems. *Chaos, Solitons & Fractals*, **7**, 877-938. http://dx.doi.org/10.1016/0960-0779(96)00002-1

[4] El Naschie, M.S. (2004) A Review of E-Infinity and the Mass Spectrum of High Energy Particle Physics. *Chaos, Solitons & Fractals*, **19**, 209-236. http://dx.doi.org/10.1016/S0960-0779(03)00278-9

[5] El Naschie, M.S. (2013) The Quantum Gravity Immirzi Parameter—A General Physical and Topological Interpretation. *Gravitation and Cosmology*, **19**, 151-155. http://dx.doi.org/10.1134/S0202289313030031

[6] May, J.P. (1977) E-Infinity Ring Spaces and E-Infinity Spectra. *Lecture Notes in Mathematics*, Springer, Berlin.

[7] Witten, E. (1998) D-Branes and K-Theory. *Journal of High Energy Physics*, **12**, 1-35. http://dx.doi.org/10.1088/1126-6708/1998/01/001

[8] Connes, A. (2000) Noncommutative Geometry Year 2000. Geometric and Functional Analysis. Special Volume, Birkhauser-Verlag, 481-599. http://dx.doi.org/10.1007/978-3-0346-0425-3_3

[9]   Penrose, R. (2004) The Road to Reality. Jonathan Cape, London.

[10] El Naschie, M.S. (2015) Kerr Black Hole Geometry Leading to Dark Matter and Dark Energy via E-Infinity Theory and the Possibility of Nano Spacetime Singularity Reactor. *Natural Science*, **7**, 210-225. http://dx.doi.org/10.4236/ns.2015.74024

[11] El Naschie, M.S. (2015) The Casimir Topological Effect and a Proposal for a Casimir-Dark Energy Nano Reactor. *World Journal of Nano Science and Engineering*, **5**, 26-33. http://dx.doi.org/10.4236/wjnse.2015.51004

[12] Schwinger, J. (1994) Cold Fusion Theory: A Brief History of Mine. A Talk Read in an Evening Session by Eugene Mallove at *the Fourth International Conference on Cold Fusion ICCF*4, Maui, 6-9 December 1994. Printed Online by Infinity Energy—The Magazine of New Energy Science & Technology (2014-2015). http://www.infinite-energy.com/iemagazine/issue1/colfusthe.html

[13] Jiang, X.L., Zhou, X.P. and Peng, W.M. (2014) Extraction of Clean and Cheap Energy from Vacuum. *Materials for Renewable Energy & Environment*, **2**, 467-471.

[14] El Naschie, M.S. (1999) From Implosion to Fractal Spheres. A Brief Account of the Historical Development of Scientific Ideas Leading to the Trinity Test and Beyond. *Chaos, Solitons & Fractals*, **10**, 1955-1965. http://dx.doi.org/10.1016/S0960-0779(99)00030-2

[15] El Naschie, M.S. (2015) Banach Spacetime-Like Dvoretzky Volume Concentration as Cosmic Holographic Dark Energy. *International Journal of High Energy Physics*, **2**, 13-21. http://dx.doi.org/10.11648/j.ijhep.20150201.12

[16] El Naschie, M.S. (2005) On 336 Kissing Spheres in 10 Dimensions, 528 P-Brane States in 11 Dimensions and the 60 Elementary Particles of the Standard Model. *Chaos, Solitons & Fractals*, **24**, 337-457. http://dx.doi.org/10.1016/j.chaos.2004.09.071

[17] El Naschie, M.S. (2014) From $E = mc^2$ to $E = mc^2/22$—A Short Account of the Most Famous Equation in Physics and Its Hidden Quantum Entanglement Origin. *Journal of Quantum Information Science*, **4**, 284-291. http://dx.doi.org/10.4236/jqis.2014.44023

[18] Schwinger, J. (1975) Casimir Effect in Source Theory. *Letters in Mathematical Physics*, **1**, 43-47. http://dx.doi.org/10.1007/BF00405585

[19] Finkelstein, D. (1996) Quantum Relativity. Springer, Berlin. http://dx.doi.org/10.1007/978-3-642-60936-7

[20] Hooft't, G. (2014) The Cellular Automata Interpretation of Quantum Mechanics. A View on the Quantum Nature of Our Universe, Compulsory or Impossible? http://arxiv.org/abs/1405.1548

[21] El Naschie, M.S. (1995) Banach-Tarski Theorem and Cantorian Micro Space-Time. *Chaos, Solitons & Fractals*, **5**, 1503-1508. http://dx.doi.org/10.1016/0960-0779(95)00052-6

Appendix

How to build spacetime from quantum particles and quantum wave and visa versa

Let us start from von Neumann-Connes' dimensional function for the x quotient noncommutative space [8] of which the Klein-Penrose fractal tiling universe [8] [9] is a generic quasi manifold

$$D = a + b\phi \tag{1}$$

where $a, b \in Z$ and $\phi = (\sqrt{5} - 1)/2$. Inserting the appropriate Z in D one finds the corresponding dimensions $D \equiv D_{(T, H)}$ where T is the Hausdorff dimension of a bijected Menger-Urysohn topological dimension T. Consequently for the zero set pre-quantum particle one finds [1] [11]

$$D_o = 0 + (1)(\phi) = \phi \equiv D(0; \phi). \tag{2}$$

On the other hand the quantum wave which is the surface of the quantum particle is given by the cobordism empty set D_{-1} of the zero set D_0. Therefore one finds [1] [11]

$$D_o = 1 + (-1)(\phi) = \phi^2 \equiv D(-1; \phi^2). \tag{3}$$

Proceeding in the same manner one finds an infinite number of successive empty and emptier still sets given by [1] [4] [11]

$$\sum_{n=2}^{\infty} \phi^n = \phi^2 + \phi^3 + \cdots = \phi^2 \left(1 + \phi + \phi^2 + \cdots\right) = \left(\phi^2\right)\left(\frac{1}{1-\phi}\right) \equiv 1. \tag{4}$$

Adding to this the quantum pre-particle ϕ we have $1 + \phi$. In addition we should include the zero dimensional state given by $\phi^0 = 1$ leading to a gross sum of $1 + 1 + \phi = 2 + \phi$. This is the basic Hausdorff dimension for three ($n = 3$) topological Menger-Urysohn dimension as given by von Neumann-Connes dimensional function [8] or alternatively by the equivalent bijection formula of E-infinity [1] [4]

$$d_c^{(n)} = (1/\phi)^{n-1} \tag{5}$$

Which for $n = 3$ gives us the same previous result, namely

$$d_c^{(3)} = (1/\phi)^{3-1} = (1/\phi)^2 = 2 + \phi = 2.618003989\cdots \tag{6}$$

Now we can contemplate two possibilities to gauge $2 + \phi$. The first is to measure it in terms of the pre-quantum particle ϕ and find the dimension of the core of E-infinity spacetime which is the Hausdorff dimension corresponding to $n = D_{(T)} = 4$, namely [4] [11]

$$D_{(\text{H-particle})} = \frac{2+\phi}{\phi} = (1/\phi)^3 = (1/\phi)^{4-1} = 4 + \phi^3 = 4 + \cfrac{1}{4 + \cfrac{1}{4 + \cdots}} \tag{7}$$

The second possibility is to gauge $2 + \phi$ in terms of the pre-quantum wave [1] [4] [11]

$$D_{(\text{H-wave})} = \frac{2+\phi}{\phi^2} = (1/\phi)^3 = (1/\phi)^{5-1} = 4 + \phi^4 = \bar{\alpha}_o/20 \tag{8}$$

where $\bar{\alpha}_o = 137 + k_o$, $k_o = \phi^5 (1 - \phi^5)$ is the inverse electromagnetic fine structure constant and ϕ^5 is Hardy's quantum entanglement. On the other hand, realizing the evident fact that a spacetime dimensionality must be consistent with the particle-wave duality then this dimensionality must be that given by the union of $4 + \phi^3$ and $\bar{\alpha}_o/20$, namely [4] [11]

$$D(\text{H-particle, wave}) = \left(4 + \phi^3\right) + \left(\bar{\alpha}_o/20\right) = 11 + \phi^5 = 11 + \cfrac{1}{11 + \cfrac{1}{11 + \cdots}} \tag{9}$$

This is simply the dimensionality of a fractal Witten's M-theory [4] [7] or a fractal super gravity spacetime. This fractal M-theory clearly supports the particle-wave duality of quantum mechanics. It is now extremely insightful to note the following: since the pre-quantum wave $(-1, \phi^2)$ is the surface of the pre-quantum particle $(0, \phi)$ and $(-2, \phi^3)$ is the surface of the pre-quantum wave, then the average ambient spacetime in the same

2D holographic boundary [20], *i.e.* $1/(4+\phi^3) = \phi^3$ is simply the same as the surface of the pre-quantum wave on average. Remembering that ϕ^3 is the topological mass of the ordinary energy as well as being the latent topological Casimir force, one can easily realize that dark energy is a Casimir global energy and that a hybrid energy form can stem from this unity making the idea of cold fusion not as outlandish as some used to think. Consequently Nobel Laureate Julian Schwinger (see **Figure A1**) was not that far off in his warm enthusiasm for cold fusion [12]. Note also that Schwinger source [18] is replaced in the present theory by our Banach-Tarski theorem on sphere decomposition [21].

**Julian Schwinger
(1918-1994)**

Figure A1. Julian Schwinger, Nobel laureate in physics and an extraordinary proponent of cold fusion. Prof. Schwinger went as far as resigning from the American Physical Society in defence of scientific freedom [12]. In the resignation letter of Prof. Schwinger he lamented what he considered censorship of science exemplified by Editors of famous mainstream journals rejecting papers based upon pretentious and unfair reports by anonymous referees. On the whole Prof. Schwinger was a careful thinker and superior mathematician compared to his colleagues and co-Nobel Prize winner of the same year, the equally extraordinary Richard Feynman. The usual cheap shots like crackpot and old wood or numerologist could never be applied to this extraordinarily intelligent and super rational thinker. By being open minded like he was, Prof. Schwinger may have provoked conventional thinkers because he did not think very highly of the extensive use of Feynman's diagrams and by introducing an alternative theory to quantum field theory known as source theory [19] which he applied similar to the present work to the Casimir effect. In fact our Banach-Tarski theorem plays in our theory the role of a Schwinger source [18] [21].

19

Angular Dependence of the Second Harmonic Generation Induced by Femtosecond Laser Irradiation in Silica-Based Glasses: Variation with Writing Speed and Pulse Energy

Jing Cao, Bertrand Poumellec, François Brisset, Anne-Laure Helbert, Matthieu Lancry

Institut de Chimie Moléculaire et des Matériaux d'Orsay (ICMMO), CNRS-Université Paris Sud, Université Paris Saclay, Bât.420, Campus Orsay, 91405 Orsay, France
Email: bertrand.poumellec@u-psud.fr

Abstract

To control second harmonic generation (SHG) in silica-based glasses is crucial for fabricating photonic devices, such as frequency doubling waveguides. Here, we investigated SHG of laser induced nonlinear optical crystals in silica-based glasses, according to writing speed and pulse energy. We observed two regions with different probing laser polarization angular dependence: a) a well-defined cosine-like curve with period of 180° at low pulse energy (0.8 µJ) whatever the writing speed or at high pulse energy (1.4 µJ) with high writing speed (25 µm/s). This is accounted for by a well-defined texture for the nano crystals with their polar axis oriented perpendicular to the writing laser polarization; and b) a double cosine-like curve revealing a second texture of the crystals at high pulse energy (1.4 µJ) with low writing speed (5 µm/s) and with the polar axis oriented closer parallel to the writing laser polarization. Therefore, a SHG dependence on probing laser polarization angle may show high contrast by a correct choice of the writing speed and pulse energy. These results pave the way for elaboration of nonlinear optical devices.

Keywords

Nonlinear Optical Crystal, Glass Ceramics, Direct Writing, Ultrafast Processes, Waveguide

1. Introduction

Second harmonic generation (SHG) is a nonlinear optical effect, in which the frequency of incident light is doubled [1] [2]. Generally, SHG requires medium without inversion symmetry. In that case, breaking the centre symmetry of glass is necessary for obtaining SHG based devices. Modification of glasses by precipitation of second order nonlinear optical materials is an effective way to reach this goal [3]-[5]. On the other hand, excellent improved mechanical, optical properties can be obtained by combining glasses and crystals [6]. Recently, this field has received more and more attention because of the excellent optical properties and potential applications such as optical memory [7] and second harmonic generation waveguides [8]-[10]. To master the orientation of nonlinear optical crystal is an important topic for obtaining highly efficient optical devices [11].

Thus far, various methods have been employed to obtain highly oriented nonlinear optical crystal [12]. Heat treatment is a traditional technique; however, the spatially precipitation of oriented crystal is limited and the process is time consuming [13]. Appling electric field, the spatial crystal orientation can be improved but it commonly needs complicated facilities and electric energy [14]. Compared with the above two methods, femtosecond laser induced crystallization in glasses is a promising tool in achieving nonlinear optical crystal orientation with three-dimensional, sub-micrometre spatial resolution [9]. Because of the nonlinear nature of light-matter interaction, pulse energy of femtosecond laser could be deposited in transparent materials [15]. At certain high repetition rate (e.g. >200 kHz), thermal accumulation effect occurs and the space-time profile can be controlled [16] [17]. In addition, it is a simple and flexible process just by adjusting the parameters of femtosecond laser [9].

Recently, the orientation of nonlinear optical crystal has been controlled by adjusting the laser parameters such as writing direction [18], pulse energy [9]. However, thus far, there is no systematic investigation of the relationship between SHG and laser parameters such as pulse energy, writing speed and writing orientation [19], in particular, how to get well contrasted angular dependence of SHG with probing polarization.

Here, we investigated thus the SHG in femtosecond laser induced $LiNbO_3$-like crystal in Li_2O-Nb_2O_5-SiO_2 system. It is a commonly used nonlinear optical crystal system due to the advantageous properties such as a wide range of chemical composition for glass formation, optically transparent and easy precipitation of crystals [20].

2. Experimental Section

2.1. Femtosecond Laser-Induced Crystallization

A silica-based glass with composition of $32.5Li_2O$-$27.5Nb_2O_5$-$40SiO_2$ (mol%) was prepared by the classical melt quenching technique. Details of the sample preparation were described elsewhere [21]. A Yb-doped fiber amplifier femtosecond laser (1030 nm, 300 fs, 250 kHz repetition rate, Satsuma, Amplitude Systèmes Ltd.) was focused 300 mm (in air) below the surface of glass using an objective (numerical aperture, NA = 0.6). Samples were mounted on a computer controlled three-dimensional stage and mechanically moved during laser fabrication process to obtain crystal lines.

2.2. Writing Configuration Definition

As a matter of fact, in previous experiments [19], the angle between writing direction and writing laser polarization direction was found insufficient to differentiate experiments made with the same angle between laser polarization direction and writing direction but with different writing direction e.g. vertical or horizontal. We detected also an asymmetrical effect when we change the sense of writing [21]. Therefore, in this paper, we refer the vector orientations to a laboratory reference described in **Figure 1(a)**, independently. In details, the geometry of the problem is denoted in terms of a set of Cartesian coordinate system along the laser propagating direction. The reference of writing configuration is based on the beam specifies at Cartesian coordinate system at the origin (in black). X_0 and Y_0 axes are defined as the horizontal and vertical directions, respectively; right (up) side of beam is donated as positive and left (down) as negative. $+Z_0$ is the laser propagation direction. It changed to Cartesian coordinate system (in green) when arrived at the sample (because of the odd number of mirrors on the optical table, the sample coordinate system changed from right-handed one to left-handed one). We investigated various configurations considering different combinations of writing and laser polarization directions. Here, we take as example for the direction of writing +45° and +225° in reference to +X (in X, Y plane) with the

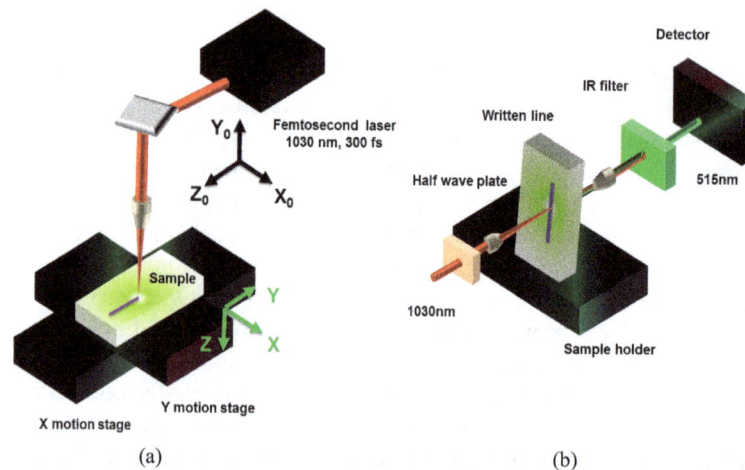

Figure 1. Schematic of experimental set up. (a) Femtosecond laser writing system, the reference of writing configuration is based on the beam specifies at Cartesian coordinate system at the origin (in black), and changed to Cartesian coordinate system when arrived at the sample (in green). N.B.: because of the odd number of mirrors on the optical table, the sample coordinate system changes from right-handed one to left-handed one; (b) Second harmonic generation measurement setup. It was measured in X, Y plane. All angles in this paper are referred to +X direction (in X, Y plane).

polarization direction parallel to Y. Femtosecond laser irradiation of the glass induced crystallization under specific conditions (repetition rate and pulse energy large enough) [9]. It is worth to note that the irradiated volume is smaller than the crystallized one. The threshold of crystallization for our glass is found at pulse energy of 0.4 μJ. Crystal seeds were produced by irradiating the sample without moving during 80 s before continuous irradiation in X, Y plane. The writing speed was varied from 1 to 25 μm/s.

2.3. Electron Backscatter Diffraction Measurement and Texture Analysis

Before to make further investigations, we observed the samples optically through a microscope with natural and polarized light and found strong index change and birefringence.

Samples were cut along the direction perpendicular to writing directions, polished and analyzed using a field-emission gun scanning electron microscope (FEG-SEM ZEISS SUPRA 55 VP) without HF etching.

Electron backscatter diffraction (EBSD) [22], is a useful tool to determine the amount of crystallized matter, the size of the crystal, their spatial distribution and their orientation if any. The first step of the measurement is the recording of the Kikuchi lines [23]. This means that the matter is crystallized at the point of electron illumination. Then, by entering space group of the expected phase, here R3c and the crystal parameters for $LiNbO_3$ [24], we can get the indexation of the Kikuchi lines using Orientation Imaging Microscopy (*OIM*™) software. This software yields several facilities for texture analysis.

The first step is to examine the Orientation Distribution Function (ODF, the angular density of crystals among the Euler space) for determining if a texture exists. Then, we can plot the Inverse Pole Figure (IPF) for displaying the preferred orientations in choosing a sample direction suitable to show the detected texture.

2.4. Second Harmonic Generation Measurement

Polarized second harmonic generation (SHG) measurement was performed in transmission mode. As illustrated **Figure 1(b)**, the fundamental beam from a Yb-doped fiber amplifier femtosecond laser system (1030 nm, 300 fs, 100 kHz repetition rate, Satsuma, Amplitude Systèmes Ltd.) was used as laser source. The polarization direction of the fundamental beam was varied by rotating a half-wave plate to obtain polarization dependent SHG signal. Sample was mounted on a stage which could be adjusted to let laser propagate perpendicular to sample X, Y plane. After passing a low-pass filter (used to block the fundamental beam), the intensity of SHG was detected by a photomultiplier. Data were recorded 5 times and we took the average value. Then intensity of data for each irradiated line was normalized at the largest value as 1. The error bars represent the standard error of mean. As a matter of fact, the absolute intensity cannot be compared because of different scattering from place to place. It is also worth to note that in a disordered material, we could expect a centre-symmetric material and from that the SHG is forbidden. This is obviously experimentally not the case and it is always the case in ceramics (a mixing between glass and nanocrystals). The explanation has been given by several authors and in particularly, Brevet *et al.* [25] have reported that the emission from a nanocrystal looses its phase relation with the other nanocrystals due to multiple scattering. Therefore, we can collect the sum of the intensity coming from each nanocrystal whatever their orientation.

3. Experimental Results

3.1. Second Harmonic Generation Properties of Irradiated Lines

For investigating the SHG properties, we recorded the SHG intensity according to two determinant laser parameters: writing speed and pulse energy.

3.1.1. Pulse Energy Effect

At low pulse energy (0.8 µJ, red curve in **Figure 2**), we obtain a well defined cosine-like curve with period of 180°. The minimum SHG intensity is obtained at an angle close to writing laser polarization direction (90°). When we increased the pulse energy to 1.4 µJ (green curve in **Figure 2**), the cosine-like curve is modified, a second maximum occurs at the place of the minimum on the simple cosine-like curve. This phenomenon becomes obvious at 1.8 µJ (blue curve in **Figure 2**). This observation is in agreement with a previous report [9].

3.1.2. Writing Speed Effect

For investigating the dependence of SHG with the writing speed, we fixed the pulse energy at 0.8 µJ and changed the writing speed. We obtained well defined cosine-like curves whatever the writing speed in **Figure 3(a)**. It is worth noting that, with the increase of writing speed, the minimum value of SHG shifts from 102° to 91°, which is closer to the writing polarization angle (90°). However, for high pulse energy (1.4 µJ), the result is quite different from the previous ones. With the decrease of the writing speed from 25 to 1 µm/s, the SHG curve changed from simple cosine-like curve (blue curve in **Figure 3(b)**) to the modified ones (green and red curves in **Figure 3(b)**).

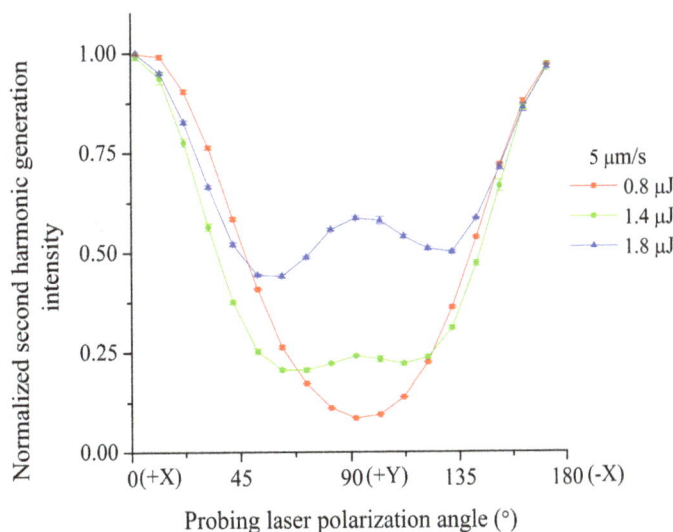

Figure 2. Normalized second harmonic generation intensity of irradiated lines as function of probing laser polarization in the X, Y plane. The variable parameter is the pulse energy. Other parameters: writing speed 5 µm/s, writing direction is along 45° direction, laser polarization is parallel to Y.

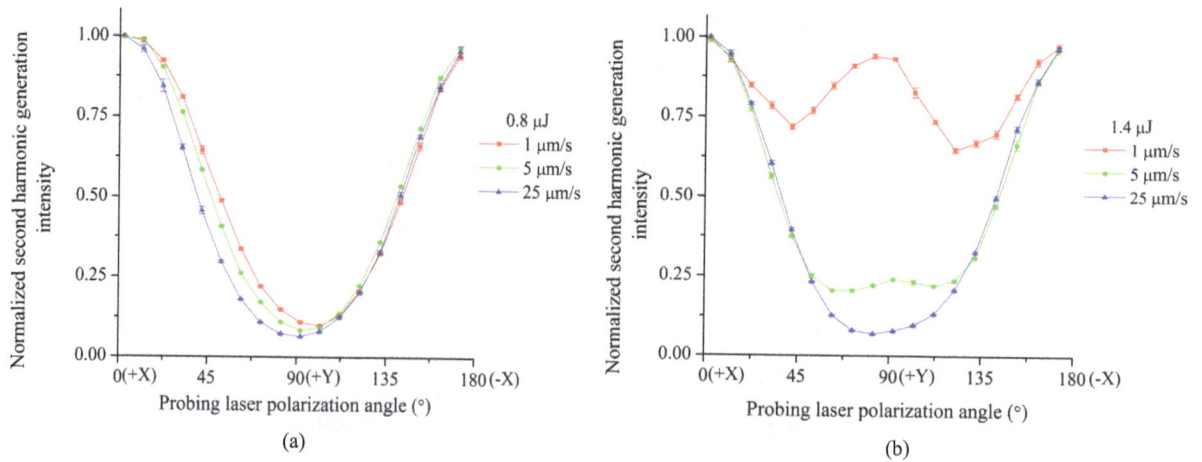

(a)

(b)

Figure 3. Normalized second harmonic generation intensity of irradiated lines as function of probing laser polarization angle. The variable parameter is the writing speed at pulse energy of 0.8 µJ for (a) and 1.4 µJ for (b). Other parameter: writing direction is along 45° direction, laser polarization is parallel to Y.

3.1.3. Writing Orientation Effect

Orientation here means the movement along a given direction. As a matter of fact, there are two orientations for a given direction. According to a previous report, the orientation of writing is peculiar for ultra-brief laser-matter interaction [26]. The glass is centre-symmetric, considering that the laser beam is XY-symmetric and the experimental geometry is also XY-symmetric; there should be no difference between the orientation of writing *i.e.* between forward direction (here 45°) and backward direction (here 225°). However, experimentally, it is not the case. In both cases, there are two textures: one with polar axis of nanocystals (<0001>) perpendicular to the writing laser polarization, X direction and one with the polar axis close to Y direction. However, the ratio between the two components and the angular position of the maxima seems slightly different. This is particularly clear in **Figure 4(a)** when writing speed is at 1 µm/s and pulse energy is 1.4 µJ. This is an Asymmetric Orientational Writing (AOW). With the increase of writing speed from 1 µm/s to 25 µm/s, the curves are cosine-like ones and the AOW effect is reduced (**Figure 4(b)**). It is worth noting that it is the first time that an AOW effect is clearly demonstrated in SHG and related to photo precipitation of crystals in a silica-based glass.

3.2. Modification of Glasses

In order to find an explanation of the above angular dependence, we have investigated the modification of glasses after irradiation. Samples were cut along the direction perpendicular to writing direction. The morphology of the cross section has been analyzed by scanning electron microscopy (SEM, **Figure 5**). At low pulse energy (0.8 µJ), a ginseng like shape laser track with a width of 1.4 µm of and length of 30 µm is obtained (**Figure 5(a)**). From the **Figure 5(b)**, a magnification of the previous figure, a rough structure with ribbon-like shape is obtained. With the increase of the pulse energy (1.4 µJ), the width of the laser trace increased to about 4 µm, another part appearing in white in **Figure 5(c)** is obtained around a rough structure.

The ODF showed that a preferential orientation has been developed. The Inverse Pole Figure (IPF, **Figure 5(d)** and **Figure 5(e)**) is used to display the crystal direction along the writing laser polarization direction. In the color coding, basic red is used for polar axis of the crystal (<0001> axis), green and blue for $\langle 0\bar{1}10 \rangle$ and $\langle 1\bar{1}00 \rangle$ axes, respectively. The inter media orientations are colored by an RGB mixture of the primary components.

The first remark we can make from IPF is that at low pulse energy (0.8 µJ, **Figure 5(d)**), nano-sized crystals have been produced. We observed that the picture is completely green and blue, indicating that $\langle 0\bar{1}10 \rangle$ and $\langle 1\bar{1}00 \rangle$ axes are parallel to the writing laser polarization direction so that the polar axis (*i.e.* <0001> axis) is perpendicular to the writing laser polarization direction. With the increase of pulse energy (1.4 µJ), sub-micro sized crystals are obtained, with nano-crystals in the core of laser track or in the tail; micro-sized crystals in the head part, especially outside of the core (**Figure 5(e)**). It should be noted that with the increase of pulse energy, the color of IPF changes from completely green to various colors, indicating that other orientations are appearing.

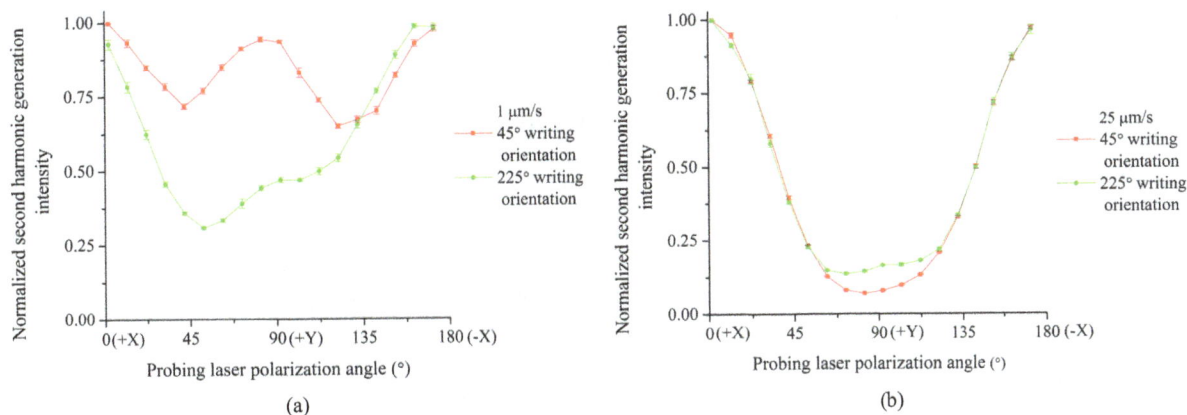

Figure 4. Normalized second harmonic generation intensity of irradiated lines as function of probing laser polarization angle. The writing orientation is the variable parameter for two writing speeds: (a) 1 µm/s and (b) 25 µm/s. Other parameters: pulse energy 1.4 µJ, laser polarization is parallel to Y.

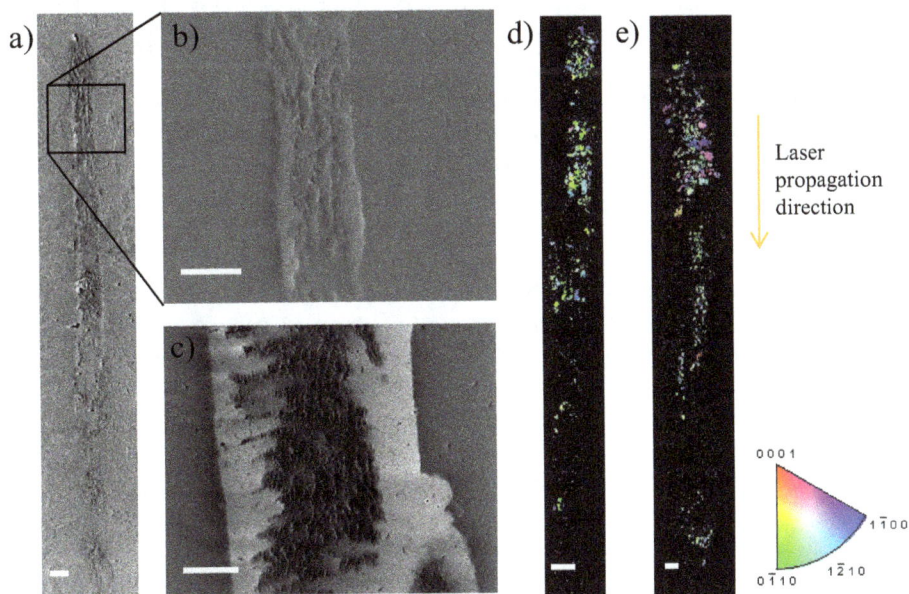

Figure 5. Structure of the laser trace by scanning electron microscopy, the section is achieved perpendicularly to the writing direction. (a) The secondary electron image for 0.8 µJ; (b) The magnification of the part framed in red in (a). (c) A similar part of the laser trace but for 1.4 µJ; (d) The corresponding Inverse Pole Figures coding crystal direction along the writing laser polarization direction for 0.8 µJ; (e) Similar information for 1.4 µJ. N.B.: the color in IPF maps (inset at the right bottom) is based on R3c space group. All scale bars are 1 µm. Other parameters: writing speed 5 µm/s, writing direction along 45° and laser polarization direction parallel to Y.

Based on the above analysis, we can defined three regions of the laser traces exhibiting different morphologies: region 1: in **Figure 5(b)** with nano-sized or sub-micro sized crystals or the core of the **Figure 5(c)**; region 2: the white part at the border of the laser trace, with micro-sized crystals (**Figure 5(c)**); region 3: for high pulse energy, the tail of the laser trace with nano-sized crystals (**Figure 5(e)**). Clearly, the texture is stronger at low pulse energy, with the acuity of the $\langle 0\bar{1}10 \rangle$ axis direction reinforced along Y.

4. Discussion

4.1. Sum up of SHG Dependence

A quantity, named anisotropy magnitude, deduced from the angular dependence of the SHG in **Figure 2** to **Figure 3** and defined as $\left(SHG_{max} - SHG_{min}\right)/\left(SHG_{max} + SHG_{min}\right)$ has been used to characterize the angular contrast. It is plotted in **Figure 6**. At the fix writing speed (5 μm/s), the anisotropy magnitude decreases with the increase of pulse energy (green line in **Figure 6**). At high speed (25 μm/s), the anisotropy magnitude is not dependent on pulse energy (blue line in **Figure 6**), whereas it is absolutely not the case when the writing speed is decreased, especially at high pulse energy (red line in **Figure 6**).

4.2. Modeling/Interpretation

We have analysed the shape of the curves in **Figure 2** on the basis of SHG angular response of the single crystal. We used a programme previously used for poling for computing the coefficients in the following expression for simulating the angular dependence of SHG intensity with the probe polarization. It deduced from the theory with the second order nonlinear tensor for the symmetry R3c attached to $LiNbO_3$ crystal, the largest SHG coefficient (d_{33} = 34.4 pm/V) is supplied in the <0001> crystal direction, *i.e.* along the polar axis. The other ones are d_{31} = 5.95 pm/V and d_{22} = 3.07 pm/V [27], $f(\theta) = a + b\cos^4(\theta) + c\sin^4(\theta)$ with a = 0.010, b = 0.033, c = −8.4 × 10^{-3}. Note that this function is maximum for θ = 0°. From this, we built another expression that may be accounted for by two populations: one randomly oriented and one textured *i.e.*

$$I_{SHG}^{Norm} = \left[\alpha + \beta f(\theta - \theta_1)\right]/\left[\alpha + \beta(a+b)\right]$$ where θ_1 is an angle shift from +X direction; α and β the weight of the two populations. We obtained a fit with α = 0.013, β = 3.2 and θ_1 = 3.4°. Result shown in **Figure 7(a)** for pulse energy of 0.8 μJ seems to indicate a large proportion of crystal with the polar axis oriented in the X direction versus non randomly oriented nanocrystals.

The same procedure has been applied to pulse energy of 1.8 μJ considering two oriented distributions. In this case, we adjusted the expression with three populations: one randomly oriented and two textured.
$$I_{SHG}^{Norm} = \left[\alpha + \beta f(\theta - \theta_2) + \gamma f(\theta - \theta_3)\right]/\left[\alpha + \beta(a+b) + \gamma(a+c)\right].$$ We obtained a reasonable fit with α = 0.0, β = 3.9, γ = 2.0, θ_2 = 1.1° and θ_3 = 94°. The result is shown in **Figure 7(b)** with one distribution with the polar axis oriented along X (at the level of 66%) and another with the polar axis oriented closely to Y (a few degrees away actually, at the level of 34%).

Because EBSD is the tool for pointing out texture, we intend to get similar information with this method.

A texture has actually been detected and is shown in **Figure 8** using pole figures. At low pulse energy (**Figure 8(a)**), we can see that the polar axis of the crystal (<0001>), is distributed perpendicular to the laser polarization direction (Y) like a fiber texture. The SHG intensity is mainly defined by the angle between the probe laser polarization direction and the polar axis. The width of this function at half maximum is just deduced from the equation $f(\theta)$ and reaches ±45°. This acceptance is represented by the insertion of a milky disk in **Figure 8(a)** meaning "not to take into account the distribution 45° around Z axis. The angular dependence plotted in **Figures 2-4** corresponds in this picture to the arc drawn from +X to −X. We see that the density shows a maximum for X but a minimum for Y as it is appearing in **Figure 2** red curve.

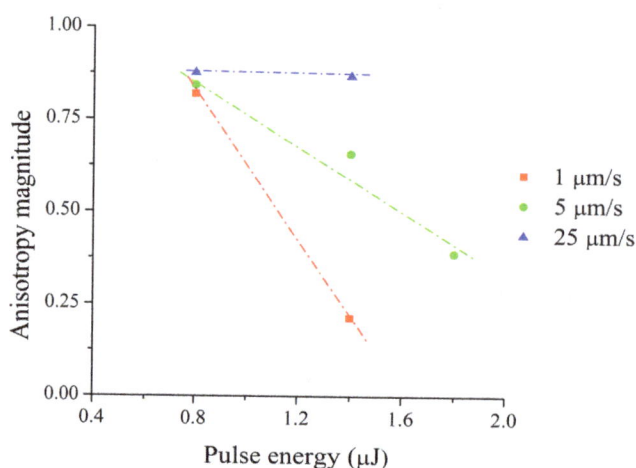

Figure 6. Anisotropy magnitude of irradiated lines computed from **Figure 2** to **Figure 3**. Here anisotropy magnitude is defined by the following ratio: $(SHG_{max} - SHG_{min})/(SHG_{max} + SHG_{min})$. Other parameter: writing direction is along 45° direction, laser polarization is parallel to Y. The lines are guides for the eyes.

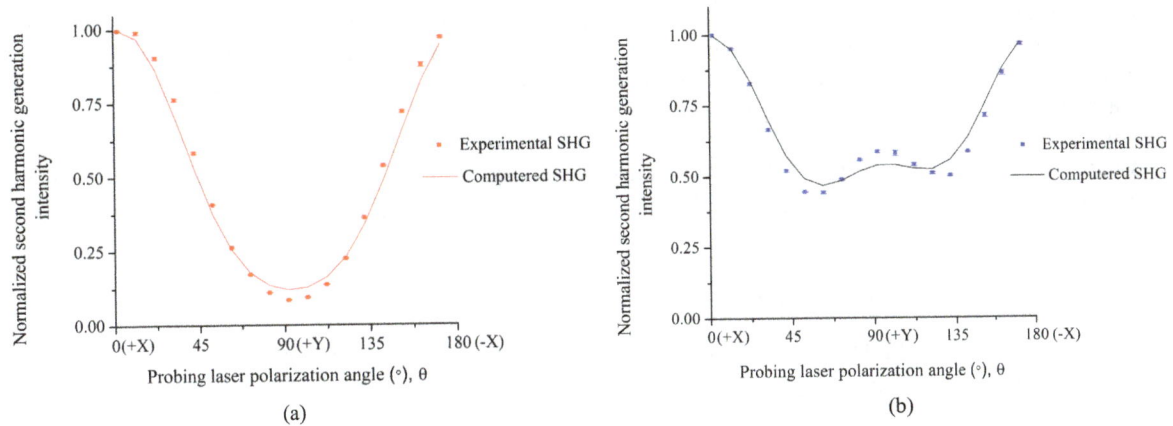

Figure 7. Experimental data and computered data of 0.8 μJ for (a) and 1.8 μJ for (b). Writing speed 5 μm/s, writing direction is along 45° direction, polarization is parallel to Y direction.

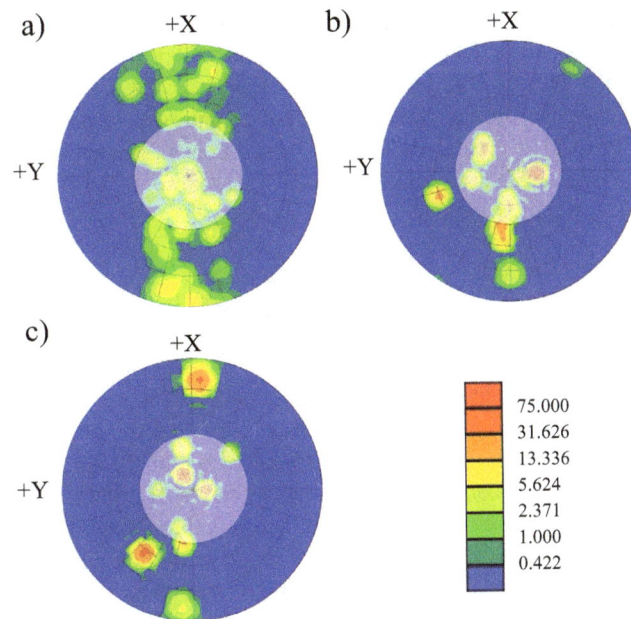

Figure 8. 0001 pole figure of texture calculated from electron backscatter diffraction (EBSD)-scans *i.e.* the polar axis from EBSD scans. At pulse energy of 0.8 μJ for (a) and 1.4 μJ for (b) and 1.8 μJ for (c), writing speed 5 μm/s, writing direction is along 45° direction, polarization is parallel to Y direction. The milky disks at the centre of the pole figures indicate the part of the distribution that should not be taken into account for SHG interpretation with probe polarization in X, Y plane (see text).

The same procedure, applied to the curves in **Figure 2** green and blue curves (1.4 and 1.8 μJ), shows two textures: one with the polar axis perpendicular to Y like for low energy and another one at another angle from +X: 112° for 1.4 μJ or 140° for 1.8 μJ. This is not exactly consistent with the fit we have obtained in angular curve fitting but we have also to note that it is not exactly the same interaction volume considered in SHG analysis (X, Y plane) and cross section analysis (135, Z plane). In the case of SHG, the measurement is performed through the sample whereas for EBSD it is performed at the surface.

From the discussion in the section above, we see that the femtosecond laser induced crystallization leads to crystals not randomly oriented but with one or two textures. This means that some forces are active for orienting the crystals during their formation. The simple preferential orientation is such that the polar axis is perpendicular to the writing laser polarization direction. We have observed this texture in many cases with several orientations of the writing or laser polarization direction providing that the pulse energy is low enough [9]. The first kind of

forces that can lead to an orientation is the thermal force due to large thermal gradient experienced in such focused irradiation. However, this can be ignored at once since in that case, there would be no effect of the laser
polarization. If we observe such an effect, this is because the light is acting. The simplest idea is a torque exerted
on a dipole of the nanocrystal during the nucleation. Assuming that the nuclei have already the structure of the
final crystals, we can notice that $LiNbO_3$ may have a spontaneous dipole (because it is ferroelectric), it has also
an induced dipole, like any crystal, but with a anisotropic susceptibility in such a way that the dipole is not al-
ways parallel to the applied electric field $\boldsymbol{E}_\omega$. In such a case, a non-oscillating torque $\breve{\Gamma}_{DC}$ is developed on the
nanocrystal (Equation (1)):

$$\breve{\Gamma}_{DC} = DCpart\left(\boldsymbol{P}_\omega \wedge \boldsymbol{E}_\omega\right), \quad \boldsymbol{P}_\omega = \varepsilon_0 \overline{\overline{\chi}} \boldsymbol{E}_\omega, \quad \overline{\overline{\chi}} \text{ is anisotropic} \tag{1}$$

with $\boldsymbol{P}_\omega$ the induced dipole, $\overline{\overline{\chi}}$ the first order susceptibility tensor, ε_0 the vacuum dielectric constant. The
relative permittivity $\overline{\overline{\varepsilon}}_r$ of a media is related to its electric susceptibility, $\overline{\overline{\varepsilon}}_r = \overline{\overline{I}} + \overline{\overline{\chi}}$, where $\overline{\overline{I}}$ is the identity
matrix.

   This torque aligns the dipole $\boldsymbol{P}_\omega$ on the electromagnetic field. But what does this alignment correspond in
the crystal to? It will be along the largest value for the susceptibility. We get these values in the literature with
some variations as $LiNbO_3$ is sensitive to non-stoichiometry [28]. They are the following that we can be deduced from the refractive index. For a crystal reference based on its principal axes, the permittivity tensor could
be written as illustrated in Equation (2) (dielectric matrix) and the principal refractive indexes n_{kk} $(k = 1, 2, 3)$
is deduced from $\varepsilon_{kk} = n_{kk}^2$. Due to its symmetry, $LiNbO_3$ is a uniaxial crystal with $\varepsilon_{11} = \varepsilon_{22} = n_o^2 \neq \varepsilon_{33} = n_e^2$
where n_o and n_e is known as ordinary and extraordinary indexes respectively. Here n_o (electric field po-
larization normal to polar axis) is greater than n_e (electric field polarization parallel to polar axis). So we can
get $\varepsilon_{\perp \text{polar axis}} > \varepsilon_{//\text{polar axis}}$. We see that the largest value is perpendicular to the polar axis of the nanocrystal and
thus this will lead to a laser polarization perpendicular to the polar axis as it is observed.

$$\varepsilon = \varepsilon_0 \begin{bmatrix} \varepsilon_{11} & 0 & 0 \\ 0 & \varepsilon_{22} & 0 \\ 0 & 0 & \varepsilon_{33} \end{bmatrix} \tag{2}$$

   The existence of a second texture in the SHG angular response is more difficult to explain. There are two possibilities: this last is not at the same location in the laser track than the previous one and at that place the predominant electromagnetic force is of different origin (non-linear), or the force is of different nature and this last
is predominant (e.g. thermal but this one is not dependent of the laser polarization). This discussion may become
speculative but in previous papers we have proposed a theoretical approach for trying to understand this effect, it
arises from the peculiarity of the femtosecond laser interaction with dielectrics [29].

   Briefly, the laser light is absorbed through multiphoton ionization or tunnelling ionization producing a
quasi-free electron plasma in the conduction band. Then, the formed plasma may be further heated by the rest of
the laser pulse through free carrier one or several-photon absorption and/or grows through avalanche ionization.
In a previous publication [19], for explaining the existence of the appearance of asymmetric effect in writing
with the femtosecond laser like it is also described in Section 3.1.3 here, we have assumed that a ponderomotive
force (the force created by the light on the electrons) increases the plasma density on a side of the beam and then
that the trapping afterwards "records" the subsequent space charge in the materials, producing a DC field and a
stress field that can act between the pulses (memory effect).

   In such a way, a second torque is appearing based on the spontaneous dipole of the nanocrystal that may have
a direction parallel to the polar axis. The direction of the polar axis is thus driven by the direction of the induced
DC electric field that is discussed elsewhere [19]. It depends on the orientation of the pulse front tilt (PFT), of
the writing laser polarization direction and of the direction of writing. The PFT has been measured here rotated
by 142° around the Z axis from X (PFT azimuth) and tilted by 67° from the Z axis (PFT colatitude). The angle
between the writing laser polarization and PFT makes 55.6°. This means from the theory developed in [19] that
the polarization can play a role. On the other hand, it is also noticed in this paper, that the stress field modify the
kinetics and the stress field is dependent of the orientation of writing. We can also remark that the angles between the direction of writing and the PFT one are 83.6° (for 45° writing direction) and 96.4° (for 225° writing
direction). Finally, the combination of a stress field differently oriented compared to the PFT vector may lead to

a variation of the crystallization, here detected on the intensity of the second texture (the one aligned with the laser polarization).

5. Conclusion

In summary, we have demonstrated that second harmonic generation (SHG) can be obtained from a glass by femtosecond laser irradiation. This is a flexible method for controlling SHG three-dimension in silica matrix. By adjusting the pulse energy and writing speed, angular dependence of SHG with the probe laser polarization can be obtained with a high contrast. In our experiment conditions: a) a well-defined cosine-like curve with period of 180° could be obtained at low pulse energy or high pulse energy with high writing speed; b) a double cosine-like curve revealing a second texture of the crystals at high pulse energy (1.4 µJ) with low writing speed (5 µm/s) and with the polar axis oriented closer parallel to the writing laser polarization. An asymmetric orientational writing was observed, especially at high pulse energy with low writing speed so when the second texture was active. A discussion has been presented, including the mechanism for tentatively explain the above observations. The main force would be the effect of the electromagnetic polarization of the writing laser on the anisotropic induced dipole. We believe that this investigation contributes to a better understanding of the mechanism of the SHG orientations and contributes to revealing the technology potential in fabricating three-dimensional nonlinear optical devices.

Acknowledgements

The authors thank Prof. T. Baudin, Dr. D. Solas and K. Verstraete in Université Paris-Sud for useful discussions. The work has been done in the frame of FLAG (Femtosecond Laser Application in Glasses) consortium project with the support of Agence Nationale pour la Recherche (ANR-09-BLAN-0172-01). The authors extend thanks to China Scholarship Council and Université Paris-Sud.

References

[1]   Franken, P.A., Hill, A.E., Peters, C.W. and Weinreich, G. (1961) Generation of Optical Harmonics. *Physical Review Letters*, **7**, 118-119. http://dx.doi.org/10.1103/PhysRevLett.7.118

[2]   Bloembergen, N. and Pershan, P.S. (1962) Light Waves at the Boundary of Nonlinear Media. *Physical Review*, **128**, 606-622. http://dx.doi.org/10.1103/PhysRev.128.606

[3]   Vigouroux, H., Fargin, E., Fargues, A., Garrec, B.L., Dussauze, M., Rodriguez, V., Adamietz, F., Mountrichas, G., Kamitsos, E., Lotarev, S. and Sigaev, V. (2011) Crystallization and Second Harmonic Generation of Lithium Niobium Silicate Glass Ceramics. *Journal of the American Ceramic Society*, **94**, 2080-2086. http://dx.doi.org/10.1111/j.1551-2916.2011.04416.x

[4]   He, X., Poumellec, B., Liu, Q., Brisset, F. and Lancry, M. (2014) One-Step Photoinscription of Asymmetrically Oriented Fresnoite-Type Crystals in Glass by Ultrafast Laser. *Optics Letters*, **39**, 5423-5426. http://dx.doi.org/10.1364/OL.39.005423

[5]   Du, X., Zhang, H., Zhou, S., Zhang, F., Dong, G. and Qiu, J. (2015) Femtosecond Laser Induced Space-Selective Precipitation of a Deep-Ultraviolet Nonlinear $BaAlBO_3F_2$ Crystal in Glass. *Journal of Non-Crystalline Solids*, **420**, 17-20. http://dx.doi.org/10.1016/j.jnoncrysol.2014.12.023

[6]   Qiu, J., Miura, K. and Hirao, K. (2008) Femtosecond Laser-Induced Microfeatures in Glasses and Their Applications. *Journal of Non-Crystalline Solids*, **354**, 1100-1111. http://dx.doi.org/10.1016/j.jnoncrysol.2007.02.092

[7]   Dai, Y., Zhu, B., Qiu, J., Ma, H., Lu, B. and Yu, B. (2007) Space-Selective Precipitation of Functional Crystals in Glass by Using a High Repetition Rate Femtosecond Laser. *Chemical Physics Letters*, **443**, 253-257. http://dx.doi.org/10.1016/j.cplett.2007.06.076

[8]   Komatsu, T., Koshiba, K. and Honma, T. (2011) Preferential Growth Orientation of Laser-Patterned $LiNbO_3$ Crystals in Lithium Niobium Silicate Glass. *Journal of Solid State Chemistry*, **184**, 411-418. http://dx.doi.org/10.1016/j.jssc.2010.12.016

[9]   He, X., Fan, C., Poumellec, B., Liu, Q., Zeng, H., Brisset, F., Chen, G., Zhao, X. and Lancry, M. (2014) Size-Controlled Oriented Crystallization in SiO_2-Based Glasses by Femtosecond Laser Irradiation. *Journal of the Optical Society of America B—Optical Physics*, **31**, 376-381. http://dx.doi.org/10.1364/JOSAB.31.000376

[10] Stone, A., Jain, H., Dierolf, V., Sakakura, M., Shimotsuma, Y., Miura, K., Hirao, K., Lapointe, J. and Kashyap, R. (2015) Direct Laser-Writing of Ferroelectric Single-Crystal Waveguide Architectures in Glass for 3D Integrated Optics.

Scientific Reports, **5**, 10391. http://dx.doi.org/10.1038/srep10391

[11] Komatsu, T., Ihara, R., Honma, T., Benino, Y., Sato, R., Kim, H.G. and Fujiwara, T. (2007) Patterning of Non-Linear Optical Crystals in Glass by Laser-Induced Crystallization. *Journal of the American Ceramic Society*, **90**, 699-705. http://dx.doi.org/10.1111/j.1551-2916.2006.01441.x

[12] Zeng, H., Poumellec, B., Fan, C., Chen, G., Erraji-Chahid, A., *et al.* (2012) Preparation of Glass-Ceramics with Oriented Nonlinear Crystals: A Review. Nova Science Publishers, New York, 89-134.

[13] Ochi, Y., Meguro, T. and Kakegawa, K. (2006) Orientated Crystallization of Fresnoite Glass-Ceramics by Using a Thermal Gradient. *Journal of the European Ceramic Society*, **26**, 627-630. http://dx.doi.org/10.1016/j.jeurceramsoc.2005.07.044

[14] Gerth, K., Rüssel, C., Keding, R., Schleevoigt, P. and Dunken, H. (1999) Oriented Crystallisation of Lithium Niobate Containing Glass Ceramic in an Electric Field and Determination of the Crystallographic Orientation by Infrared Spectroscopy. *Physics and Chemistry of Glasses*, **40**, 135-139.

[15] Stuart, B., Feit, M., Herman, S., Rubenchik, A., Shore, B. and Perry, M. (1996) Nanosecond-to-Femtosecond Laser-Induced Breakdown in Dielectrics. *Physical Review B*, **53**, 1749-1761. http://dx.doi.org/10.1103/PhysRevB.53.1749

[16] Eaton, S., Zhang, H., Herman, P., Yoshino, F., Shah, L., Bovatsek, J. and Arai, A. (2005) Heat Accumulation Effects in Femtosecond Laser-Written Waveguides with Variable Repetition Rate. *Optics Express*, **13**, 4708-4716. http://dx.doi.org/10.1364/OPEX.13.004708

[17] Gattass, R., Cerami, L. and Mazur, E. (2006) Micromachining of Bulk Glass with Bursts of Femtosecond Laser Pulses at Variable Repetition Rates. *Optics Express*, **14**, 5279-5284. http://dx.doi.org/10.1364/OE.14.005279

[18] Stone, A., Sakakura, M., Shimotsuma, Y., Stone, G., Gupta, P., Miura, K., Hirao, K., Dierolf, V. and Jain, H. (2009) Directionally Controlled 3D Ferroelectric Single Crystal Growth in $LaBGeO_5$ Glass by Femtosecond Laser Irradiation. *Optics Express*, **17**, 23284-23289. http://dx.doi.org/10.1364/OE.17.023284

[19] Poumellec, B., Lancry, M., Desmarchelier, R., Hervé, E., Brisset, F. and Poulin, J.C. (2013) Asymmetric Orientational Writing in Glass with Femtosecond Laser Irradiation. *Optical Materials Express*, **3**, 1586-1599. http://dx.doi.org/10.1364/OME.3.001586

[20] Todorović, M. and Radonjić, L. (1997) Lithium-Niobate Ferroelectric Material Obtained by Glass Crystallization. *Ceramics International*, **23**, 55-60. http://dx.doi.org/10.1016/0272-8842(95)00140-9

[21] Fan, C. (2011) Directional Writing Dependence of Birefringence in Multicomponent Silica-Based Glasses with Ultrashort Laser Irradiation. *Journal of Laser Micro/Nanoengineering*, **6**, 158-163.

[22] Dingley, D.J. and Randle, V. (1992) Microtexture Determination by Electron Back-Scatter Diffraction. *Journal of Materials Science*, **27**, 4545-4566. http://dx.doi.org/10.1364/OL.37.002955

[23] Fan, C., Poumellec, B., Lancry, M., He, X., Zeng, H., Erraji-Chahid, A., Liu, Q. and Chen, G. (2012) Three-Dimensional Photoprecipitation of Oriented $LiNbO_3$-Like Crystals in Silica-Based Glass with Femtosecond Laser Irradiation. *Optics Letters*, **37**, 2955-2957. http://dx.doi.org/10.1364/OL.37.002955

[24] Weis, R.S. and Gaylord, T.K. (1985) Lithium Niobate: Summary of Physical Properties and Crystal Structure. *Applied Physics A*, **37**, 191-203. http://dx.doi.org/10.1007/BF00614817

[25] Butet, J., Russier-Antoine, I., Jonin, C., Lascoux, N., Benichou, E. and Brevet, P.-F. (2013) Effect of the Dielectric Core and Embedding Medium on the Second Harmonic Generation from Plasmonic Nanoshells: Tunability and Sensing. *The Journal of Physical Chemistry C*, **117**, 1172-1177. http://dx.doi.org/10.1021/jp310169u

[26] Yang, W., Kazansky, P.G. and Svirko, Y.P. (2008) Non-Reciprocal Ultrafast Laser Writing. *Nature Photonics*, **2**, 99-104. http://dx.doi.org/10.1038/nphoton.2007.276

[27] Träger, F. (2007) Springer Handbook of Lasers and Optics. Springer, New York. http://dx.doi.org/10.1007/978-0-387-30420-5

[28] Zelmon, D.E., Small, D.L. and Jundt, D. (1997) Infrared Corrected Sellmeier Coefficients for Congruently Grown Lithium Niobate and 5 mol.% Magnesium Oxide—Doped Lithium Niobate. *Journal of the Optical Society of America B*, **14**, 3319-3322. http://dx.doi.org/10.1364/JOSAB.14.003319

[29] Mao, S.S., Quéré, F., Guizard, S., Mao, X., Russo, R.E., Petite, G. and Martin, P. (2004) Dynamics of Femtosecond Laser Interactions with Dielectrics. *Applied Physics A*, **79**, 1695-1709. http://dx.doi.org/10.1007/s00339-004-2684-0

20

Effect of TiO$_2$ Thin Film Morphology on Polyaniline/TiO$_2$ Solar Cell Efficiency

Amer N. J. Al-Daghman[1], K. Ibrahim[1], Naser M. Ahmed[1], Kareema M. Zaidan[2]

[1]Nano-Optoelectronic Research and Technology Laboratory, School of Physics, University Sains Malaysia, Pulau Pinang, Malaysia
[2]Physics Department, Collage of Science, University of Basrah, Basrah, Iraq
Email: amer78malay@yahoo.com.my

Abstract

Nanocrystalline titanium dioxide (TiO$_2$) thin films were prepared by using sol-gel through spin-coating method. An assembly of indium tin oxide (ITO)/TiO$_2$/polyaniline (PANI)/Ag was made in a sandwich panel structure. The obtained junction shows rectifying behavior. Additionally, the I/V characteristic indicates that a P-N junction at nanocrystalline PANI/TiO$_2$ interface has been created. In this experimental study, we depended only on the ratio between titanium and PANI in the process of preparing sol-gel (PANi/TiO$_2$ at 20% wt). The largest open circuit voltage of 656 mV and short current density of 0.00315 mA/cm^2 produce 0.0004% power conversion solar cell (η) under simulated solar radiation (50 mW/cm^2). The thin films of PANI and titanium oxide (TiO$_2$)/PANI composites were synthesized by sol-gel technique. Pure TiO$_2$ powder with nanoparticle size of less than 25 nm and PANI were synthesized through chemical oxidative polymerization of aniline monomers. The composite films were characterized by high resolution X-ray diffraction, Fourier transform infrared spectroscopy, field effect scanning electron microscopy, and UV-vis spectroscopy. The results were compared with the corresponding data on pure PANI films. The intensity of diffraction peaks for PANI/TiO$_2$ composites is lower than that for TiO$_2$. The characteristic of the FTIR peaks of pure PANI shifts to a higher wave number in TiO$_2$/PANI composite, which is attributed to the interaction of TiO$_2$ nanoparticles with PANI molecular chains.

Keywords

TiO$_2$, Polyaniline, Crystal Structure, Solar Cells

1. Introduction

Photovoltaics has received increasing attention over the past decades as a feasible way to replace the diminishing fossil fuels and reduce environmental damage.

Inorganic-organic heterojunction photovoltaic devices have been elicited because of their advantages, such as low cost and light weight. Inorganic semiconductor particles, such as TiO_2 [1] [2], have been used as electron acceptor in solar cells. The sol-gel method has been selected to allow the sample preparation of high-purity films at low cost. Conducting polymers used as hole transporting layers have been recently applied on photovoltaic (PV) cells. We have investigated the effect of TiO_2 nanoparticle concentration on thin film morphology and the performance of PANI/TiO_2 solar cells.

Conducting PANI is important and has exhibited great potential for commercial applications because of its unique electrical, optical, and photoelectrical properties, as well as its easy preparation and excellent environmental stability [3] [4]. Nanocrystalline TiO_2 has also been frequently used for preparing various nanocomposites with conducting polymers because of its excellent physical and chemical properties and promising applications in advanced coatings, solar cells, gas sensors, and photo catalysts [5]. Therefore, PANI/TiO_2 nanocomposites have been the most intensively studied among various nanocomposites, because they combine the merits of PANI and nanocrystalline titanium dioxide (TiO_2) particles within a single material and could be applied in electronic devices, nonlinear optical system, gas sensors, and photoelectrochemical devices [6] [7]. Most of the properties of these materials are based on the synergy between the properties of the components, which are a direct result of their chemical and structural compositions, and thus, can be tailored. For instance, coatings based on organic-inorganic hybrid materials have the capability to combine the flexibility and easy processing of polymers with the interesting properties of the inorganic part: hardness, thermal stability, as well as electrical and electrochemical distinguished properties. The combination of nanocrystalline titanium dioxide (TiO_2) and polyaniline (PANI) is attractive because the combination of PANI and metal oxide exhibits excellent electrical, mechanical, and optical properties, such as surface hardness, modulus, strength, transparency, high refractive index, and acids; their derivatives are highly promising coupling molecules that allow the anchoring of organic groups to inorganic solids [8] [9]. Thus, the preparation of PANI-nano-TiO_2 has been a subject of interest in many studies. Feng *et al.* synthesized a composite of PANI encapsulating TiO_2 nanoparticles through *in situ* emulsion polymerization [10].

The authors explained the nature of chain growth and interaction between PANI and nano-TiO_2 particles by Fourier transform infrared (FTIR) spectroscopic analyses [10]. Xia and Wang prepared PANI nanocrystalline titanium dioxide (TiO_2) composite through ultrasonic irradiation, which is a novel method for the preparation of 1D to 3D conducting polymer nanocrystalline composites [7]. Somani *et al.* reported the preparation of highly piezoresistive conducting PANI-TiO_2 composite through *in situ* deposition technique at low temperature (0°C) [9]. The technological relevance of both conducting PANI and semiconducting material TiO_2 in nano form leads to the preparation of a composite of PANi and TiO_2 at molecular-level interaction. Such molecular-level interaction may lead to novel properties in these two dissimilar chemical components [11]-[13]. In this paper, we report the synthesis of PANi/TiO_2 composite by sol-gel method. Their morphological, structural, electrical, and optical properties are also studied.

2. Materials and Methods

2.1. Materials

Aniline (C_6H_5NH) from Merck (Schuchardt, Germany) was purified through distillation at reduced pressure before it was used. Ammonium peroxydisulfate (APS) was purchased from Merck (KGaA, Germany). Nanodimensional titanium dioxide (TiO_2, 99.7%) and anatase nanoparticles with size <25 nm were also used.

2.2. Synthesis of Polyaniline

PANI was synthesized through the polymerization of aniline in the presence of hydrochloric acid as a catalyst and ammonium peroxydisulfate as an oxidant by chemical oxidative polymerization method. For the synthesis, 50 ml of 1 M HCl was taken, and 2 ml of aniline was added together into a 250 ml equipped with electromagnetic stirrer. Then, 5 mg of ammonium peroxydisulfate (($NH_4)_2S_2O_8$) in 50 ml and 1 M HCl were suddenly added into the above solution. The polymerization temperature at 0°C was maintained for 5 h to complete the polymerization reaction. Then, the obtained precipitate was filtered.

The product was washed successively by 1 M HCl followed by distilled water and washed until the solution turned colorless. Then, the product was re-filtered and thoroughly washed once again by distilled water to obtain the emeraldine salt (ES) form of PANI. To obtain the emeraldine base (EB) form of PANI, the ES form of PANI with 0.1 M NH_4OH solution was dried at 60°C in vacuum oven for 24 h. Thus, the powder of insulating PANI EB polymer was obtained [9].

2.3. Synthesis of (TiO₂/PANI) Nanocomposite

Figure 1 schematically shows that the photovoltaic device structure is ITO glass/TiO₂/PANI/Ag.

The device dimension for this measurement was 1 cm². Titanium dioxide was used as a material for the thin films of the nanocomposite. About 4 ml of m-cresol (C_7H_8O) 97.7% from Acros, USA was added to the TiO₂ nanoparticle powder under vigorous stirring for 12 h for peptization. The sols were deposited on ITO conducting glass through spin-coating method at 1500 rpm for 60 s. Then, the sols were annealed at 450°C for 2 h in a tube furnace (Model: LENTON VTF/12/60/700). The TiO₂ nanoparticle with thickness of 120 was prepared through sol-gel method and annealed at 450°C and had a perfect crystalline structure. The formation of the Ti-O-Ti bonds in the films was observed after thermal treatment. However, the film became crystalline at anatase phase after annealing at 450°C.

PANI (emeraldine base, EB) powder was dissolved in 1:1 m-cresol deposit on the obtained TiO₂ thin films through spin-coating method at 3000 rpm for 60 s, and then dried at 100°C for 10 min. Afterward, the film was dried at 60°C for 24 h in an oven vacuum. Ag, an electrode, was evaporated in high vacuum with 10^{-4} Pa pressure during evaporation.

2.4. Characterization and Measurement Methods

X-ray diffraction (XRD) studies were carried out using high resolution X-ray diffractometer (Model: PANalytical X pert Pro MRD PW3040). The XRD patterns were recorded in the 2θ range of 20° - 70° with step width of 0.02° and step time of 1.25 s by using $CuK\alpha$ radiation ($\lambda = 1.5406$ A°). XRD patterns were analyzed by matching the observed peaks with the standard pattern provided by JCPDS file. FTIR spectroscopy (Model: Perkin Elmer Spectrum Gx) of iO₂, PANi, and PANi:TiO₂ (20%) composite was studied in the frequency range of 400 - 4000 cm^{-1}. Morphological study of the films of PANi and PANi:TiO₂ composite was carried out using field effect scanning electron microscopy (Model: FEI Nova NanoSEM 450) operated at 20 kV. UV-vis spectra of the samples, which were dispersed in deionized water under ultrasonication, were recorded on a Shimadzu 1800 UV-vis spectrophotometer.

The I/V characteristic measured by a Keithley 2400 current-voltage source in the dark indicated that no barrier was apparent at the Ag/PANI or ITO/TiO₂ interface.

3. Result and Discussion

Figures 2(a)-(c) show the XRD patterns of pure PANI in EB form, TiO₂, and PANI:TiO₂ (20%) composite. The XRD pattern of PANI shows a broad peak at $2\theta = 22.68°$, which corresponds to 112 planes of PANI [10]. In **Figure 2(b)** and **Figure 2(c)**, the patterns show sharp and well-defined peaks, indicating the crystallinity of the synthesized materials. The observed 2θ values were consistent with the standard values and showed the tetragonal structure of TiO₂ [8]. **Figure 2(b)** shows that a = 3.78 A° and c = 9.51 A° [14].

The intensity of the diffraction peaks for PANI:TiO₂ composites was lower than that for TiO₂ (**Figure 2(c)**). Noncrystalline PANI reduced the volume fraction percentage of TiO₂, and thus, weakened the diffraction peaks of TiO₂ in the composite.

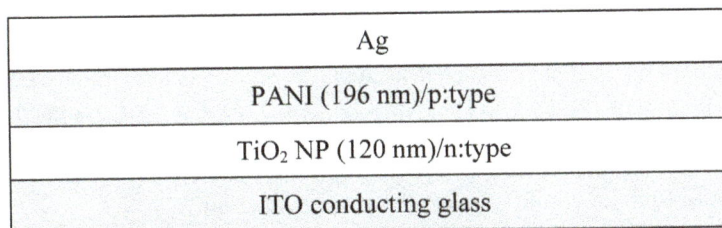

Ag
PANI (196 nm)/p:type
TiO₂ NP (120 nm)/n:type
ITO conducting glass

Figure 1. Structure of the (1 cm × 1 cm) ITO glass/TiO₂/PANI/Ag.

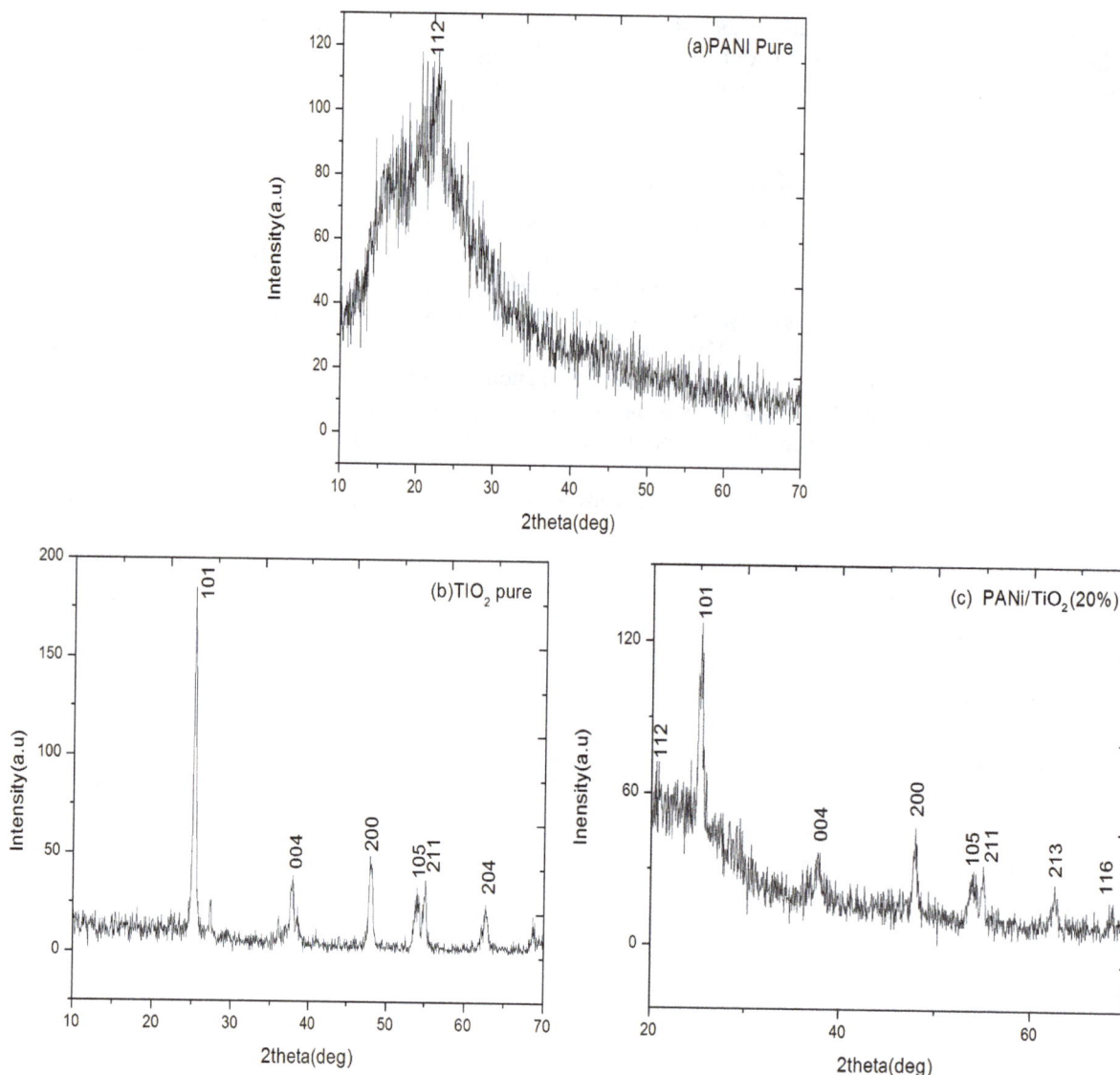

Figure 2. X-ray diffraction (a) PANi (EB), (b) TiO$_2$, and (c) PANI:TiO$_2$ nanocomposite.

Figures 3(a)-(c) show the FESEM of pure PANI, pure TiO$_2$, PANI:TiO$_2$ (20%), and nanocomposite. FESEM image of the composite shows a uniform distribution of the TiO$_2$ particles in the PANI chains without any agglomeration. According to the FESEM images, the nanostructure TiO$_2$ particles are embedded within the netlike structure built by PANI chains. The composite is highly microporous and is able to increase the liquid-solid interfacial area [15] [16].

Figures 4(a)-(c) show the FTIR spectra of the undoped PANi, PANi-TiO$_2$ composite, and TiO$_2$ nanoparticles. The origins of the vibration bands are as follows: 3365, 2922, and 622 cm^{-1}, which are caused by the NH stretching of aromatic amine, CH-stretching, and CH out-of-plane bending vibration, respectively. The CH out-of-plane bending mode has been used as a key to identify the type of substituted benzene.

The bands at 1665 and 1489 cm^{-1} are attributed to the C=N and C=C stretching mode of vibration for the quinonoid and benzenoid units of PANI. The peaks at 1296 and 1155 cm^{-1} are assigned to the C-N stretching mode of benzenoid ring.

The bands in the region 1000 - 1115 cm^{-1} are caused by the in-plane bending vibration of C-H mode. The bend at 850 cm^{-1} originates from the out-of-plane C-H bending vibration.

The low wavenumber region exhibits a strong vibration around 621 cm^{-1}, which corresponds to the antisymmetric Ti-O-Ti mode of the titanium oxide [8].

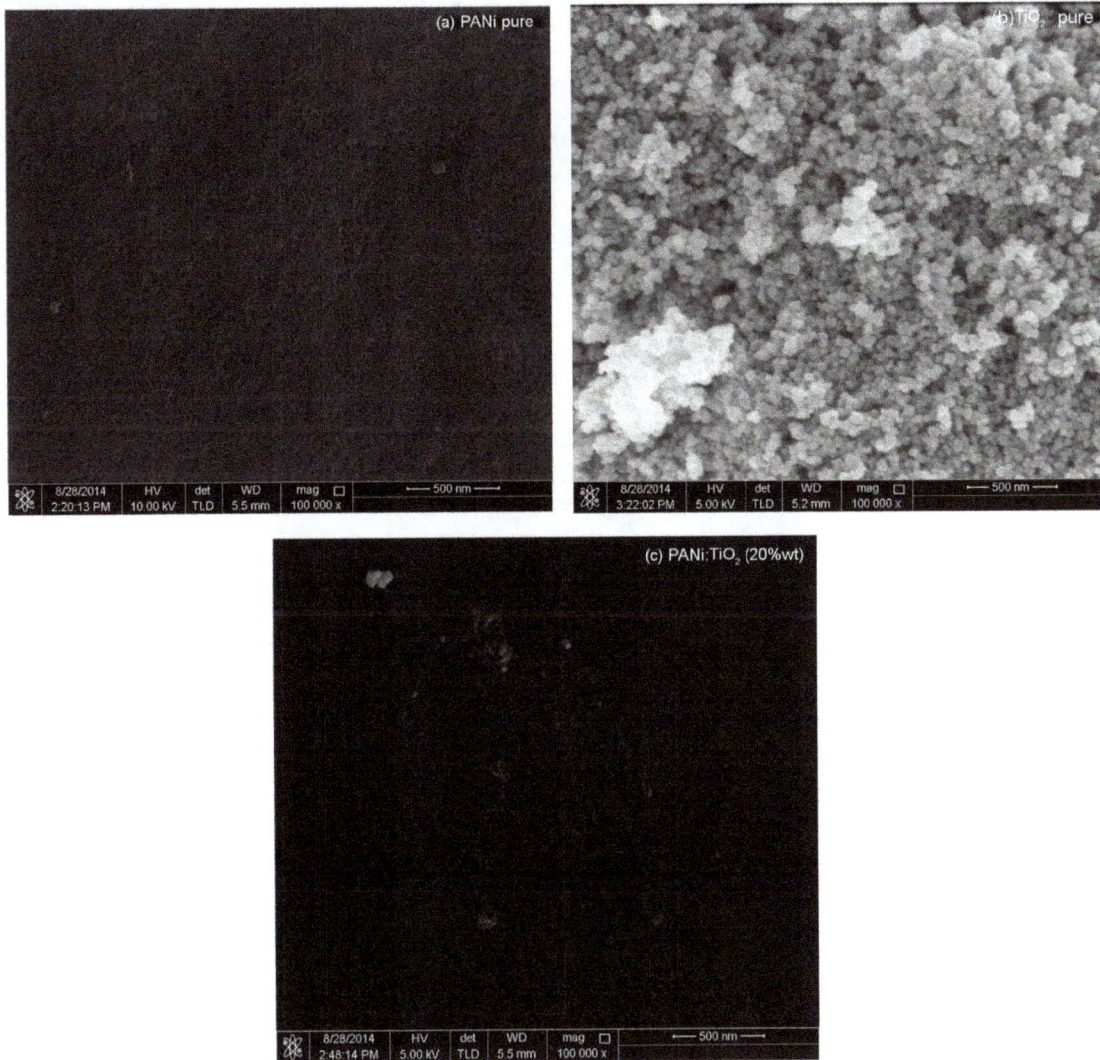

Figure 3. FESEM morphology of (a) pure PANI, (b) pure TiO_2, and (c) PANI:TiO_2 (20% wt).

The absorption of PANI pure film at the visible spectrum, which was measured on a Shimadzu UV1700 ultraviolet visible spectrophotometer, is shown in **Figure 5**.

Notably, two peaks lie at about 426 and 805 nm. These peaks indicate that the insertion of nanoparticles TiO_2 has the effect of doping the conducting PANI, and hence, should lead to an interaction at the interface of PANI and nanoparticles TiO_2. Strong terrestrial solar photon flux between 400 - 900 nm was noted. A primary factor influencing the photo-induced carrier mechanism of solar cells should also be considered. Therefore, the limiting factor of TiO_2/PANI and solar cell devices is the low absorption of photons. We could solve this issue by increasing the absorption spectrum of polyaniline in the visible zone using suitable dopants.

A built-in electric field at the nanocrystalline TiO_2/polyaniline interface has been created. **Figure 6** shows the I/V characteristics obtained from the devices under 50 mW/cm^2. A short-circuit current density of 3.15 mA/cm^2 and an open-circuit voltage of 0.656 V were obtained from device. The efficiency of the solar cell was very minimal ($\eta = 0.0004\%$) because of the increased resistance of the device, leading to the reduction of the open-circuit voltage. **Figure 6** shows the absorption spectrum of the polyaniline in the visible spectrum. The strong terrestrial solar photon flux between 400 and 900 nm should be considered a primary factor influencing the photo-induced carrier mechanism of a solar cell. This result suggests that the low absorption of photons, which is the limiting factors of TiO_2/polyaniline solar cells, might be solved by the increase of the absorption spectrum of polyaniline in the visible region by using suitable dopants.

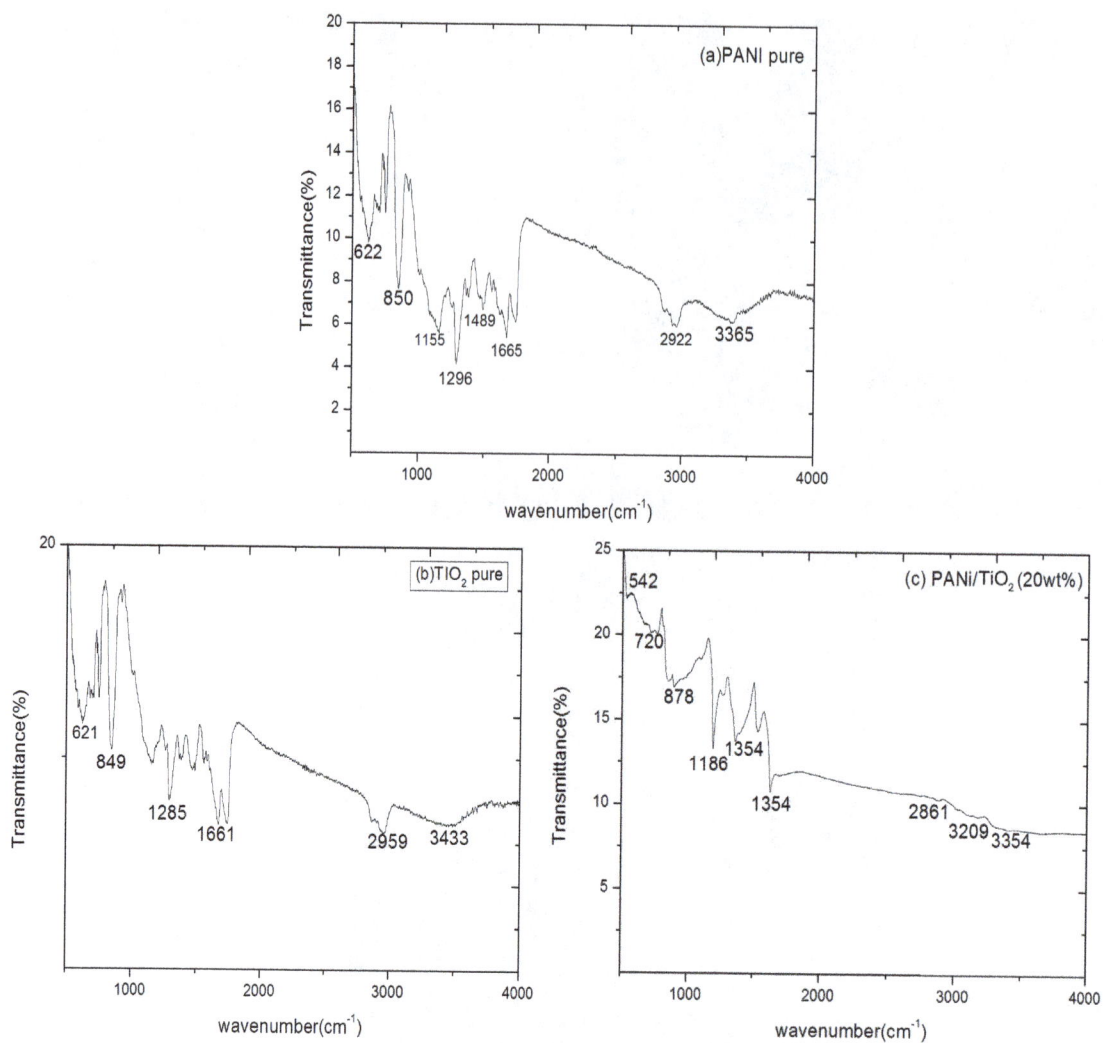

Figure 4. FTIR spectra of (a) PANI (EB), (b) PANI:TiO$_2$, and (c) TiO$_2$.

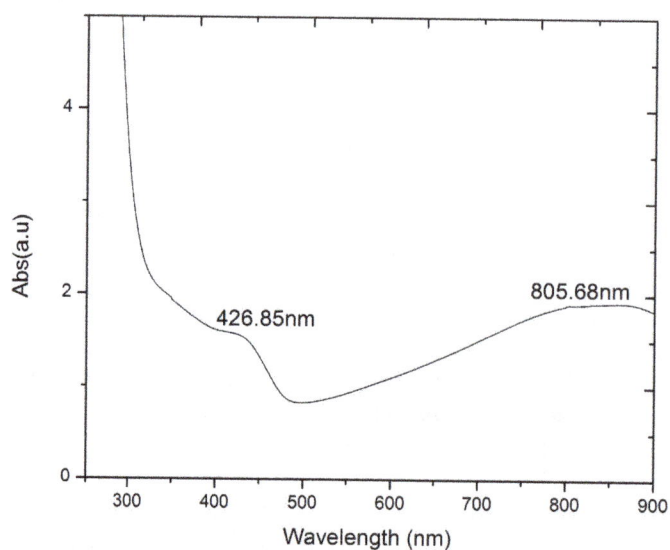

Figure 5. The absorption spectrum of polyaniline (EB) in the visible spectrum.

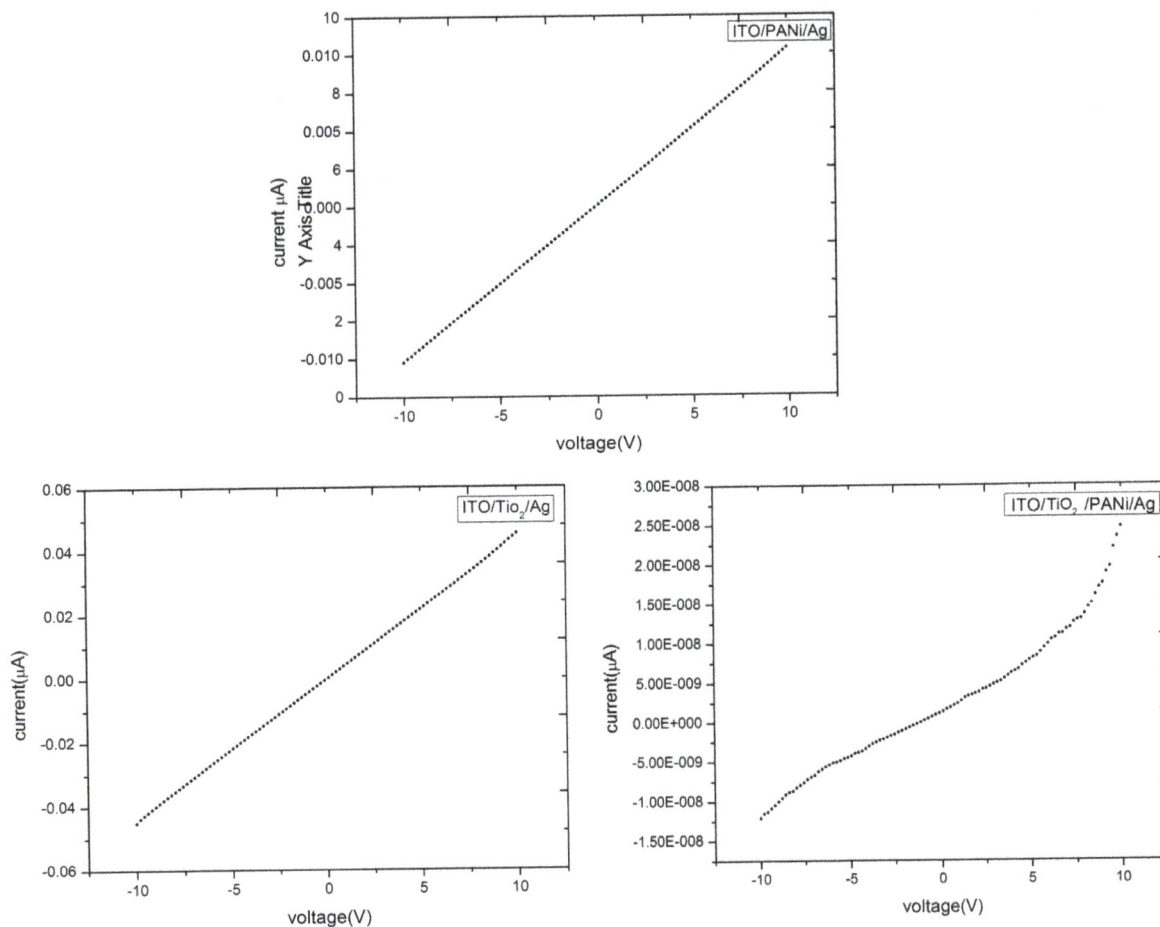

Figure 6. I/V characteristic of the sandwich-type structure of PANI, TIO₂, and ITO/TiO₂/PANI/Ag.

4. Conclusion

Thin films of conducting polymer (PANI), TiO₂ nanoparticles, and PANI/TiO₂ nanocomposites were synthesized through sol-gel method. The absorption peaks in the FTIR and UV-vis spectra of PANI/TiO₂ composite films were found to shift around higher wavenumber compared with those in pure PANI. The observed shifts were attributed to the interaction between the TiO₂ particles and polymer molecular chains PANI. A change in the value of the lattice parameter of TiO₂ in the PANi/TiO₂ composite was observed, which also indicated the presence of interaction between TiO₂ particles and PANI matrix polymer. FESEM analysis of PANI/TiO₂ composite films revealed uniform distribution of TiO₂ particles in the PANI matrix. The I/V characteristic for the device under simulated solar radiation (50 mw/cm^2) has the largest open-circuit voltage of 0.656 V and short-circuit current density of 315 mA/cm^2.

Acknowledgements

We gratefully acknowledge the support of the School of Physics, USM, under the short-term grant nos. 203.PSF.6721001 and 304/PFIZIK/6312076.

References

[1]   Regan, B.O. and Gratzel, M. (1991) Low-Cost, High Efficiency Solar Cell Based on Dye Sensitized Colloidal TiO₂ Film. *Nature*, **353**, 737-739. http://dx.doi.org/10.1038/353737a0

[2]   Senadeera, G.K.R., Nakamura, K., Kitamura, T., Wada, Y. and Yanagida, S. (2003) Fabrication of Highly Efficient Polythiophene-Sensitized Metal Oxide Photovoltaic Cells. *Applied Physics Letters*, **83**, 5470-5472.

http://dx.doi.org/10.1063/1.1633673

[3]   Huang, J., Virji, S., Weiller, B.H. and Kaner, R.B. (2003) Polyaniline Nanofibers: Facile Synthesis and Chemical Sensors. *Journal of the American Chemical Society*, **125**, 314-315. http://dx.doi.org/10.1021/ja028371y

[4]   Huang, J.X. and Kaner, R.B. (2004) A General Chemical Route to Polyaniline Nanofibers. *Journal of the American Chemical Society*, **126**, 851-855. http://dx.doi.org/10.1021/ja0371754

[5]   Deore, B.A., Yu, I. and Freund, M.S. (2004) A Switchable Self-Doped Polyaniline: Interconversion between Self-Doped and Non-Self-Doped Forms. *Journal of the American Chemical Society*, **126**, 52-53. http://dx.doi.org/10.1021/ja038499v

[6]   Tiwari, A. (2007) Gum Arabic-Graft-Polyaniline: An Electrically Active Redox Biomaterial for Sensor Applications. *Journal of Macromolecular Science, Part A*, **44**, 735-745. http://dx.doi.org/10.1080/10601320701353116

[7]   Roy, A.S., Anilkumar, K.R. and Ambika Prasad, M.V.N. (2011) Studies of AC Conductivity and Dielectric Relaxation Behavior of CdO-Doped Nanometric Polyaniline. *Journal of Applied Polymer Science*, **123**, 1928-1934. http://dx.doi.org/10.1002/app.34696

[8]   Tiwari, A., Sen, V., Dhakate, S.R., Mishra, A.P. and Singh, V. (2008) Synthesis, Characterization, and Hoping Transport Properties of HCl Doped Conducting Biopolymer-Co-Polyaniline Zwitterion Hybrids. *Polymers for Advanced Technologies*, **19**, 909-914. http://dx.doi.org/10.1002/pat.1058

[9]   Zhang, L.J., Wan, M.X. and Wei, Y. (2005) Polyaniline/TiO_2 Microspheres Prepared by a Template-Free Method. *Synthetic Metals*, **151**, 1-5. http://dx.doi.org/10.1016/j.synthmet.2004.12.021

[10] Feng, W., Sun, E.H., Fujii, A., Wu, H.C., Niihara, K. and Yoshino, K. (2000) Synthesis and Characterization of Photoconducting Polyaniline-TiO_2 Nanocomposite. *Bulletin of the Chemical Society of Japan*, **73**, 2627-2633.

[11] Xia, H.S. and Wang, Q. (2002) Ultrasonic Irradiation: A Novel Approach to Prepare Conductive Polyaniline/Nanocrystalline Titanium Oxide Composites. *Chemistry of Materials*, **14**, 2158-2165. http://dx.doi.org/10.1021/cm0109591

[12] Somani, P.R., Marimuthu, R., Mulik, U.P., Sainkar, S.R. and Amalnerkar, D.P. (1999) High Piezoresistivity and Its Origin in Conducting Polyaniline/TiO_2 Composites. *Synthetic Metals*, **106**, 45-52. http://dx.doi.org/10.1016/S0379-6779(99)00081-8

[13] Matsumura, M. and Ohno, T. (1997) Concerted Transport of Electrons and Protons across Conducting Polymer Membranes. *Advanced Materials*, **9**, 357-359. http://dx.doi.org/10.1002/adma.19970090416

[14] Yoneyama, H., Takahashi, N. and Kuwabata, S. (1999) Catalytic Asymmetric Reaction of Lithium Ester Enolates with Imines. *Journal of the Chemical Society, Chemical Communications*, **2**, 716-719.

[15] Pawar, S.G., Patil, S.L., Chougule, M.A., Jundale, D.M. and Patil, V.B. (2010) Microstructural, Optical and Electrical Studies on Sol Gel Derived TiO_2 Thin Films. *Archives of Physics Research*, **1**, 57-66.

[16] Gospodinova, N. and Terlemezyan, L. (1998) Conducting Polymers Prepared by Oxidative Polymerization: Polyaniline. *Progress in Polymer Science*, **23**, 1443-1484. http://dx.doi.org/10.1016/S0079-6700(98)00008-2

21

Synthesis and Properties of Fumed Silicas Modified with Mixtures of Poly(methylphenylsiloxane) and Dimethyl Carbonate

Iryna S. Protsak[1]*, Valentyn A. Tertykh[1], Yulia M. Bolbukh[1], Dariusz Sternik[2], Anna Derylo-Marczewska[2]

[1]Department of Chemisorption, Chuiko Institute of Surface Chemistry of National Academy of Sciences of Ukraine, Kiev, Ukraine
[2]Faculty of Chemistry, Maria Curie-Sklodowska University, Lublin, Poland
Email: *iryna_protsak@yahoo.com

Abstract

Effect of the concentration ratios of organosiloxane/initiator and treatment temperature on the characteristics of hydrophobic products obtained by modification of surface of fumed silica with poly(methylphenylsiloxane) (PMPS) in the presence of dimethyl carbonate has been studied. Morphology, particle size, surface area and coating microstructure of modified silicas were analyzed by methods of transmission electron and atomic force microscopies, nitrogen adsorption-desorption data. Carbon contents in the grafted modifying layer of organosilicas were determined using IR spectroscopy and elemental analysis. Hydrophilic-hydrophobic properties of surface of the obtained modified silicas were estimated by measurements of contact angles of wetting. It was shown that modification of pyrogenic silicas with mixtures of poly(methylphenylsiloxane) and dimethyl carbonate allows to obtain the homogeneous hydrophobic products and serve their nanodispersity.

Keywords

Fumed Silica, Poly(methylphenylsiloxane), Dimethyl Carbonate, Surface Modification, Hydrophilic-Hydrophobic Properties

1. Introduction

Modified disperse silicas with grafted phenyl groups are applied as the stationary phases and carriers in chro-

---

*Corresponding author.

matography, and as the fillers for manufacturing adhesives, coatings and materials with high heat resistance [1]-[3]. Appropriate phenyl-containing alkoxy and chlorosilanes are mainly used for preparation of such type of modified silicas. However, the case of phenyl-containing alkoxysilanes chemisorption of alcohol molecules is observed as a side process at high temperatures of silica treatment. Using the phenyl-containing chlorosilanes for surface modification is not always suitable because of the elimination of hydrogen chloride from reaction with silanol groups of silica. It may result in the presence of potential electrolytes in the obtained modified products. Therefore, application of the phenyl-containing siloxane oligomers for modification of silica surface is undoubtedly interesting. However, the processes of chemisorption involving silanol groups of the silica surface and oligomeric phenylsiloxanes proceed at high temperatures (>350°C).

It is known from the experiment [4] [5] that due to interaction between dimethyl carbonate and polydimethylsiloxane, a reaction product is formed with viscosity considerably lesser than those of parent substances, so a conclusion can be made on the rupture of chemical bonds in polydimethylsiloxane chains, and it is established [6] that in the presence of dimethyl carbonate, chemisorption of poly(dimethylsiloxane) on the silica surface is carried out at lower temperatures.

Dimethyl and diethyl carbonates as it was earlier established are effective reagents in the reaction of siloxane bond cleavage in poly(dimethylsiloxanes) [4] [7] [8]. Thus, it can be expected that the simultaneous use of dimethyl carbonate, which belongs to "green chemistry" solvents and reagents [9] [10], and phenyl-containing polysiloxane in the processes of chemical modification of silica will provide deoligomerization of siloxane and promote to cleavage of siloxane bridges on the silica surface.

In this work, such approach was tested in the simultaneous application of dimethyl carbonate and poly(methylphenylsiloxane) (PMPS) for modification of the silica surface. Special attention was paid for investigation of structural and surface characteristics of the products obtained during modification of fumed silica with the mixtures of different composition.

2. Experimental

2.1. Reagents

Fumed silica characterized by a specific surface area of 260 m^2/g (A-300, Kalush, Ukraine), synthesized via high-temperature hydrolysis of silicon tetrachloride was taken for surface modification. As a modifying reagent poly(methylphenylsiloxane) fluid PMPS-4 (Zaporizhzhya, Ukraine, GOST 15866-70) with degree of polymerization n = 8 - 10 was used. The applied organosilicon polymer is characterized by a wide temperature (from −20°C to +350°C) range of stability, chemical inertness and hydrophobic properties. Dimethyl carbonate (DMC) was supplied by Sigma Aldrich.

Synthesis of Composite Materials

Modification of fumed silica surface with PMPS was performed at the different temperatures for 2 hrs with or without the addition of dimethyl carbonate (DMC). The modification process was performed in a glass reactor with a stirrer at rotational speed from 20 to 300 ppm. After loading with fumed silica, all air volume in the reactor was filled with nitrogen and the reactor was heated up to a defined temperature. Then, nitrogen supply was interrupted and the modifying reagent was added into reactor by means of its aerosol spraying through a nozzle.

Four series of the samples were synthesized.10 g of fumed silica were treated with 2.2 g of PMPS in the first series. In these second series the surface treatment was carried out with the mixture of PMPS and DMC at the weight ratios SiO_2:PMPS:DMC = 10:2.2:1.5. The weight ratios SiO_2:PMPS:DMC = 10:2.2:1.0 were used in the third series. Modification of the surface was carried out with the mixture of the PMPS and DMC in the weight ratios SiO_2:PMPS:DMC = 10:2.2:0.5 in the fourth series. Modification of the fumed silica surface was performed at three different temperatures: 200°C, 250°C and 300°C. Removal of the physically adsorbed reactants was carried out in a Soxhlet apparatus with n-hexane as a solvent at 68°C for 1 h. Then, the washed sample was dried at 80°C for 2 hrs.

2.2. Methods

2.2.1. Infrared Spectra

In order to control the flow of surface reactions, IR spectra were recorded using a Specord M-80 spectrophotometer in a range of wave numbers 4000 - 200 cm^{-1}. The silica samples were pressed into rectangular 28 × 8 mm plates of 25 mg weight.

2.2.2. Nanostructural Characterization

Surface microstructure and morphology as well as coating homogeneity for the modified samples were analyzed by transmission electron microscopy, TEM (Tecnai G2T20 X-TWIN, USA) and atomic force microscopy, AFM (Nanoscope V Digital Instruments, USA, with a Tapping Mode technique). AFM data processing was performed using the SPIP program (version 5.0.6).

2.2.3. Nitrogen Adsorption Measurements

Porous structure (specific surface area, pore volume and pore size distribution function) was characterized using nitrogen adsorption-desorption data measured using Accelerated Surface Area and Porosimetry analyzers ASAP 2020 and 2420 (Micromeritics, USA). Before the measurements, the samples were outgassed at 110°C. The specific surface area, S_{BET}, was calculated using the BET method [11]. The total pore volume V_p was evaluated using the nitrogen adsorption data at the relative pressure p/p_o~0.98 - 0.99, where p and p_o are equilibrium pressure and vapor pressure at the temperature 77.4 K, respectively. Pore size distribution (PSD) was calculated by employing the regularization approach at the fixed regularization parameter $\alpha = 0.01$ according to the Nguyen-Do method for the bimodal PSD [12]-[14]. This method was used previously to describe the structural properties of carbonaceous materials [15]-[18], composites [14] [19] [20] and aluminosilicates [21]. In the paper the model of the pores as voids between spherical particles and the model of the cylindrical pores were used.

2.2.4. Contact Angle Measurements

Hydrophilic-hydrophobic properties of the surface of obtained modified silicas were estimated by measurements of contact angles of water drops. The contact angle data were measured using a commercial Contact Angle Meter (GBX Scientific Instruments, France) equipped with a temperature and humidity controlled measuring chamber and a digital camera (T = 20°C; RH = 50%).

2.2.5. Elemental Analysis

To measure the content of grafted organic groups in the synthesized samples, the Perkin-Elmer 2400 CHN-analyzer (USA) was used. The modifying layer was oxidized to produce H_2O and CO_2 during the samples heating in the oxygen flow at 750°C.

3. Results and Discussion

Control of the surface reactions under the surface modification of the fumed silica with poly(methylphenylsiloxane) and dimethyl carbonate was performed by IR spectroscopy. IR spectra of the silicas modified with PMPS are shown in **Figure 1**. Surface treatment was performed at 200°C, 250°C and 300°C for 2 hrs. High intensity bands at 2900 - 3100 cm^{-1} (asymmetric stretching C–H vibrations in methyl and phenyl groups) and accompanying band at 2910 cm^{-1} (symmetric stretching C–H vibrations) are observed in the IR spectra of the samples of fumed silica modified with poly(methylphenylsyloxane) at 200°C, 250°C and 300°C (spectra 3 - 5). The samples of silicas modified at 200°C and 250°C (spectra 3, 4) are characterized by the highest content of methyl groups and it is correlated with data of elemental analysis (**Table 1**).

IR spectra are also characterized by the presence of a broad absorption band at 3600 - 3000 cm^{-1} corresponding to O–H vibrations in adsorbed water molecules and in silanol groups which formed hydrogen bonds with molecules of adsorbate. The band at 3750 cm^{-1} corresponding to O–H stretching vibrations of the free silanol groups [22] [23] is intense in the spectrum of the original silica, but this band is not seen in the spectra of the modified silicas. The band of deformation vibrations of the adsorbed water (at 1630 cm^{-1}) is observed in all three spectra of the modified samples (spectra 3-5), but this band is the most intensive in the spectrum of the pristine silica (spectrum 1).

Carrying out the process of modification in the simultaneous presence of PMPS and dimethyl carbonate in the reaction mixture results in both the increase of intensity of the bands of stretching vibrations of C–H bonds in the methyl groups in the wave number range 3000 - 2900 cm^{-1} [24] and the increase of intensity of the bands of stretching vibrations of C–H bonds in the phenyl groups manifested in the frequency range of 3100 - 3000 cm^{-1}

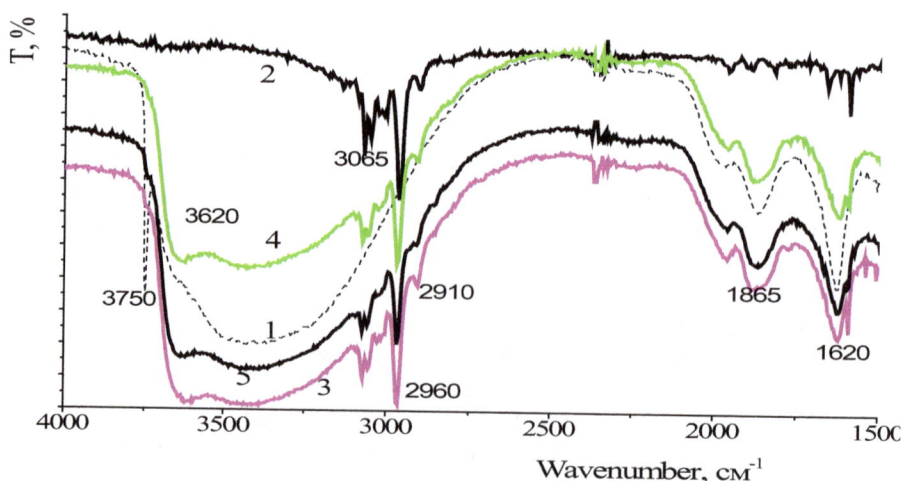

Figure 1. IR spectra of the original fumed silica A-300 (1), PMPS (2), and the fumed silicas modified with PMPS for 2 hrs at 200°C, 250°C and 300°C (3 - 5, respectively).

Table 1. Structural characteristics of fumed silicas modified with PMPS and its mixtures with DMC determined from nitrogen adsorption, and hydrophilic-hydrophobic properties of the surface determined on the basis of elemental analysis and values of contact angles of wetting.

T, °C	Component ratios	S_{BET}, m^2/g	V_p, cm^3/g	R_{ave} nm	Carbon content, wt.%	Contact angle of wetting Θ, degrees
	Pristine SiO$_2$	260	0.539	11	0	0
	SiO$_2$ + PMPS (10 g + 2 ml)	102	0.024	40	7.3	133
	SiO$_2$ + PMPS + 0.5 ml DMC	151	0.021	15	6.0	130
200	SiO$_2$ + PMPS + 1.0 ml DMC	158	0.021	15	6.0	130
	SiO$_2$ + PMPS + 1.5 ml DMC	171	0.027	31	7.4	130
	SiO$_2$ + PMPS	168	0.027	31	7.9	133
250	SiO$_2$ + PMPS + 0.5 ml DMC	168	0.022	16	7.3	130
	SiO$_2$ + PMPS + 1.0 ml DMC	173	0.021	14	6.6	129
	SiO$_2$ + PMPS + 1.5 ml DMC	191	0.024	17	6.5	130
	SiO$_2$ + PMPS	190	0.028	28	6.0	134
	SiO$_2$ + PMPS + 0.5 ml DMC	160	0.023	17	7.0	130
300	SiO$_2$ + PMPS + 1.0 ml DMC	170	0.022	15	6.7	130
	SiO$_2$ + PMPS + 1.5 ml DMC	178	0.022	16	7.0	130

(**Figure 2**). The content of adsorbed water in the samples synthesized in the presence of alkyl carbonate is smaller, unlike the silica modified in the absence of alkyl carbonate for which the band intensity with maximum at 3450 cm^{-1} is much higher.

It should be noted that there is no absorption band O–H in the free silanol groups at 3750 cm^{-1} in the spectra of the modified products, indicating their full participation in the reactions with the mixture of PMPS and DMC.

Figure 3(a) shows the adsorption-desorption isotherms of nitrogen for pristine silica and for silica modified with pure PMPS and its mixtures with DMC. The type of isotherm is common for the all modified silica, namely they are characterized by the fourth type according to the Brunauer's classification (a certain superposition of isotherms of II and IV types of the IUPAC classification without plateau adsorption [25] [26]) and the third type of the hysteresis loop in the range of high values of relative pressures [27]. Within the range of relative pressures $p/p_0 = 0.4 - 0.9$ one can observe lowering of all adsorption-desorption isotherms of nitrogen for modified silicas

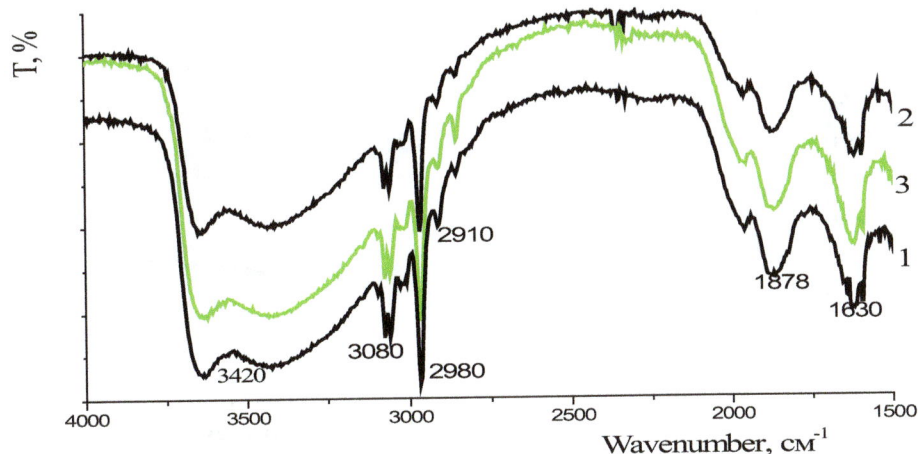

Figure 2. IR spectra of fumed silicas modified with mixtures of PMPS (2 ml) and DMC (0.5 ml) for 2 hrs at 200°C, 250°C and 300°C (1 - 3, respectively).

in comparison with the pristine silica.

Structural and surface characteristics of the pristine silica and the samples modified with poly(methylpheny-lsiloxane) are given in **Table 1**.

The samples synthesized by modification of the silica surface with PMPS have large number of voids with the average radius $R_{ave} > 28$ nm. For the samples modified with the mixture PMPS and DMC size-range of voids is $R_{ave} < 17$ nm.

Materials obtained by treatment with poly(methylphenylsiloxane) and mixtures of PMPS and DMC with different ratios of the components have similar contact angles of wettings (from 129° to 134°). It indicates the high hydrophobic properties of surface of the synthesized organosilicas (**Table 1**). As stated before, hydrophobicity of the modified materials is largely associated with the structure of the surface layer and determined by the concentration of the grafted organic groups.

According to the elemental analysis data (**Table 1**) one can find that using pure poly(methylphenylsiloxane) and mixtures of PMPS with dimethyl carbonate for modification of silica at 250°C, allows to reach the high concentrations of grafted organic groups in the synthesized samples (the average of carbon content ~7 wt.%). It can be explained by the fact that at higher temperature the depolymerisation process becomes more intense, *i.e.* molecular weight of oligomers is decreased and number of contacts of the modifying reagent with the surface sites is increased. The average of carbon content is smaller (≥ 6.6 carbon) for the samples synthesized at 200°C and 300°C. It can be assumed that at 200°C the depolymerisation degree is low what promotes chemisorption of the oligomers with relatively higher molecular weight and with weaker contacts with the surface sites. At 300°C an increased number of low-weight oligomers with lower carbon content is formed that affect the value of chemisorption and the structure of the grafted surface layer.

Extreme character of dependence of the carbon content on the modification process temperature is a result both the chemisorption kinetics and the depth of thermal depolymerisation. During modification with the mixture of PMPS and DMC two processes are competing: a thermal and chemical (with participation of DMC) depolymerisation. The degree of modification depends on kinetic features of chemisorption and the process of splitting of polysiloxane chains. At 200°C the rate-limiting step of modification is splitting the siloxane bonds in the reaction with dimethyl carbonate with following attachment of oligomers to the silica surface. However, at 250°C and 300°C the thermal depolymerisation prevails, the quantity of initiator slightly affects the surface carbon content that controls the degree of the siloxane depolymerisation. Generally, with increasing amounts of initiator the degree of the polysiloxane grafting increases, however the temperature 250°C is optimal for maximum anchoring of the functional groups to the silica surface.

In **Figure 4**, the TEM images of fumed silica modified with PMPS and its mixture with DMC are presented. The data of the transmission electron microscopy showed that particles of silica modified with PMPS formed the aggregates with size of 20 - 40 nm (**Figure 4(a)**). The structure of fumed silica modified with the mixture of DMC and PMPS consists mostly of smaller particles of 10 - 12 nm (**Figure 4(b)**).

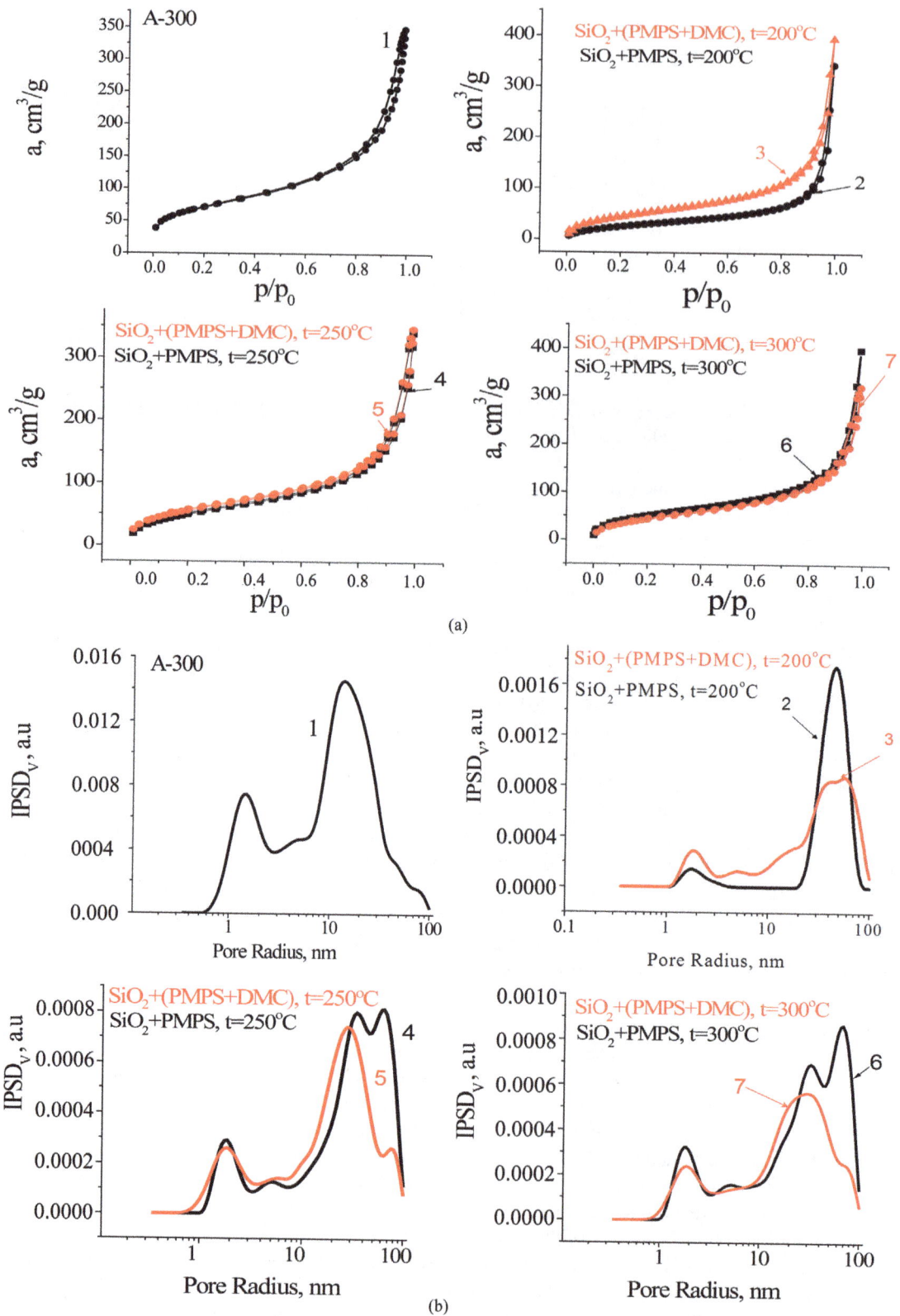

Figure 3. Adsorption-desorption isotherms of nitrogen at 77 K (a) and pore size distribution (b) for the pristine silica A-300 (1) and fumed silicas modified with pure PMPS (2, 4, 6) and with mixture of PMPS and DMC (3, 5, 7) for 2 hrs at 200°C (2, 3), 250°C (4, 5) and 300°C (6, 7).

It is known that the atomic force microscopy is crucial in measuring the characteristics of surface topography. In this paper, the microstructure of the surface layer of modified silica was studied by applying the AFM technique. The AFM images obtained for the pristine silica and after its modification with PMPS and PMPS/DMC mixture are presented in **Figure 5**. On the AFM images it can be observed that for the initial fumed silica (**Figure 5(a)**)

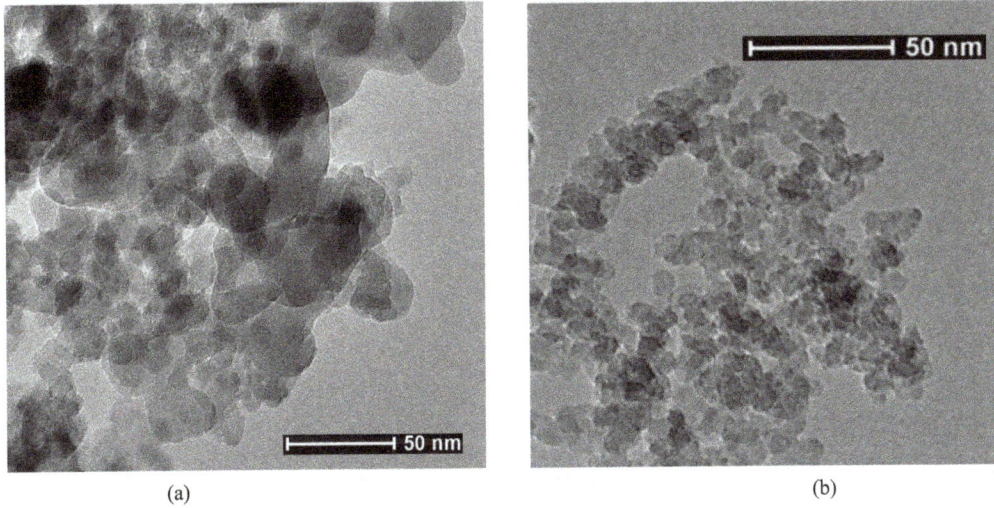

(a)   (b)

Figure 4. TEM images of the fumed silicas modified with PMPS (a) or with PMPS/DMC mixture (b) at 250°C for 2 hrs.

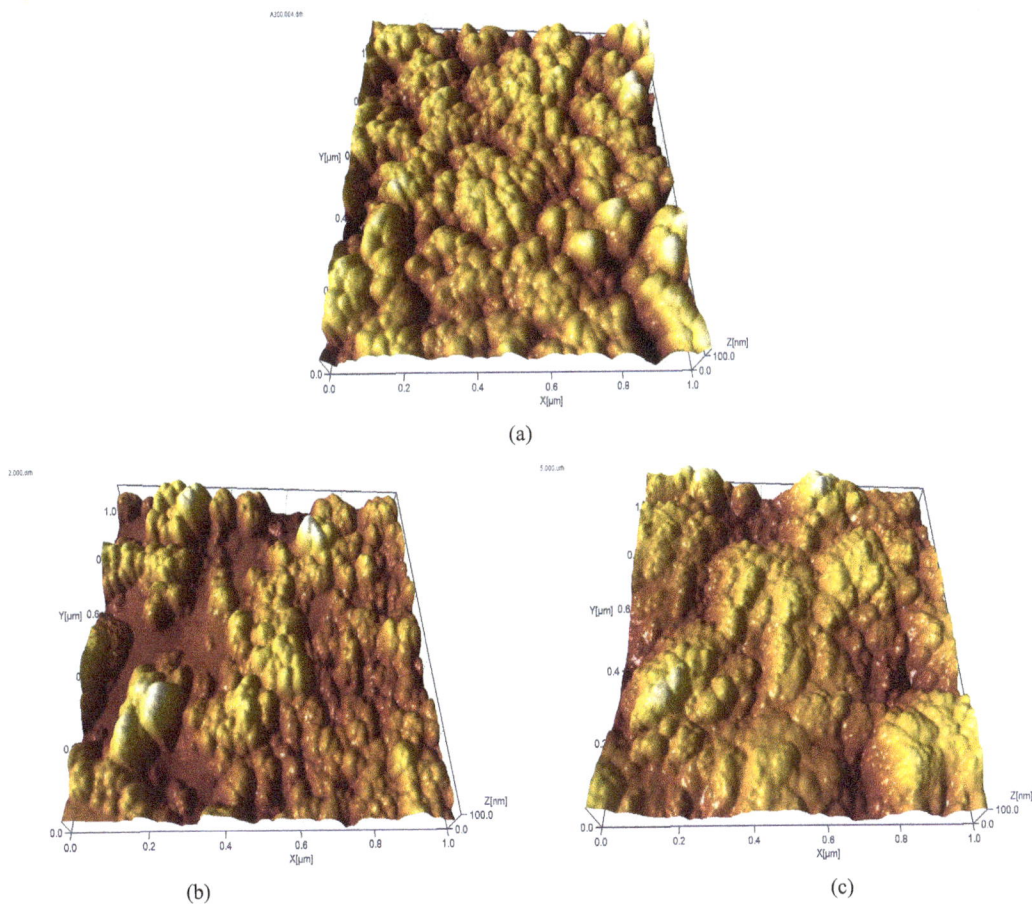

(a)

(b)   (c)

Figure 5. AFM images of pristine silica A-300 (a) after treatment at 250°C for 2 hrs with pure PMPS (b) and with PMPS/DMC mixture (c).

the structure of the surface layer is relatively homogeneous. The silica modified with the mixture of DMC/PMPS is characterized by a narrow size distribution and smallest grain size (**Figure 5(c)**). The coating microstructure for this sample is rather smooth in comparison to other coatings. After the modification of silica with pure PMPS a wider size distribution (**Figure 5(b)**) is observed. The particles show tendency to form bigger agglomerates than SiO_2 modified with the mixture of DMC/PMPS.

4. Conclusions

The properties of modified fumed silicas were studied using the methods of IR spectroscopy, elemental analysis, transmission electron microscopy, nitrogen adsorption-desorption data, atomic force microscopy and measurements of contact angles of wetting. According to IR spectroscopy data, mixtures of PMPS with DMC were shown to provide an increase of the concentration of grafted organic groups in the surface layer and full participation of the free silanol groups in the chemisorptions process at the moderate temperatures (200°C). At 250°C, the highest concentrations of grafted organic groups in the synthesized samples (the average of carbon content ~7 wt.%) were reached. The average of carbon content is smaller ($\geq$6.6 wt.% carbon) for the samples synthesized at 200 and 300°C. These data indicate that the temperature 250°C is optimal for maximum attachment of the functional groups on the silica surface. The quantity of initiator slightly influenced the surface carbon content in the surface modifying layer.

The type of nitrogen adsorption isotherm is common for all modified silica, namely they are characterized by the fourth type according to the Brunauer's classification and the third type of the hysteresis loop in the range of high values of relative pressures. Coating microstructure and morpholology of the modified samples is relatively homogeneous under modification of the silica surface with the mixture of PMPS and DMC. Their structure consists of particles of size 10 - 12 nm. After the modification of silica with pure PMPS a wider size distribution is observed. Particles of silica modified with PMPS form the aggregates of the size 20 - 40 nm.

Acknowledgements

The research leading to these results has received funding from the People Programme (Marie Curie Actions) of the European Union's Seventh Framework Programme FP7/2007-2013/under REA grant agreement No PIRSES-GA-2013-612484.

References

[1] Zenkevich, I.G., Makarov, A.A. and Ivanova, K.V. (2014) Characteristic Variations of Gas-Chromatographic Retention Indices for Phases of Variable Composition. *Analytical Chemistry*, **69**, 1089-1095. http://dx.doi.org/10.1134/S1061934814110148

[2] Yang, M.H., Chen, I.L. and Wu, D.H. (1996) Chemically Bonded Phenylsilicone Stationary Phases for the Liquid-Chromatographic Separation of Polycyclic Aromatic-Hydrocarbons and Cyclosiloxanes. *Journal of Chromatography A*, **722**, 97-105. http://dx.doi.org/10.1016/0021-9673(95)00577-3

[3] Lisichkin, G.V. (1986) Modified Silicas in Sorption, Catalysis and Chromatography. Khimiya, Moscow. (In Russian)

[4] Protsak, I.S., Kozakevich, R.B., Bolbukh, Yu.M. and Tertykh, V.A. (2013) Viscosimetric Study of Polydimethylsiloxane Depolymerization with Dimethyl Carbonate. *Chemical Industry*, **117**, 58-62. (In Ukrainian)

[5] Demianenko, E.M., Grebenyuk, A.G., Lobanov, V.V., Tertykh, V.A., Protsak, I.S., Bolbukh, Yu.M. and Kozakevych, R.B. (2014) Quantum Chemical Study on Interaction of Dimethyl Carbonate with Polydimethylsiloxane. *Chemistry, Physics and Technology of Surface*, **5**, 473-479.

[6] Protsak, I.S., Tertykh, V.A., Goncharuk, O.V., Bolbukh, Yu.M. and Kozakevich, R.B. (2014) Hydrophobization of the Fumed Silica Surface with Polydimethylsiloxanes in the Presence of Alkyl Carbonates. *Chemistry, Physics and Technology of Surface*, **5**, 226-235. (In Ukrainian)

[7] Okamoto, M., Suzuki, S. and Suzuki, E. (2004) Polysiloxane Depolymerization with Dimethyl Carbonate Using Alkali Metal Halide Catalysts. *Applied Catalysis A: General*, **261**, 239-245. http://dx.doi.org/10.1016/j.apcata.2003.11.005

[8] Okamoto, M., Miyazaki, K., Kado, A. and Suzuki, E. (2001) Deoligomerization of Siloxanes with Dimethyl Carbonate over Solid-Base Catalysts. *Chemical Communication*, **18**, 1838-1839. http://dx.doi.org/10.1039/b104371b

[9] Tundo, P., Arico, F., Rosamilia, A.E., Grego, S. and Rossi, L. (2008) Dimethyl Carbonate: Green Solvent and Ambident Reagent. *Green Chemical Reactions*, 213-232.

[10] Arico, F. and Tundo, P. (2010) Dimethyl Carbonate: A Modern Green Reagent and Solvent. *Russian Chemical Reviews*, **79**, 479-489. http://dx.doi.org/10.1070/RC2010v079n06ABEH004113

[11] Brunauer, S., Emmett, P.H. and Teller, E. (1938) Adsorption of Gases in Multimolecular Layers. *Journal of the American Chemical Society*, **60**, 309-319. http://dx.doi.org/10.1021/ja01269a023

[12] Nguyen, C. and Do, D.D. (1999) A New Method for the Characterization of Porous Materials, *Langmuir*, **15**, 3608-3615.

[13] Gun'ko, V.M. (2014) Composite Materials: Textural Characteristics. *Applied Surface Science*, **307**, 444-454. http://dx.doi.org/10.1016/j.apsusc.2014.04.055

[14] Gun'ko, V.M. (2000) Consideration of the Multicomponent Nature of Adsorbents during Analysis of Their Structural and Energy Parameters. *Theoretical and Experimental Chemistry*, **36**, 319-324. http://dx.doi.org/10.1023/A:1005264427135

[15] Gun'ko, V.M., Meikle, S.T., Kozynchenko, O.P., Tennison, S.R., Ehrburger-Dolle, F., Morfin, I. and Mikhalovsky, S.V. (2011) Comparative Characterization of Carbon Adsorbents and Polymer Precursors by Small-Angle X-Ray Scattering and Nitrogen Adsorption Methods. *The Journal of Physical Chemistry A*, **115**, 10727-10735. http://dx.doi.org/10.1021/jp201835r

[16] Do, D.D., Nguyen, C. and Do, H.D. (2001) Characterization of Micromesoporous Carbon Media. *Colloids and Surfaces A: Physicochemical and Engineering*, **187**, 51-71. http://dx.doi.org/10.1016/S0927-7757(01)00621-5

[17] Gun'ko, V.M. and Mikhalovsky, S.V. (2004) Evaluation of Slitlike Porosity of Carbon Adsorbents. *Carbon*, **42**, 843-849. http://dx.doi.org/10.1016/j.carbon.2004.01.059

[18] Toth, A., Voitko, K.V., Bakalinska, O., Prykhod'ko, G.P., Bertóti, I., Martiinez-Alonso, A., Tascon, J.M.D., Gun'ko, V.M. and Laszloa, K. (2012) Morphology and Adsorption Properties of Chemically Modified MWCNT Probed by Nitrogen, *n*-Propane and Water Vapor. *Carbon*, **50**, 577-585. http://dx.doi.org/10.1016/j.carbon.2011.09.016

[19] Myronyuk, L.I., Myronyuk, I.F., Chelyadyn, V.L., Sachko, V.M., Nazarkovsky, M.A., Leboda, R., Skubiszewska-Zięba, J. and Gun'ko, V.M. (2013) Structural and Morphological Features of Crystalline Nanotitania Synthesized in Different Aqueous Media. *Chemical Physics Letters*, **583**, 103-108. http://dx.doi.org/10.1016/j.cplett.2013.07.068

[20] Gun'ko, V.M., Voronin, E.F., Nosach, L.V., Turov, V.V., Wang, Z., Vasilenko, A.P., Leboda, R., Skubiszewska-Zięba, J., Janusz, W. and Mikhalovsky, S.V. (2011) Structural, Textural and Adsorption Characteristics of Nano-Silica Mechanochemically Activated in Different Media. *Journal of Colloid and Interface Science*, **355**, 300-311. http://dx.doi.org/10.1016/j.jcis.2010.12.008

[21] Sternik, D., Majdan, M., Deryło-Marczewska, A., Żukociński, G., Gładysz-Płaska, A., Gun'ko, V.M. and Mikhalovsky, S.V. (2011) Influence of Basic Red 1 Dye Adsorption on Thermal Stability of Na-Clinoptilolite and Na-Bentonite. *Journal of Thermal Analysis and Calorimetry*, **103**, 607-615. http://dx.doi.org/10.1007/s10973-010-1014-3

[22] Davydov, V.Y. (2000) Adsorption on Silica Surface. In: Papirer, E., Ed., *Surfactant Science Series*, Marcel Dekker, New York, 63-118.

[23] Gun'ko, V.M., Borysenko, M.V., Pissis, P., Spanoudaki, A., Shinyashiki, N., Sulim, I.Y., Kulik, T.V. and Palyanytsya, B.B. (2007) Polydimethylsiloxane at the Interfaces of Fumed Silica and Zirconia/Fumed Silica. *Applied Surface Science*, **253**, 7143-7156. http://dx.doi.org/10.1016/j.apsusc.2007.02.185

[24] Gao, D., Jia, M. and Luo, Y. (2013) Crosslinked Organosiloxane Hybrid Materials Prepared by Condensation of Silanol and Modified Silica: Synthesis and Characterization. *Chinese Journal of Polymer Science*, **31**, 974-983. http://dx.doi.org/10.1007/s10118-013-1289-5

[25] Condon, J.B. (2006) Surface, Area and Porosity Determinations by Physisorption, Measurements and Theory. Elsevier, Amsterdam.

[26] Sing, K.S.W. (1982) Reporting Physisorption Data for Gas/Solid Systems with Special Reference to the Determination of Surface Area and Porosity. *Pure and Applied Chemistry*, **54**, 2201-2218. http://dx.doi.org/10.1351/pac198254112201

[27] Gregg, S.J. and Sing, K.S.W. (1991) Adsorption, Surface Area and Porosity. 2nd Edition, Academic Press, London.

Permissions

List of Contributors

Sabri M. Husssein and Sattar S. Ibrahim
Department of Chemistry, College of Science, University of Anbar, Anbar, Iraq

Omar H. Shihab
Department of Chemistry, College of Women Education, University of Anbar, Anbar, Iraq

Naser M. Ahmed
Nano-Optoelectronics Research and Technology Laboratory, School of Physics, University Sains Malaysia, Penang, Malaysia

Mohamed S. El Naschie
Department of Physics, University of Alexandria, Alexandria, Egypt

K. Vijaya Kumar
Department of Physics, Jawaharlal Nehru Technological University Hyderabad College of Engineering, Nachupally (Kondagattu), Karimnagar-Dist, Telangana State, India

D. Paramesh and P. Venkat Reddy
Sreenidhi Institute of Science and Technology (Autonomous), Hyderabad, India

Chernet Amente
Physics Department, Addis Ababa Science and Technology University, Addis Ababa, Ethiopia

Keya Dharamvir
Physics Department, Panjab University, Chandigarh, India

Vânia M. Dias, Inês Portugal and Dmitry V. Evtuguin
CICECO/Department of Chemistry, University of Aveiro, Aveiro, Portugal

Alena Kuznetsova, João Tedim, Aleksey A. Yaremchenko and Mikhail L. Zheludkevich
CICECO/Department of Materials and Ceramic Engineering, University of Aveiro, Aveiro, Portugal

Md. Khalid Hossain
Institute of Electronics, Atomic Energy Research Establishment, Savar, Dhaka, Bangladesh

Jannatul Ferdous and Md. Manjurul Haque
Department of Applied Physics, Electronics & Communication Engineering, Islamic University, Kushtia, Bangladesh

A. K. M. Abdul Hakim
Department of Glass and Ceramic Engineering, Bangladesh University of Engineering and Technology, Dhaka, Bangladesh

Rojalin Sadual and Sushanta K. Badamali
Department of Chemistry, Utkal University, Bhubaneswar, India

Sudhir E. Dapurkar
Tata Chemicals Limited, Pune, India

Rajesh K. Singh
Department of Chemistry, North Orissa University, Baripada, India

Emad H. M. Zahran
Department of Mathematical and Physical Engineering, College of Engineering, University of Benha, Shubra, Egypt

Naraavula Suresh Kumar
Mallareddy Institute of Engineering & Technology, Secunderabad, India

Katrapally Vijaya Kumar
Department of Physics, JNTUH College of Engineering Jagtial, Nachupally (Kondagattu), Karimnagar-Dist, TS, India

R. Desmarchelier, B. Poumellec, F. Brisset, S. Mazerat and M. Lancry
ICMMO, UMR CNRS-UPSud 8182, Université Paris Sud (in Université Paris Saclay), Orsay, France

M. S. El Naschie
Department of Physics, University of Alexandria, Alexandria, Egypt

Ludovic Richert
Centre National de la Recherche Scientifique (CNRS), UMR, Faculté de Pharmacie de l'Université de Strasbourg (UdS), Illkirch, France

Laetitia Keller and Quentin Wagner
Institut National de la Santé et de la Recherche Médicale (INSERM), Osteoarticular and Dental Regenerative Nanomedicine, UMR, Faculté de Médecine de l'Université de Strasbourg and FMTS, Strasbourg, France
Faculté de Chirurgie Dentaire de l'Université de Strasbourg (UdS), Strasbourg, France

Fabien Bornert, Catherine Gros, Sophie Bahi, François Clauss, William Bacon, Nadia Benkirane-Jessel and Florence Fioretti
Institut National de la Santé et de la Recherche Médicale (INSERM), Osteoarticular and Dental Regenerative Nanomedicine, UMR, Faculté de Médecine de l'Université de Strasbourg and FMTS, Strasbourg, France
Faculté de Chirurgie Dentaire de l'Université de Strasbourg (UdS), Strasbourg, France
Hôpitaux Universitaires de Strasbourg, Strasbourg, France

Philippe Clézardin
Institut National de la Santé et de la Recherche Médicale (INSERM), UMR, Faculté de Médecine Laënnec de l'Université Claude Bernard Lyon 1, Lyon, France

Ashok K. Bharimalla, Prashant G. Patil and Nadanathangam Vigneshwaran
ICAR-Central Institute for Research on Cotton Technology, Mumbai, India

Suresh P. Deshmukh
Department of General Engineering, Institute of Chemical Technology, Mumbai, India

Mahsa Rezaei
Faculty of Food Industries, Ayatollah Amoli Branch, Islamic Azad University, Amol, Iran

Ali Motamedzadegan
Faculty of Food Industries, Ayatollah Amoli Branch, Islamic Azad University, Amol, Iran
Department of Food Science, Sari University of Agricultural Sciences and Natural Resources, Sari, Iran

Valentin Tsvetkov and Sergei Pasechnik
Problem Laboratory of Molecular Acoustics, Moscow State University of Instrument Engineering and Computer Sciences, Moscow, Russia

Aleksei Dronov
National Research University of Electronic Technology, Moscow, Zelenograd, Russia

Jacob Y. L. Ho, Vladimir Chigrinov and Hoi-Sei Kwok
State Key Laboratory on Advanced Displays and Optoelectronics Technologies, Clear Water Bay, Kowloon, Hong Kong, China

Eman Alzahrani
Chemistry Department, Faculty of Science, Deanship of Scientific Research, Taif University, Taif, KSA

Ahmed H. Kurda and Yousif M. Hassan
Physics Department, College of Science, University of Salahaddin, Erbil, Kurdistan of Iraq

Naser M. Ahmed
Nano-Optoelectronic Research & Technology Laboratory, School of Physics, Universiti Sains Malaysia, Penang, Malaysia

Jing Cao, Bertrand Poumellec, François Brisset, Anne-Laure Helbert and Matthieu Lancry
Institut de Chimie Moléculaire et des Matériaux d'Orsay (ICMMO), CNRS-Université Paris Sud, Université Paris Saclay, Bât.420, Campus Orsay, 91405 Orsay, France

Amer N. J. Al-Daghman, K. Ibrahim and Naser M. Ahmed
Nano-Optoelectronic Research and Technology Laboratory, School of Physics, University Sains Malaysia, Pulau Pinang, Malaysia

Kareema M. Zaidan
Physics Department, Collage of Science, University of Basrah, Basrah, Iraq

Iryna S. Protsak, Valentyn A. Tertykh and Yulia M. Bolbukh
Department of Chemisorption, Chuiko Institute of Surface Chemistry of National Academy of Sciences of Ukraine, Kiev, Ukraine

Dariusz Sternik and Anna Derylo-Marczewska
Faculty of Chemistry, Maria Curie-Sklodowska University, Lublin, Poland